Simulation, Experiment and Modeling of Coal Fires

Simulation, Experiment and Modeling of Coal Fires

Guest Editors

Wei Liu
Zeyang Song
Caiping Wang
Bobo Shi

MDPI

Basel • Beijing • Wuhan • Barcelona • Belgrade • Novi Sad • Cluj • Manchester

Guest Editors

Wei Liu
School of Emergency Management and Safety Engineering
China University of Mining & Technology
Beijing
China

Zeyang Song
College of Safety Science and Engineering
Xi'an University of Science and Technology
Xi'an
China

Caiping Wang
College of Safety Science and Engineering
Xi'an University of Science and Technology
Xi'an
China

Bobo Shi
School of Safety Engineering
China University of Mining & Technology
Xuzhou
China

Editorial Office
MDPI AG
Grosspeteranlage 5
4052 Basel, Switzerland

This is a reprint of the Special Issue, published open access by the journal *Fire* (ISSN 2571-6255), freely accessible at: https://www.mdpi.com/journal/fire/special_issues/OOR1CVZ56X.

For citation purposes, cite each article independently as indicated on the article page online and as indicated below:

Lastname, A.A.; Lastname, B.B. Article Title. *Journal Name* **Year**, *Volume Number*, Page Range.

ISBN 978-3-7258-6167-5 (Hbk)
ISBN 978-3-7258-6168-2 (PDF)
https://doi.org/10.3390/books978-3-7258-6168-2

Contents

About the Editors

Wei Liu

Wei Liu, Ph.D., is a Professor and doctoral supervisor in the School of Emergency Management and Safety Engineering at China University of Mining and Technology (Beijing). His research focuses on the prevention and control of mine fires, gas disasters, and risk assessment in complex industrial systems. He has led over 20 national and provincial-level research projects, including initiatives from the National Natural Science Foundation of China. His work has contributed significantly to the understanding of coal spontaneous combustion mechanisms and the development of intelligent fire prevention technologies. He obtained his Ph.D. in Safety Technology and Engineering from the China University of Mining and Technology (Beijing). He has published more than 80 academic papers, holds over 20 invention patents, and has received multiple provincial and ministerial-level awards for scientific and technological progress.

Zeyang Song

Zeyang Song is a Professor in the College of Safety Science and Engineering at Xi'an University of Science and Technology. He received his Ph.D. in Safety Technology and Engineering from China University of Mining and Technology (Beijing) in 2015. His research interests are focused on fire modelling, fire remote sensing, smoldering combustion, heat and mass transfer in porous media, and AI-powered safety engineering. As PI, he led more than 10 research projects funded by the National Natural Science Foundation of China and provincial natural science foundation in China. He has co-authored over 60 papers, and been honored with the Youth Talents of Science and Technology in Shaanxi Province and Top 2% Scientists in 2024 and 2025. He served as editor, editorial board, and topical advisory panel for five peer-reviewed journals.

Caiping Wang

Caiping Wang, Ph.D., is a Professor and Deputy Dean of the College of Safety Science and Engineering at Xi'an University of Science and Technology. Her research is primarily focused on the mechanisms of coal spontaneous combustion, early warning systems for mine fire disasters, and comprehensive disaster prevention and control in deep mining environments. She has led numerous research projects supported by the National Natural Science Foundation of China, Shaanxi Province, and major enterprises. Her work aims to develop precise and intelligent monitoring technologies for fire hazards and to establish multi-parameter early warning indicator systems. She obtained her Ph.D. from Xi'an University of Science and Technology, where she also completed her postdoctoral research. She has published over 50 research papers and holds numerous authorized invention patents.

Bobo Shi

Bobo Shi, Ph.D., is a Professor in the School of Safety Engineering at China University of Mining and Technology. His main research interests include the intelligent prevention and control of coal mine fires, the coupled dynamics of coal and gas, and multi-disaster prevention in deep mines. He has been the principal investigator for several projects funded by the National Natural Science Foundation of China. His work explores the fundamental mechanisms of coal oxidation and heating, as well as developing advanced early warning and emergency response technologies for major mining disasters. He received his Ph.D. in Safety Technology and Engineering from China University of Mining and Technology. He has authored or co-authored over 40 SCI-indexed papers and has received awards for scientific and technological innovation from national industry associations.

Preface

Coal fires, as a major disaster in the mining industry, pose a persistent threat to global energy and production security. Underground coal fires, typically initiated by spontaneous combustion, are a worldwide challenge to control effectively due to their concealed sources, dynamic propagation, and high susceptibility to reignition. The core obstacle lies in the fact that a coal fire is fundamentally a complex dynamic process involving coupled multi-physics, including chemical reactions, heat and mass transfer, and geomechanics. Therefore, employing a combined approach of experiments, numerical simulations, and theoretical modeling to investigate the ignition mechanisms of coal and the propagation patterns of underground fire zones is of critical scientific importance for achieving their precise prevention and control.

This Special Issue, entitled "Simulation, Experiment and Modeling of Coal Fires", aims to bring together the latest research findings in this field. Our goal is to reveal the disaster-causing mechanisms of coal fires from the multi-dimensional perspectives of simulation, experimentation, and modeling, to elucidate the spatio-temporal evolution of their occurrence and development, and to provide a solid theoretical basis for their precise prevention and control. The 12 articles in this Special Issue collectively offer a multi-faceted, in-depth examination of the frontiers of coal fire research: from the micro-molecular mechanisms of coal spontaneous combustion to macroscopic mine prevention and control engineering; from intelligent monitoring and early warning systems based on optical fibers and machine learning to the precise prediction of secondary disasters such as gas outbursts; and, more forward-looking, the exploration of new pathways for resource utilization, turning the waste heat from coal fires from a hazard into a benefit.

We extend our heartfelt gratitude to all the authors for their outstanding contributions, as their rigorous academic work forms the core of this Special Issue. We would also like to express our sincere thanks to the expert reviewers, whose insightful comments have significantly enhanced the quality of the papers. Finally, we thank the editorial team of the journal *Fire* for their professional support and hard work throughout the preparation and publication process. We believe that this Special Issue will provide valuable references and inspiration for researchers, engineers, and decision-makers in the fields of mining safety, fire engineering, and environmental science.

Wei Liu, Zeyang Song, Caiping Wang, and Bobo Shi
Guest Editors

fire

MDPI

Review

Advances in Detection and Monitoring of Coal Spontaneous Combustion: Techniques, Challenges, and Future Directions

Lucica Anghelescu and Bogdan Marian Diaconu *

Faculty of Engineering, "Constantin Brancusi" University of Targu Jiu, Calea Eroilor 30, 210135 Targu Jiu, Romania; lucica.anghelescu@e-ucb.ro
* Correspondence: bogdan.diaconu@e-ucb.ro

Abstract: Coal spontaneous combustion (CSC) is a multifaceted research domain that has been widely explored in the literature, ranging from analytical and numerical modeling to the development of fire suppression materials and methods. A comprehensive review of the literature has revealed several distinct research trajectories, or "roadmaps", identified through criteria such as the volume of studies addressing each theme, the presence of review papers dedicated to a specific roadmap, and the explicit mention of coal spontaneous combustion in the title or keywords. This classification framework has outlined six primary roadmaps: (1) spread, quantification, and impact; (2) mechanisms, models, factors, and parameters; (3) experimental studies and models; (4) detection, monitoring, and prediction; (5) prevention and control; and (6) applications. While interconnections exist between these roadmaps, and all ultimately converge towards roadmap 5 (prevention and control), each roadmap constitutes a distinct research cluster. The focus of this review is on roadmap 4, specifically addressing the methods and technologies for detection, monitoring, and prediction of CSC events. This review encompasses studies published from 2010 to the present, providing a thorough examination of the various detection techniques employed, with particular emphasis on their limitations and the strategies proposed to overcome these challenges. A critical analysis highlights the key advantages and disadvantages of each category of techniques, offering insights into their practical applications and the potential for future advancements in this field. The present review aims to contribute to the refinement of detection and monitoring methods for CSC, with the goal of enhancing early detection capabilities and improving fire management strategies.

Keywords: coal spontaneous combustion; detection techniques; gas monitoring; ground temperature measurement; thermal infrared imaging; remote sensing

1. Introduction

Spontaneous combustion is a class of combustion processes triggered by a temperature increase caused by exothermic chemical and physical phenomena. In general, it is agreed that spontaneous combustion has two phases: (i) The slow phase, during which the reaction rate is low and the whole mass of material heats up almost uniformly. The heat is dissipated at the interface with the environment by various heat transfer mechanisms, mainly convection. Depending on the thermal equilibria conditions, the process can be slowed down or it can evolve and reach the second phase. (ii) The thermal runaway phase, in which the exothermic internal reaction speeds up due to the temperature increase and create an avalanche effect. The temperature increase rate becomes extremely high and eventually autoignition occurs.

Both inorganic and organic materials can undergo spontaneous combustion, but with different triggering mechanisms. Coal is an organic material with a significant spontaneous combustion potential, which is a key hazard in coal mines. Spontaneous combustion of coal is a complex process influenced by many factors, either directly or through a chain of triggering processes. Coal spontaneous combustion occurs in coal stockpiles, coalfields, underground mines, and, in general, whenever coal is stockpiled for long periods. The spontaneous combustion of coal causes economic losses, generates greenhouse gases and poses risks to property through the spread of fire. Coal fires caused by spontaneous combustion have two distinctive features compared to conventional fires:

1. Coal fires can develop in underground seams as smoldering combustion, representing some of the most persistent fires on Earth, with occurrences dating back several million years, as noted by Heffern and Coates [1]. These fires are notoriously difficult or even impossible to extinguish, leading to both immediate, observable consequences such as wildfires, and longer-term, less predictable impacts like the alteration of adjacent geological formations. The terrain-shaping effects of coal fires have been documented in a study by Heffern and Coates [1].
2. Coal fires do not always require an ignition source. Once the thermal runaway threshold temperature is reached, the coal mass ignites spontaneously. The thermal runaway threshold can be anywhere between 80 and 120 °C (Song et al. [2]) depending on many factors.

A bibliometric analysis on the topic of coal spontaneous combustion was carried out by Liu et al. [3]. The analysis included metrics such as the number of publications over time, the countries/regions and organizations from where the publications originated, the journals, and various domain maps discussing the keywords spontaneous combustion, spontaneous ignition, self-ignition, self-heating and pyrophoric. A key conclusion derived from this study is that spontaneous combustion research is a multidisciplinary domain, integrating knowledge from areas such as fuel science, chemistry, thermodynamics, heat and mass transfer.

Another bibliometric analysis was conducted by Yang and Qiu [4]. The following list of terms contained in either title, abstracts, and keywords list, was used to extract from Science Citation Index Expanded publications from 1984 to 2018: coal spontaneous combustion, coal self-heating, coal spontaneous heating, coal low-temperature oxidation. A number of 829 publications were selected, out of which 91.31% were journal articles. The bibliometric analysis reported in this paper focused mainly on the following issues:

➢ Originating geographic area of the publications. The top four positions were occupied by China with 348, followed by Australia with 96, USA with 92 and India with 45 publications. China is far ahead of the pack, which is explained by the high prevalence of coal fires.
Note: A significant number of coal seams, particularly in the regions Xinjiang and Inner Mongolia, meet the geological conditions and mining practices to render them highly susceptible to fire, Huang et al. [5]. Coal fire cause annual losses exceeding 100 million ¥, spontaneous combustion fires accounting for 90–94% of the total fire events in mines [5]. Huang et al. [5] present a statistic (without mentioning the source of the figures) stating that 32 underground coal spontaneous combustion or gas explosion accidents occurred in China between 2001 and 2014, causing 614 deaths.
➢ Journals having published papers on the topic of CSC.
➢ Keyword occurrence. The top four keywords and their respective occurrence frequency values were low temperature oxidation (353), coal spontaneous combustion (302), coal (169) and mine fire (137). The keyword with the highest occurrence frequency can be explained by the fact that low temperature coal oxidation is the triggering factor that ultimately results in coal spontaneous combustion. Other keywords related to chemical interaction between coal and molecular oxygen were encountered, such as kinetics, mechanism, model and particle size. Keywords related to prevention and control such as prevention (37), prediction (26) and inhibitor suggest the existence of a consolidated research direction.

Onifade and Genc [6] conducted a review on the topic of "spontaneous combustion of coal". The content of this review is limited to three sections, mathematical modelling, experimental methods and statistical methods. The article has a narrative structure summarizing several works, selected on unspecified criteria.

Zhang et al. [7] conducted a review on numerical studies on the topic of self-heating of a coal stockpile. A bottom-up approach has been employed in structuring the article discussing first the mechanisms of coal stockpile self-heating, the transfer equations and the constants used by various authors, a section dedicated to the variables (coal and stockpile characteristics, moisture and wind) and concluding with a section describing the main findings: the simple Arrhenius dependence on defining the coal oxidation rate is an acceptable approach after the occurrence of the thermal runaway but not at low temperatures; the Arrhenius parameters change with the particle; the reviewed studies are not consistent in assessing the effect of the coal particle size on the dynamics of the coal stockpile self-heating; three distinct stages were identified in the temperature profile: an initial slow rise due to the self-heating, a plateau phase which was explained by the energy expenditure to support the phase transition of water, and a third, steep-gradient stage, explained by the fast oxidation of dry coal. Song and Kuenzer [8] reviewed the literature on the coal fires occurring in China a decade back (from the date the article was published). It was reported that written evidence of coal fires dates as early as 386–534 AD. Coal fire distribution and development were discussed and some theories were put forward to explain the occurrence of coal fires in the north of China. A significant section of the article discusses issues related to detection and monitoring of coal fires. The detection and monitoring methods were grouped into four categories, as follows: underground, ground, airborne and space remote sensing methods. Three other sections were included discussing modelling of underground fires, environmental impact and fire control techniques.

In general, reviews covering the larger topic of CSC cannot thoroughly discuss all the components of the problem. For this reason, recent reviews focus on more narrow issues connected to the ultimate goal of the CSC research, that is prevention and control. Two sub-topics were identified where significant research has been published and several reviews were conducted:

1. Theoretical background, analytical models and numerical studies of the self-heating coal stockpiles and self-ignition are a set of issues of importance in designing prevention and control techniques. A comprehensive review on the topic of CSC hazard assessment based on thermal-kinetic and heat and mass transfer was carried out by Lu et al. [9]. The large volume of literature on this topic was divided into several sections: propensity rating for coal spontaneous combustion with small loading, fire potential for coal spontaneous combustion based on large-scale approach, and field investigation. This division is significant in understanding the main research directions pertaining to the CSC. Several key conclusions defining the need for more in-depth research were formulated: extrapolating small-scale indices, such as self-heating rate, heat release intensity, oxygen consumption, as well as dimensionless parameters, to large and/or full-scale models has not been fully validated; the macroscopic approach used in current mathematical modelling of CSC are not always consistent with the measured data, the development of a multiscale modelling methodology is highly desirable; although the existence of a threshold between self-sustaining and extinction conditions for coalfield fire has been established, more research is required to define quantitatively the threshold conditions. Studies employing numerical solutions for the self-heating of the coal stockpiles and spontaneous combustion were reviewed by Zhang et al. [7]; a short discussion on the topic of mathematical models and factors that determine the dynamics of the self-heating process was included in the general review conducted by Onifade and Genc [6].
2. Mechanisms and methods for prevention and control of CSC is actually the ultimate goal of research in this direction. Li et al. [10] conducted a review discussing the most prevalent materials used in prevention of CSC. The mechanisms by which the inhibitors slow down or suppress the chemical and physical processes leading ultimately to CSC were discussed and a classification of materials was carried out

based on the inhibition mechanism. The availability, cost and environmental impact of inhibitors were assessed. The techniques, methods, and materials applied in the prevention of coal spontaneous combustion were reviewed in Onifade [11] (measures), Kong et al. [12] (methods to prevent CSC), Han et al. [13], and Lu et al. [14] (fire prevention materials), and Dai et al. [15] (heat pipe systems to prevent CSC in coal storage piles).

2. Research Methodology

A Scopus search from 2010 to date including document type article, conference paper, review, conference review resulted in 1176 documents using the keyword "coal spontaneous combustion". The Scopus search was limited to the "Keyword" field. A bibliometric analysis by means of VOS Viewer reported an author keywords number of 2119. We set the following filters as follows: Document type to Article, Conference paper and Review, Language to English, and Source type to Journal and Conference proceeding, the total number of entries was reduced to 928. In the VOS Viewer tool, the number of keyword occurrence was set to the default value 5, resulting a number of 91 keywords. The list of keywords contained entries with high semantic similarity (e.g., "coal spontaneous combustion" and "spontaneous combustion of coal") which were not removed since the main purpose of this analysis was to identify/define the main research directions in CSC. The top five keywords semantically distinct were "coal spontaneous combustion" with 576 occurrences, "activation energy"—38, "characteristic temperature"—34, "coal"—44, and "apparent activation energy"—29. An initial manual screening was performed, first based on titles and then on abstracts, to define research roadmaps. The criteria for defining a roadmap topic were as follows:

1.1 A minimum number of 20 studies broadly discussing a topic define a research roadmap.
OR

1.2 At least one review article was identified.
AND

2. The coal spontaneous combustion is mentioned in the title, keywords, or abstract.

In defining the roadmaps, some degree of topic disjunction was sought, in the sense that the roadmap classification would become less meaningful if two or more topics would merge.

Six roadmaps were identified as follows:

1. Spread, quantification and impact
   These are mainly experimental studies and field reports discussing:
   1.1 General characteristics, specific features, morphology and large-scale effects of coal fires occurring throughout the world: Song and Kuenzer [8], Kuenzer et al. [16], Saini et al. [17], Kuenzer and Stracher [18], Kus [19], Xu et al. [20], etc.
   1.2 Coal fires as a source of pollution and environmental impact: Li et al. [21], Carroll et al. [22], Oliveira et al. [23], Deng et al. [24], etc.
2. Mechanisms, models, factor and parameters
   This category includes
   2.1 Studies advancing analytical models for self-heating of coal and spontaneous combustion: Lu et al. [9], Xia et al. [25], Rua et al. [26], Xi et al. [27], Xiao et al. [28], etc.
   2.2 CFD studies presenting numerical simulation of the coal self-heating and CSC as well as the influence of physical factors: Yuan et al. [29], Taraba et al. [30], Chen et al. [31], Zhuo et al. [32], Xiaomeng et al. [33], etc.
   2.3 Parametric studies investigating the influence of physical factors (wind, moisture content, particle size) and chemical (coal grade, presence of foreign com-

pounds, etc.): Zhang et al. [34], Yu et al. [35], Plakunov et al. [36], Qiao et al. [37], Deng et al. [38], Zhang et al. [39], Zhang et al. [40], Li et al. [41], etc.

2.4 Studies relying on artificial intelligence models and nature-inspired algorithms, primarily aimed at assessing the propensity of coal spontaneous combustion: Said et al. [42], Lei et al. [43], Sahu et al. [44], Xie et al. [45], Li et al. [46], Sahu et al. [47], Wang et al. [48], etc.

3. Experimental studies and models
A class of studies attempting to investigate experimentally the CSC and validate numerical models:
3.1 The dynamics and susceptibility of CSC: Li et al. [49], Onifade et al. [50], Wen et al. [51], Zhang et al. [52], Yan et al. [53], Zhao et al. [54], etc.
3.2 Fire control methods and materials: Fan et al. [55], Zhang et al. [56], Zhang et al. [57], Cheng et al. [58], etc.
3.3 In situ measurements and tests: Liu et al. [59] and Tang et al. [60].

4. Detection, monitoring and prediction
This category includes studies from the following sub-categories:
4.1 Methods for predicting or assessing the risk of coal self-combustion (CSC) in various systems, including coal stockpiles, coal seams, coal goafs, and coal gobs, both in underground and surface mining: Hu et al. [61], Song et al. [2], Du et al. [62], Wei et al. [63], Ma et al. [64], etc.
4.2 Detection of coal fires after the ignition phase using various methods, both direct and indirect, including secondary effects such as gas emissions: Guo et al. [65], Guo et al. [66], Gao et al. [67], Zhou et al. [68], etc.
4.3 Monitoring of existing and developing fires: Biswal and Gorai [69], Wang et al. [70], Hu et al. [71], Kong et al. [72], etc.

5. Prevention and control.
5.1 Techniques for CSC prevention: Liu et al. [73], Liu et al. [59], Shi et al. [74], Zhai et al. [75], etc.
5.2 Materials and methods for extinguishing the coal fires: Cheng et al. [76], Cheng et al. [77], Fan et al. [78], Li et al. [79], etc.

6. Applications

This research direction is relatively new, arising from the observation that certain coal fires are inextinguishable. Consequently, technologies have been developed to harness the thermal energy released by these fires: Shi et al. [80], Su et al. [81], Xiao et al. [82], Xiao et al. [83], etc.

It is important to observe that numerous studies span multiple research roadmaps. This overlap is expected, as the criteria for categorizing a study under a single roadmap can be inherently subjective. Review studies that specifically address or include at least one roadmap topic in a dedicated section were identified accordingly are presented in Table 1 (comprehensive overviews of coal spontaneous combustion and brief reviews are not included).

The detection, monitoring, and prediction of coal spontaneous combustion (CSC) are essential for managing and mitigating the risks posed by coal fires, which can cause extensive environmental damage, release hazardous emissions, and disrupt mining operations. Detection methods primarily aim to identify early-stage combustion events, where low-temperature oxidation occurs below the surface, often without visible signs. Techniques such as thermal infrared (TIR) imaging, gas emissions monitoring, and multispectral and hyperspectral sensors are widely employed to detect heat anomalies, volatile gases, and changes in the physical properties of coal and surrounding materials. Monitoring methods, on the other hand, focus on tracking the progression and spread of established fires, often using aerial and satellite-based remote sensing platforms to map fire boundaries, assess subsurface smoldering, and evaluate the effectiveness of fire suppression efforts.

Table 1. Review papers addressing fully or partially the roadmap themes.

Roadmap Theme	Author and Ref.	Year	Estimated Percentage Covering the Roadmap Theme
Spread quantification and impact	Song et al. [8]	2014	25
Mechanisms, models, factor and parameters	Song et al. [8] Lu et al. [9] Zhang et al. [7]	2014 2022 2016	12 100 100
Experimental studies and models	None identified	Not applicable	
Detection, monitoring and prediction	Song et al. [8] Liang et al. [84]	2014 2019	12 100
Prevention and control	Li et al. [10]	2020	100
Applications	Xiao et al. [82]	2023	100

Predictive modeling is also a critical component of CSC management, as it helps anticipate fire behavior by simulating parameters such as combustion depth, temperature gradients, and fire propagation rates. These models rely on field data obtained from detection and monitoring efforts, enabling the identification of high-risk areas and informing targeted intervention strategies. Despite advancements in these technologies, challenges remain, particularly in adapting detection techniques to varying geological and environmental conditions, ensuring continuous monitoring, and improving the accuracy of early-stage CSC detection. This review focuses on recent developments in CSC detection, monitoring, and prediction methods, evaluating their practical applications and highlighting key limitations and areas for further research.

3. Coal Spontaneous Combustion Detection

Coal fires, whether fully developed or in the developing phase, can be detected using a wide variety of methods and sensory systems. Accurate and timely detection is essential for any coal spontaneous combustion (CSC) detection technique. The selection of an appropriate technique depends primarily on the general layout and specific site conditions. More precise detection can be achieved by combining multiple techniques, leveraging the strengths of each while compensating for their limitations.

Three comprehensive CSC detection categories will be introduced in this review, as follows:

- Gas emission assessment
- Ground measurements
- Remote sensing

3.1. Gas Emission Assessment

3.1.1. Index Gases

Self-ignition of coal, either in coal stockpiles, seams or gobs, is the result of the oxidation exothermic process. Three stages were identified in the kinetics of the CSC by Dong et al. [85]: slow oxidation, accelerated oxidation and intense oxidation. Different kinds of gas products with different concentrations are released during each phase. The slow oxidation stage is the most complicated process, according to Xueqiu et al. [86], with low to moderate CO and CO_2 emissions at a constant rate until 150 °C, as presented in an experimental study performed by Dong et al. [85]. Between 150 and 180 °C, the emission rate starts to increase. At 180 °C, the concentrations of CO and CO_2 rise rapidly, indicating that the coal oxidation reaction has entered an accelerated stage. Guo et al. [66] reported slightly different values: the CO emission rate started to increase rapidly above 60 °C. This value creates the conditions for the formation of coal–oxygen complexes. After 130 °C, the CO formation rate is even higher. The temperature threshold above which the rate of CO formation increases rapidly can be used as a marker of the rapid oxidation process and so to predict the inception of the spontaneous combustion. The emission of hydrocarbons is another important marker of ongoing oxidation processes. In an experimental study, Guo et al. [66] reported that the CH_4 emission is minimal below 100 °C (for long flame coal) and 80 °C (for anthracite). In fact, the CH_4 emission rate decreases slowly as the temperature increases. This is explained by the fact that the CH_4 emission consists of desorption of the CH_4 adsorbed in the coal pores. At temperature values varying between 100 and 140 °C, the CH_4 emission rate starts to increase, this being caused by the break of the side chains in the coal molecules. The temperature at which the methane production rate reverses indicates the onset of an oxidative process according to Zhao et al. [87]. In the same experimental study on CSC gas emissions conducted by Zhao et al. [87], it was reported that C_2H_4 was not present in coal samples below 100 °C. This is an indication of the fact that C_2H_4 does not exist in the raw coal and it is actually a result of the oxidative processes, according to Zhao et al. [87]. Another hydrocarbon commonly associated with coal oxidative processes is C_2H_6. Guo et al. [66] reported that C_2H_6 can be detected at temperatures as low as 20–30 °C, suggesting that trace amounts are adsorbed by the raw coal. Unlike C_2H_4, which is an unsaturated compound (Guo et al. [66]), C_2H_6 can be deposited in geological formations like coal seams, where it is more stable than C_2H_4, according to Liang et al. [84]. For this reason, the C_2H_6 presence cannot be used to predict the slow oxidation phase.

For any coal rank, the oxidative processes result in the production of CO, CO_2, CH_4, C_2H_4 and C_2H_6. The generation paths for hydrocarbons are complex (Liang et al. [84]) consisting of oxidation of functional groups and decomposition of stable complexes under the effect of the heat. It must be noted though that details of these reactions are not fully understood (Liang et al. [84]). Given the complex nature of gas emission dynamics during oxidative processes leading to coal spontaneous combustion (CSC), selecting an index gas to indicate key milestones in the process chain must consider sensitivity, detection feasibility, and consistency, according to Zhao et al. [87]. A summary of gas kinds used as indicators of CSC has been compiled by Liang et al. [84]. Detection of CSC based on index gases can be categorized into single gas and composite index detection techniques. Single gas techniques detect CO, which is the most important product of coal low-temperature oxidation. Other techniques are based on detection of hydrogen, Cliff et al. [88], or ethylene, Xie et al. [89]. Composite index gas techniques for detecting CSC involve analyzing a combination of gas species emitted during the low-temperature oxidation and combustion processes. These techniques are based on the principle that monitoring the ratios or composite indices of multiple gases enhances detection accuracy by integrating the concentration patterns of several species, rather than relying on a single indicator. This approach allows for more sensitive and reliable detection, as it accounts for variables such as coal rank, temperature, and environmental factors that may influence gas emissions. A summary of the single gas and composite gas index methods is presented in Table 2.

In practical applications, several factors inherent to the mining environment can compromise the performance of CSC detection systems, leading to either false positives (erroneous alarms) or undetected CSC events. Table 3 provides an overview of the common conditions that impact gas emission-based CSC detection systems and the methodologies employed to address these limitations.

Table 2. CSC detection based on gas concentration measurement. Single gas and composite index methods.

Single Gas	Advantages/Specific Features	Problems and Limitations	Used in
CO C_2H_4 C_2H_2 H_2	CO is a very effective indicator of early stages of low temperature oxidation. Monitoring techniques are available that can detect trace amounts of CO. CO has a density lower than the dry air, which stimulates the diffusion into the surrounding gases. C_2H_4 can serve as a useful signature gas for detecting CSC, as its presence is not associated with other known processes.	The absolute concentration of detected CO provides limited information about the fire status, whereas an increasing trend in CO production typically indicates a worsening fire condition. However, trend analysis alone cannot distinguish whether the CO is generated from widespread low-temperature coal oxidation or intense heating in a localized area. Air or other gases can dilute the index gas below the detection limit. In a less severe scenario, dilution can result in an underestimation of the heating state. Other gas sources (thermal engines emissions, seam gases) can contaminate the index gas. The intensity of heating cannot be determined solely based on the concentration of a single gas. H_2 may also be the product of the reaction of galvanized steel and acidic water during the sampling process. C_2H_4 is not an early indicator as it does not occur before 1500 °C.	Wen et al. [51] Qing et al. [90] Hu et al. [61] Ma et al. [91] Yang et al. [92]
Composite index:			
CO Make [93]. Defined as the CO volume flowing past a fixed point per unit time: $\text{COMake} = K \times CO \times Q\left[\frac{\text{liters}}{\text{min}}\right]$ K = 0.06 if CO is measured in ppm K = 600 if CO is measured in % $Q\left[\frac{m^3}{s}\right]$ airflow rate CO monoxide carbon concentration Typical values: >10 liters/min: investigation required >20 liters/min: significant fire danger >30 liters/min: extreme fire danger	Removes the effect of air dilution.	It cannot be used behind seals or in closed boreholes since it requires a continuous airflow. This limits significantly its applicability.	Liu et al. [94]
Graham's Ratio [93]. The percent change in CO concentration to change in O_2 concentration: $GR = \frac{100 \times CO}{0.256 \times N_2 - O_2}$ Typical values: <0.4: Normal 0.4–1.0: Uncertain (investigation required) 1.0–2.0: Heating >2: Intense heating or fire	Graham's Ratio is primarily utilized for detecting heating events or fires that might be obscured by variations in ventilation and for monitoring their progression over time. The trend in the ratio readings is of greater significance than the absolute values. An upward trend indicates a rise in temperature.	May underestimate the state of progression.	Xu et al. [95]
Young's ratio [93]: $YR = \frac{CO_2}{0.256 \times N_2 - O_2}$	No universal trigger levels can be established due to the significant variation in CO_2 generation with temperature across different coal ranks. The trend in the CO_2 ratio is more informative than the absolute values.	Other CO_2 sources—seam gas or thermal engines or CO_2 loss (dissolved in water) may influence the value.	Danish and Onder [96] Singh et al. [97]
CO/CO_2 ratio [93]: <0.02: Normal <0.05: coal temperature <60 °C <0.10: coal temperature <80 °C <0.15: coal temperature <100 °C <0.35: coal temperature <150 °C	Independent of oxygen deficiency. Increases rapidly during the early stages of heating. At high temperatures, the rate of increase reduces significantly.	Only intended as an early warning for heating. Other CO_2 sources from seam gas or vehicle exhaust may influence the value The potential loss of CO_2 as it readily dissolves in water. Can only be employed where no CO_2 emissions from other sources exist.	Yan et al. [98] Lu et al. [99] Wang et al. [100] Li et al. [101]

Table 2. *Cont.*

Single Gas	Advantages/Specific Features	Problems and Limitations	Used in
Morris Ratio [93]: $MR = \frac{N_2 - 3.774 \times O_2}{CO + CO_2}$	Valid in early stages of heating when increasing trend indicates increasing heating activity.	MR increases to a maximum at approximately 120 °C then decreases. The size of the peak varies with the coal rank.	Singh et al. [97]
Jones-Trickett Ratio [93]: $JTR = \frac{CO_2 + 0.75 \times CO - 0.25 \times H_2}{0.265 \times N_2 - O_2}$ Milestone values: <0.4 Normal conditions <0.5 Methane fire indication <1.0 Coal fire possible >1.6 Impossible	Increasing ratio indicates intensifying heating/temperature increase. Milestone values allow objective conclusions. Dilution with fresh air of the combustion products has no effect on the ratio.	Invalid if the intake air is oxygen deficient through the injection of nitrogen or carbon dioxide.	Singh et al. [97]
Litton Ratio [93]: $LR = \frac{1}{3} \times CO \times (\%Rg)^{-1.5} \times (\%O_2)^{-0.5}$ CO Carbon monoxide concentration in ppm $\%O_2$—Oxygen concentration in percent %Rg—residual gas in percent, defined as: $\%Rg = 100 - 4.774 \times O_2 - \text{seamgas}$ >1 Combustion process <1 and stable: Safe conditions <1 and decreasing: Equilibrium conditions not reached	Milestone values allow objective conclusions.	Can only detect the combustion but it is not sensitive enough to identify the initial phase of low-temperature oxidation.	Singh et al. [97]
Willet Ratio [93]: $WR = \frac{CO_2 \text{ produced}}{\text{Blackdamp} + \text{Combustibles}}$ No specific/milestone values defined	A falling trend suggests decreasing activity. Stable values may indicate no activity. More effective than Graham's Ratio in determining the state of spontaneous combustion activity behind sealed areas.		Singh et al. [97]
H_2/CO Ratio [93]	An increasing ratio indicates intensifying heating. Independent of dilution with fresh air, seam gas or oxygen deficiency.	CO can be depleted by bacterial activity. Thermal engine emissions can modify the trend. Rate of change slowed in sealed areas resulting in averaged values instead of instantaneous. Inaccurate for low H_2 values due to detection limitations.	
Ratio between hydrocarbons [93]: C_2H_6/C_2H_4 C_2H_6/CH_4 C_3H_8/CH_4 C_4H_{10}/CH_4 C_3H_8/C_2H_6 C_4H_{10}/C_2H_6		Hydrocarbons are produced at higher temperature and cannot be used to indicate the early stages of low-temperature oxidation.	

Table 3. Common issues in CSC detection via gas concentration measurement and proposed solutions.

Issue	Reference	Proposed Approach and Solution
Influence of the goaf atmosphere composition on the formation of index gases/composite index.	Liu et al. [102] 2021	The low-temperature oxidative process in a lean-oxygen atmosphere caused by the presence of CH_4 delays the formation of CO, CO_2, H_2 and C_2H_6. The temperature milestones for the occurrence of the gaseous components indicating CSC were modified in the presence of methane. For temperature values below 170 °C the composite index CO/CO_2 was not significantly influenced by the O_2 concentration and it can be determined with: $\frac{CO}{CO_2} = -0.675 + 0.333 \times e^{0.0257T}$ For temperature values above the 170 °C threshold, the presence of CH_4 and N_2 in the goaf atmosphere rendered the CO/CO_2 irrelevant for the indication of CSC

Table 3. *Cont.*

Issue	Reference	Proposed Approach and Solution
Ventilation air dilution lowers the index gas concentration (C_2H_4, CO) below the detection limits of the instruments.	Xie et al. [89] 2011	A C_2H_4 enriching system was proposed consisting of a gas flow system, a C_2H_4 enriching and adsorption system, a temperature control system and a gas detection system. It was reported that C_2H_4 detection sensitivity increased by approximately 10 times. Field application of the EES at two coal mines demonstrated that the detection of ethylene was accelerated by 2 to 3 weeks.
Threshold concentration values may be relative and misleading in some environments and conditions.	Liu et al. [94] 2024	Temperature and CO concentration was measured in the air return roadway in two points: close to the working surface and at a distance of 16 m. A relationship between the temperature and CO concentration at the two points was established to determine the statistical patterns between the two variables. Eventually, a relationship was established between the measured CO concentration in the air return roadway and the critical concentration in the goaf.
CSC stages cannot be precisely and timely identified using the standard index gas or composite index.	Ma et al. [64] 2020	A controlled CSC experiment was designed using coal samples from Huainan mining area, China. A new index—the oxidation index—was proposed for predicting coal spontaneous combustion, defined as the ratio of oxygen consumption per unit time to the total oxygen consumption. Threshold values for the oxidation index were identified to indicate the transition between stages.
	Zhang et al. [103] 2021	Six characteristic temperatures and their corresponding threshold values were determined, allowing the coal spontaneous combustion process to be divided into seven stages: the latent stage, recombination stage, self-heating stage, activation stage, thermal decomposition stage, fission stage, and combustion stage.
	Cheng et al. [104] 2022	A critical CO concentration value was defined based on temperature, gob geometry and gob ventilation conditions.
	Yan et al. [98] 2023	Optimized indexes for temperature ranges consisting of ratios of different alkanes were defined.
Mining environment factors alter the gas sensor signal or sensor failures occur.	Chang and Chang [105] 2023	Gas concentration data were converted into recurrence plots [106]. By means of deep learning-based image feature engineering (VGG16, VGG19, and ResNet18), the temporal features of gas concentration time series were converted into image format. This approach enables the representation and comparative analysis of sensor monitoring data from different locations, facilitating the identification of operational status and anomalies.

3.1.2. Radon Gas Emission

Radon exhalation of coal is a complex process, which is has been observed and documented but no analytical model has been proposed to date. The energy recoil generated by radium (the parent nuclide of radon) represents the acting force that causes the migration of radon. The recoil energy causes the radon atoms to come loose from the shackles of the mineral lattice and move freely. Some of the newly freed radon atoms penetrate the pore space of the coal, through which a migration process begins. The migration is favored by the cracks in the coal structure, connected to the surface. The diffusion process is driven by the concentration and pressure gradients.

The radon release has two distinct stages, Lu et al. [107], Zhou et al. [108], Wen et al. [109]:

1. Some radon atoms are absorbed and captured in the internal pores of the coal and others are held captive in the water-filled pores because the transport range of radon in the water is considerably smaller than in air. The exothermic oxidation reaction causes the vaporization of porous water and the radon atoms are released into the external environment.
2. As the temperature increases, the pyrolysis process causes the closed pore of the coal to collapse and connect and in the same time new crack paths to the environment are created. The thermal diffusion and convection cause the radon atoms to migrate through these paths.

Observations show that the radon concertation increases linearly until 50–70 °C. After these values, a sharp increase in the exhalation rate occurs. Between 100 and 120 °C, the radon exhalation rate remains steady. A numerical study modelling the radon exhalation process occurring was carried out by Lu et al. [107] considering the permeability of the goaf (and its dependence on the crushing coefficient), the momentum and energy conservation equation, and species conservation equation, considering the radon decay term. It was reported that CSC in the goaf induces a porous chimney effect, characterized by thermal buoyancy and vortex formation in the high-temperature region. The thermal buoyancy effect arises from the increase in coal temperature, which reduces gas density in the affected area. This density reduction generates a vertical upward buoyant flow. Concurrently, the decreased gas density creates an apparent local negative pressure in the combustion zone, leading to vortex formation around the area due to the suction effect of the negative pressure. The porous chimney effect offers a theoretical framework for understanding radon source formation in the goaf. The distribution of radon in the goaf results from the two concurrent processes, air leakage and the porous chimney effect. During the low-temperature stage of coal, air leakage predominates, causing radon to accumulate near the air return side. As coal temperature rises, the porous chimney effect becomes dominant. The updraft generated by thermal buoyancy rapidly transporting radon to the top of the goaf. The vortex-induced fresh airflow supplies both oxygen and radon to the high-temperature zone of coal spontaneous combustion. Once heated, this airflow re-establishes an upward buoyant flow, further intensifying radon accumulation at the top of the goaf.

Chan et al. [110] conducted a controlled experiment in which long-flame coal samples with particle size ~30 mm and porosity 0.3 were placed in a heating tank under a constant rate airflow of 1 Liter/min. The samples were subject to heating at eight temperature values, from 30 to 350 °C and the radon concentration exhaled from coal was measured after the thermal oxidation of coal. The dependence of radon concentration on temperature followed a pattern consistent with that described by other authors, showing a peak around 150 °C. However, unlike previous experimental studies, a second peak corresponding to coal pyrolysis was not observed. Instead, the radon concentration decreased continuously up to 350 °C. In order to gain further insight, a numerical simulation was conducted on a model presented in Figure 1 (left, the physical system and right, the mesh and the dimensions).

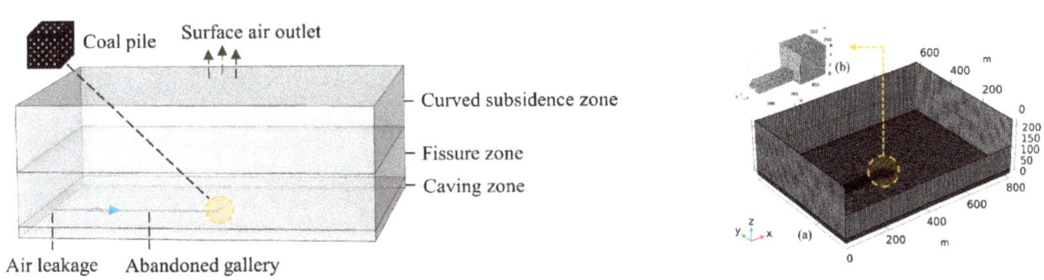

Figure 1. The physical model considered in [110] (**left**) and the mesh with dimensions (**right**): (a) overall view and (b) detail view of the coal pile. Reproduced from [110] with permission from Elsevier.

This study is interesting as it investigates an important characteristic of the geological strata, that is the air circulation inside the stock pile, as shown in Figure 2. A high value of the velocity gradient can be observed in the entrance region of the coal stockpile.

The air flow is one of the key factors that controls the CSC in the coal stockpile. In the entrance region, on the path denoted a in Figure 2-left, the air velocity has the highest values, which intensifies the convective heat transfer. As a result, high temperature regions cannot develop in this region since heat is rapidly removed. The moderate air velocity on the paths denoted b and c and the longer trajectory compared to the path a create the conditions for the coal oxidation and heat accumulation. The radon concentration profile in the coal stockpile is presented in Figure 3. In Figure 3a the radon concentration on the sides of the stockpile is plotted after 15, 35 and 48 days. A steady increase in the radon concentration on the upper side is consistent with the field and controlled experiments reported by other authors. In the early stages of the CSC, radon accumulations occur at the base of the coal stockpile. As the CSC progresses in time, the radon concentration distribution increases in a non-uniform manner due to the effects of airflow and temperature gradients. On the windward side, the temperature is lower, leading to reduced radon emission. Additionally, the higher airflow velocity in this area facilitates the transport of radon to other locations, resulting in a lower radon concentration on the windward side of the coal pile. At the early stage of coal spontaneous combustion, the radon concentration is higher on the leeward side at the base of the coal pile due to reduced gas flow and initial oxidation reactions. However, in the later stages, the radon concentration decreases as the self-heating process in the coal is inhibited. A key conclusion from this study is that the porosity discontinuity at the interface between the caving zone and fissure zone (Figure 1) acts as an obstacle for the radon flow, resulting in a radon accumulation and a continuous increase in the concentration at this interface. This provides mass and pressure gradient to support the long-distance migration of radon, which can pass through geological structures and reach the surface. Thus, the position of the radon anomaly peak is longitudinally consistent with the position of the fire. Moreover, the peak value of the radon concentration is positively correlated to the fire source temperature.

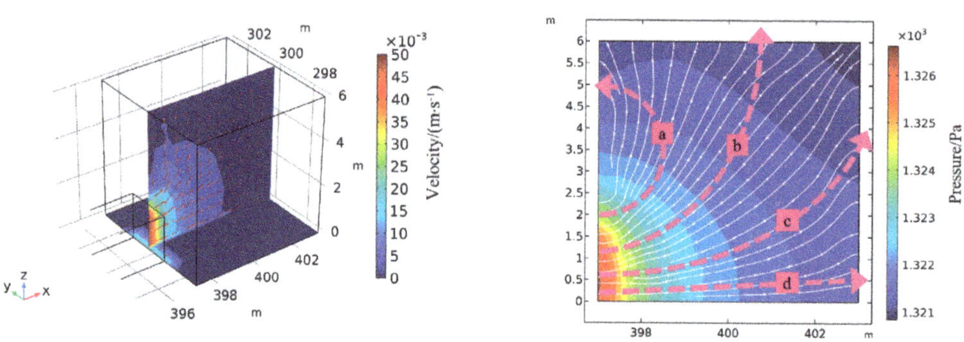

Figure 2. (**left**) Air velocity profile (and streamlines) inside the coal stockpile. (**right**) The pressure distribution and the potential lines inside the stockpile: a—high-velocity, short-length air path; b, c—moderate-velocity, long air paths; d—moderate-velocity, medium-length air path Reproduced from [110] with permission from Elsevier.

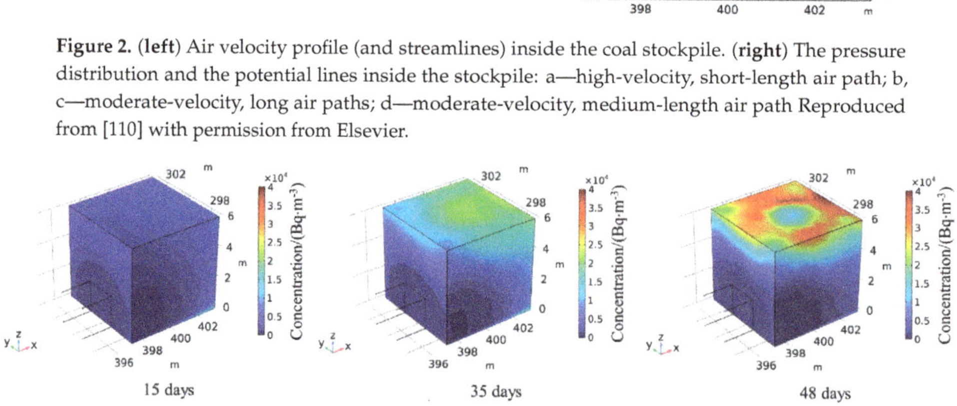

(a)

Figure 3. *Cont.*

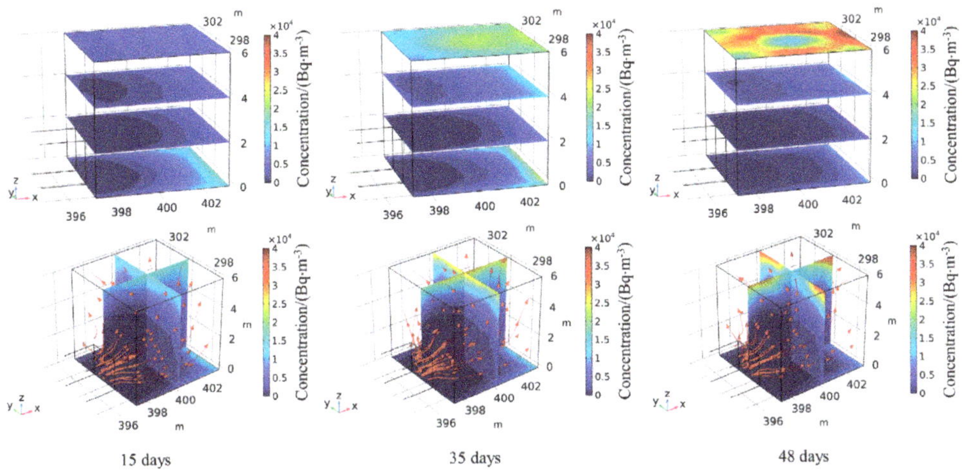

(b)

Figure 3. (**a**) The radon concentration on the faces of the coal stockpile. (**b**) The radon concentration in horizontal and vertical planes. The streamlines are depicted as red three-dimensional curves, with arrows illustrating the direction of velocity. Reproduced from [110] with permission from Elsevier.

A selection of experimental studies relating the CSC to the radon gas emission is presented in Table 4.

Table 4. Laboratory and field studies investigating the radon emission mechanisms during CSC.

Ref.	Experiment	Results
Wen et al. [111] 2020	Controlled experiment: Coal samples from the underground working face CJG (China). Coal particles with maximum size 5–7 mm. Experiments performed in a spontaneous combustion experimental furnace.	The experimental data were modeled by fitting an exponential function to the observed values. A robust correlation was identified between radon concentration and temperature within the thermal range of 50–350 °C, as evidenced by a coefficient of determination (R^2) of 0.94, indicating a strong goodness of fit. This suggests that radon emission is highly sensitive to temperature variations within this interval. However, at temperatures exceeding 350 °C, a pronounced increase in data variability was observed, likely due to complex underlying physical processes not accounted for by the exponential model. This scatter indicates a potential departure from the initial correlation, suggesting that the mechanisms governing radon release may change significantly at higher temperatures.
Du et al. [62] 2021	Field measurements: A sampling system consisting of a negative pressure air sampling pump and a sampling pipe inserted in the soil at approximately 1 m. Surface temperature measurement conducted by IR imaging on a 1600 × 450 m area known for underground coal fires. Boreholes were drilled in high surface temperature areas to measure temperature and gas concentration.	No match was observed between IR imaging high temperature regions and radon local concentration. This was attributed to factors such as • The hot gas flow generated during the CSC causes the heating of the geological structures along the air leakage path. Different physical characteristics of the geological strata cause different temperature levels. • The detection depth of radon is higher than the IR imaging. • The radon migration speed in soil is different depending on the soil permeability, being much higher in fissures and cracks than solid rock.

Table 4. *Cont.*

Ref.	Experiment	Results
Zhou et al. [108] 2021	Controlled experiment: Four coal ranks (with the corresponding Radium nuclide content in Bq/kg) were investigated: lignite (54.2), long-flame coal (44.4), coking coal (31.2) and lean coal (37.7). 1. The sealed tank was heated to 30, 50, 100, 150, 200, 250 and 300 °C. An air pump was simultaneously connected to the inlet gas pipe to oxidize the samples in the sealed tank for 48 h. 2. When the coal samples oxidation was completed at a constant temperature, the inlet and outlet gas pipes, dryer, and Rad7 radon detector were connected in a closed loop to measure the radon concentration in the tank. A total of two-hour 24 cycles was set for radon measurement at each temperature. A low-temperature nitrogen adsorption method was used to measure the specific surface area of the coal samples at different temperatures.	Radon exhalation of the coal increases rapidly with increasing oxidation temperature. The temperature corresponding to the peak radon concentration depends on the coal rank. The two-stage coal exhalation mechanism was confirmed by the radon concentration–temperature curve. Two characteristic vertices were identified on this curve: (a) The absolute maximum value, corresponding to the water evaporation temperature at approximately 100 °C. (b) The coal pyrolysis temperature, corresponding to the second stage of the exhalation process, occurring at approximately 250 °C. The specific surface area (m^2/g) first decreases with the temperature reaching a minimum value at roughly 200 °C and then it increases abruptly after 250 °C as the coal undergoes the pyrolysis process. The downward trend in specific surface area up to 200 °C suggests that it is not the primary factor contributing to the increase in radon concentration.
Zhou et al. [68] 2018	Field measurements conducted on the gob of a small abandoned coal mine in the region Bao Shan Yao Zhai (China). Radon concentration in the soil was determined by means of an alpha-cup emanometer (Ding et al. [112]). The procedure consisted of burying the frustum-shaped alpha cup for 4 h at 30–40 cm. Two measurement fields were defined: field I with an area of 396,000 m^2 and field II with an area of 89,600 m^2. The basic point distance was 20 × 20 m with a total of 1530 measurement points: 1050 in field I and 480 in field II.	A 2D color map of the radon concentration was created, showing the coordinates of the high concentration areas. Drilling verification was performed in the radon high value zone of area A confirming an abnormal temperature area.
Hu et al. [113] 2023	Controlled experiment with long-flame coal disk-shaped (Φ50 × 20 mm) samples collected from Yunlin region (China) heated at 5 °C/min up to the set point and then maintained for 1h. Two sets of experiments were conducted, aerobic and anaerobic (with samples wrapped and sealed to prevent oxygen exposure). The radon emission concentration after cooling down to the ambient temperature was measured.	Temperatures values of 300 °C and 500 °C are pivotal in altering the properties the investigated coal. At these values, there is a significant increase in the mass loss ratio, specific surface area, pore volume, and fracture ratio. The expansion and propagation of pores and fractures, along with the release of substantial amounts of pyrolysis gases, result in a massive desorption and dispersion of radon. The radon release rate from pyrolysis products is inversely correlated with specific surface area, pore volume, and fracture ratio. The development of fissures and the reduction in radon emission rates are more pronounced in the aerobic environments. Elevated temperatures promote the transformation of dissolved, adsorbed, and trapped radon into free radon within the coal, thereby accelerating its migration rate. This results in a decrease in the residual radon content in pyrolysis products and a marked reduction in the radon exhalation rate. There is a negative correlation between the heating temperature and the radon emission rate.

3.1.3. Gas Emission Assessment—Insights and Unexplored Areas

The gas analysis with the purpose of detecting the CSC in various phases is a feasible approach considering the relative simplicity and cost effectiveness of the technique. The gas sensor systems commercially available are precise enough and the measurement chains are in general fast and accurate. Rugged, fault-tolerant gas measurement systems were designed for the harsh mining environment. These systems are in general more expensive and they still require periodic calibration and verification, as they are critical sensor applications. For such applications, parameters such as precision (the number of true positives divided by the total number of positive predictions) and recall (the number of true positives flagged by the system divided by the total number of fire incidents) are of extreme importance. No relevant studies on these important issues were identified, although there are many incident reports where gas emission-based CSC detection techniques failed to detect the fire [93].

The usual gas monitoring systems are discussed comparatively from the advantages/disadvantages point of view in Table 5.

Table 5. Gas monitoring techniques currently integrated in CSC detection systems.

Gas Monitoring Technique	Advantages	Disadvantages
Gas analyzers located at the surface with a tube bundle system. Consist of PET tubes extended from the surface to selected locations underground. The tubes are general high-grade quality non permeable materials with a variable diameter ranging from 6 mm to 20 mm (depending on the length) and lengths of up to several km. Negative pressure pumps located on the surface ensure the gas circulation.	With integration of flame traps, no explosion-proof devices are required. Most of the primary components are positioned on the surface, which simplifies maintenance. Calibration of the analyzers can be conducted on the surface. No power required for underground components. Analyzers available for a wide range of gases.	Leakage and infiltrations in the tube system cannot be easily identified. Results are not in real time. Moisture removal systems failure results in formation of liquid droplets, which hinder the gas circulation or cause erroneous readings. Tubes may be easily damaged by fire, explosion or earth works.
Real-time telemetry systems. Consist of fixed sensors are generally installed where real-time data are required. Sensors suitable for this type of systems: catalytic combustion (CH_4), electrochemical (CO and O_2), and IR detectors (CO and CH_4).	Real-time indication Since gas sensors generate an electrical signal, they can be positioned at significant distances from the analyzer. A sensor failure cannot go unnoticed.	High maintenance. Limited sensor life. Not suitable for oxygen deficient atmosphere. Some sensors exhibit cross sensitivity. Catalytic sensors may undergo poisoning.
Periodic inspection using portable devices.	Diversity and flexibility in selection the sampling location. Detection accuracy depends on the accuracy of the analyzer device.	Not possible to use behind sealed boreholes or closed seams. Continuous monitoring not possible. Personnel cost.
Sampling using gas bags subsequently analyzed by a gas analysis service provider.	High precision. A wide range of gases can be detected.	Results are not in real time. Expensive. Not possible to use behind sealed boreholes or closed seams. Continuous monitoring not possible. Personnel cost.
Gas chromatographic systems. Ultra-fast micro gas chromatographs.	A wide range of gases can be detected. Fast analysis (in minutes). Simple to operate. The ability to detect and analyze key components of spontaneous combustion, such as H_2, CO, ethylene, ethane, and propylene, at concentrations ranging from parts per million (ppm) to percent levels. This requires a single type of detector, specifically a thermal conductivity detector for analyzing the mine atmosphere.	Expensive. High maintenance.

Accurate and timely detection of CSC in its early stages using index gas methods presents a complex challenge that extends beyond sensor accuracy. A significant influence is exerted by the specific conditions and characteristics of the mining facility (such as coal gob or coal seam) and the properties of the coal itself. Modifications in the local atmosphere (e.g., CO and CO_2 emissions from thermal engines, CO_2 absorption in water, CO consumption due to bacterial activity) and transient environmental conditions can cause interferences, leading to deviations in sensor measurements from actual conditions. While CO can serve as a reliable indicator of CSC initiation under controlled laboratory conditions, in real-world industrial settings, numerous and unpredictable interferences render CO-based CSC detection systems unreliable. To address this issue, the concept of composite index gases was introduced, as outlined in Table 2. However, composite indices may also be subject to similar limitations as single gas indices due to their dependence on gas sensors. Although composite indices may offer greater robustness in detecting CSC under certain conditions compared to single gas indices, their effectiveness remains influenced by the coal rank and the specific conditions of the mining environment, Gbadamosi et al. [114].

Radon gas detection methods rely on the exhalation of radon driven by temperature increases. Due to the strong correlation between radon exhalation rates and temperature, these methods exhibit robustness and are relatively less affected by specific mining factors. However, the coal rank significantly influences both the exhalation rate profile and the observed local maxima. A critical challenge for radon-based methods lies in the radon migration from subsurface to surface levels. The local geological morphology, characteristics of geological formations, and the presence of fractures and fissures can lead to substantial deviations between the surface detection points of peak radon concentration and the actual underground locations of CSC development. Another notable limitation of radon-based methods, in comparison to index gas methods, is their inability to be integrated into real-time measurement systems. Consequently, these methods do not provide instantaneous detection.

Most of the experimental studies identified and included in this review are conducted in controlled laboratory conditions, which might limit the generality of the conclusions.

3.2. *Electromagnetic, Acoustic and Optic Techniques*

3.2.1. Electromagnetic Effects

Coal heating resulting from low-temperature oxidation and combustion induces modifications in the surrounding magnetic and electric fields (de Boer et al. [115]). The generation of electromagnetic radiation (EMR) signals in response to coal heating is a well-established phenomenon (Kong et al. [116], 2018), though the underlying mechanisms are complex and not yet fully elucidated. The sources of EMR charges during coal deformation and fracturing include (Wang et al. [117]): (1) transients of electric dipoles (electric dipole layers) induced by applied stress, (2) variable motion of separated charges due to crack propagation and frictional interactions, (3) energy dissipation associated with oscillations in RC circuits within fractured coal walls, and (4) relaxation of separated charges coupled with electromagnetic radiation resulting from high-velocity particle collisions with wall cracks.

The volumetric thermal expansion induced by heating leads to deformation and detachment of the morphological structures within the coal matrix. As temperature rises, the thermal swelling stress within the coal increases proportionally, thereby enhancing the likelihood of coal deformation and fracturing. Upon reaching a critical temperature, the tensile strain experienced by the particles within the coal exceeds their ultimate tensile strength, resulting in the disruption of mineral particle interfaces. This process generates microcracks and facilitates the propagation of slip within the coal structure. (Kong et al. [118], 2016). The non-uniform thermal expansion and deformation of particles within the coal matrix induces the migration of free charges from regions of high concentration to regions of lower concentration. This differential expansion creates various stress and density zones within the coal. Consequently, charge clusters with varying densities accumulate in these distinct stress and density regions, leading to the formation of numerous dipoles, which subsequently emit electromagnetic waves (Kong et al. [119], 2017). During the processes of coal deformation and failure, non-uniform strain induced by charge transfer, piezoelectric effects, frictional interactions, chemical bond rupture, and dislocation slip under thermal stress can result in charge separation. When the internal stress in the coal exceeds the critical stress threshold for crack initiation, cracks develop and propagate. This process generates free electrons that were previously bound at the crack tips, converting them into free electrons. The movement of these free electrons generates EMR signals due to their displacement and interaction with the surrounding field (Kong et al. [120], 2018).

Electric properties of coal play a pivotal role in the characteristics of the EMR signal generated during the coal heating. Zhu et al. [121] investigated in a laboratory-controlled study the dielectric spectrum of coal during oxidation, reporting that the dielectric constant initially decreased up to 150 °C due to reduced dipole polarizability from moisture release, then it increased between 150 °C and 275 °C. Up to 225 °C, the increase was attributed to the formation of oxygen-containing bonds outpacing their breakage, enhancing dipole polarization. Beyond 225 °C, changes in oxygen bond quantity and molecular chain deformation were the main causes of the increase. The dielectric constant then dropped sharply until 400 °C which was attributed to the rupture of polar bonds and volatile release, before increasing again beyond 400 °C, likely from enhanced electron migration.

The variation in the coal electrical resistivity with the temperature was investigated in a laboratory-controlled experiment by Zhu et al. [122]. Coal samples from five different mining regions were considered. Coal from two of these regions was known for having a high spontaneous combustion tendency while coal from one region was known to be stable to low-temperature oxidation. An impedance analyzer was employed to determine the real and imaginary parts of the impedance. Under an alternating current ranging from 20 Hz to several MHz, the volume resistivity of coal samples exhibited a consistent pattern with frequency: it decreased rapidly at mid to low frequencies and more gradually at mid to high frequencies. At a test frequency of 100 kHz, the resistivity initially increased with temperature and then decreased. At the same frequency, both the real and imaginary components of the coal's complex relative permittivity decreased with increasing temperature, with a progressively lower slope. Due to the segmented variation in the resistivity-temperature relationship, the same resistivity value can correspond to multiple temperatures. As a result, resistivity alone cannot accurately determine the coal's temperature, and the imaginary part of the complex permittivity shows a similar limitation. Therefore, when using resistivity to infer the temperature in a fire zone, the geological conditions and other characteristic parameters must be considered for accurate interpretation.

In a laboratory-controlled experiment, Kong et al. [116] (2018) measured the variations in the EMR field produced during the low-temperature oxidation and combustion of coal. During the spontaneous combustion of coal, distinct electromagnetic radiation (EMR) signals are observed, with their intensity increasing in direct correlation with rising temperatures. The EMR signals measured at frequencies of 100 kHz and 1 MHz demonstrate nearly identical variation patterns, indicating a consistent response across different frequency ranges. Quantitatively, the EMR signals exhibit a positive correlation with temperature, characterized by a correlation coefficient greater than 0.73, suggesting a strong linear relationship. Moreover, the EMR signal intensity showed a substantial correlation with the evolution of CO volume, reflecting a parallel change trend that underscores the potential of EMR monitoring in tracking combustion progression and gas release dynamics. An experimental field study conducted by Kong et al. [72] (2019) attempted to correlate the EMR spectrum with the CSC in a section of a mining area where IR imaging revealed several hot-spots (~55 °C). The underground temperature measurements showed an ongoing spontaneous combustion process, with temperatures exceeding 300 °C. The correlation between the heatmap and the EMR spectrum is presented in Figure 4 using the data obtained by applying the Kriging interpolation method to compute the variogram of the EMR data collected from 20 measuring points across 5 measurement lines within the high-temperature region.

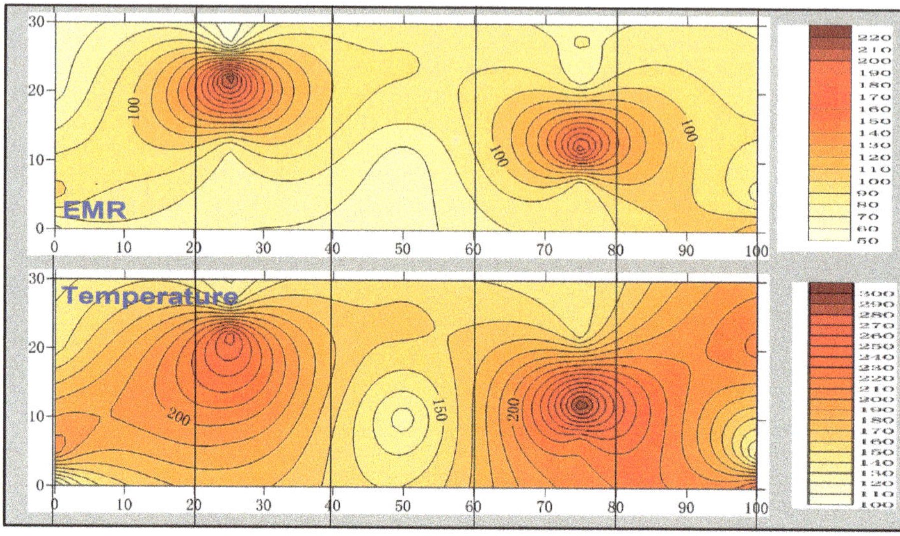

Figure 4. EMR spectrum and surface heat map. Reproduced from [72] with permission from Elsevier.

3.2.2. Electric Properties

Karaoulis et al. [123] proposed a combined method consisting of simultaneous inversion of self-potential and resistivity data, employing joint attribute analysis to delineate and localize the burning front of a coal seam fire. The experimental study was applied to a subbituminous coal formation fire at depth value ranging from 10 to 15 m. The self-potential survey systems comprised 160 measurement stations distributed across 5 profiles, with 32 stations per profile, positioned at the ground surface. The observed self-potential anomalies ranged from +70 m to −50 mV relative to a reference point located distal to the burning front. The resistivity survey included 5 profiles, each with 118 measurements, totaling 590 resistivity readings. The burning front was characterized by extremely low electrical resistivity values (<15Ω • m) and a source current density likely attributable to thermoelectric effects. The integrated inversion and interpretation of self-potential and resistivity data distinctly delineated the position of the coal seam fire. This finding was further validated by the detection of a thermal anomaly at a depth of 30 cm and observation of sporadic hot steam, suggesting a minimum depth for the thermal source of approximately 9 m. The high-resolution earth resistivity method, initially developed to detect underground cavities (bunkers, tunnels, etc.) was applied to detect underground burning areas by Shi and Wu [124]. The method consists of placing the potential electrode at one end of the survey line. A current is injected into the ground through a series of current electrodes, and the corresponding potential differences are measured at the initial position of the potential electrode. The potential electrode is then moved to the next measurement point along the survey line, and the process is repeated: current is supplied sequentially through each current electrode, and the resulting potential differences are recorded. This procedure continues until the potential electrode reaches the opposite end of the survey line. By following this sequence, a complete two-dimensional profile of electrical potential differences along the survey line is obtained. Uneven electrical properties of the Earth's surface may cause serious errors for this method. The second method described in Shi and Wu [124], which overcomes the main drawback of the high-resolution earth resistivity method, is the transient electromagnetic method. An ungrounded loop or grounded line source is used to send a primary electromagnetic pulse of a specific waveform into the ground. Between these pulses, a coil or grounded electrode is used to measure the secondary magnetic field (caused by eddy currents) or the space-time distribution of the induced electric field. This method provides information about the electrical properties of underground geological structures. Compared to the frequency-domain method, the transient electromagnetic method offers higher detection accuracy, as measurements are taken from the same location, reducing issues related to recording points. Additionally, the use of a loop device minimizes the effects of static shifts caused by surface electrical property variations and is less affected by terrain changes. Therefore, this method is particularly suitable for detecting burned areas and high-temperature centers in combustion zones, making it a critical component of the comprehensive electromagnetic approach discussed in this study. The results of the two methods are presented in Figure 5 (left, high-resolution earth resistivity and right, transient electromagnetic method).

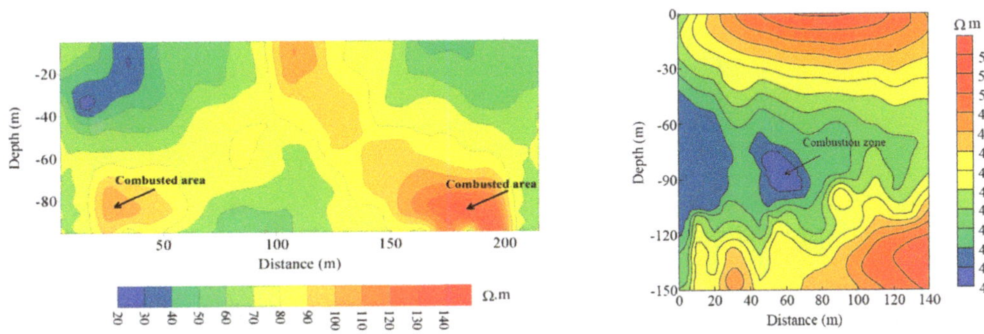

Figure 5. Apparent resistivity profile determined through the high– resolution earth resistivity method (**left**) and transient electromagnetic method (**right**). Reproduced from [124] with permission from Elsevier.

The modification of the acoustic properties of solids with temperature is another method that allow for identification of areas where CSC exists. Two methods that rely on temperature-induced modification of the acoustic properties exist: (i) methods that are based on the change in speed of sound and (ii) methods that determine the change in frequency of the acoustic wave. Acoustic velocity thermometry is a commonly employed technique for measuring temperature. This method involves determining the propagation velocity of an acoustic wave by measuring its travel time over a known distance within the medium of interest. The temperature of the medium is then inferred from the established relationship between acoustic wave velocity and temperature. The sound source signals used in acoustic thermometry typically encompass both low- and medium-frequency acoustic waves, as well as ultrasound. Ren et al. [125] conducted acoustic attenuation tests in an experimental study in a laboratory-controlled environment with four types of coal (lean coal, coking coal, non-stick coal and long flame coal) prepared into seven particle size distribution ranges (>10 mm, 7–10 mm, 5–7 mm, 3–5 mm, 0.9–3 mm and <0.9 mm). An experimental apparatus consisting of an impedance tube, acoustic driver (loudspeaker), microphone array, and data acquisition system was employed to quantify the attenuation coefficient over a specified range of frequencies and temperatures. For all coal samples and temperature conditions, the attenuation coefficient demonstrated a non-linear, oscillatory behavior with local maxima superimposed upon a generally increasing trend. A modest increase in the attenuation coefficient was observed with rising temperatures. Attenuation peaks were detected at frequencies of 400, 700, 1100, and 1600 Hz, which can be attributed to mechanisms of acoustic absorption and scattering resulting from the granular and porous nature of the coal. Deng et al. [126] used a similar experimental setup to investigate the acoustic attenuation characteristics of anthracite, using coal samples collected from Si he coal mine (China). Pseudo-random, pulse, and swept-frequency acoustic signals were utilized in this study to evaluate the attenuation characteristics of anthracite coal. The findings indicated that the pseudo-random signal exhibited the highest attenuation, whereas the pulse signal demonstrated the greatest sensitivity to temperature variations.

When estimating the temperature of anthracite coal using an optimized acoustic signal composed of a linear frequency sweep between 600 and 1000 Hz with a 0.1 s duration, the maximum absolute temperature error was determined to be 4.22%. In a laboratory-controlled study, Kong et al. [127] (2024) built a test system for exploring the sound wave kinematics parameters in the process of CSC. The elastic wave computed tomography (CT) inversion algorithm was employed to reconstruct the acoustic wave velocity field, which was subsequently used to infer the temperature distribution within the coal sample. Results indicated that regions of elevated sound wave velocity corresponded to areas exhibiting temperature anomalies, particularly in zones with higher coal temperatures. The accuracy of the reconstructed sound velocity field was validated through correlation with the results obtained from single-path acoustic velocity measurements. This study also proposes a hypothesis to account for the variability of sound velocity in regions affected by CSC. Scanning electron microscopy (SEM) analysis revealed that as the temperature rises, the extent of damage and fracturing of coal particles increases, resulting in the expansion of inter-particle gaps. As sound waves propagate through these gaps, the elevated air temperature within them contributes to an increase in sound velocity. Consequently, the velocity of sound waves rises in tandem with the heating of loose coal.

Liu et al. [128] developed a laboratory-scale experimental setup to investigate infrasound signals generated during the CSC process. Their study aimed to establish the frequency domain characteristics and the correlation between infrasound signals and CSC temperature. The analysis focused on the frequency domain and energy characteristics of infrasound waves across different stages of the CSC process. Two key conclusions were derived from this study: (i) The time-domain analysis revealed that the infrasound waves generated during CSC can be classified into three distinct stages: a stable stage, a development stage, and a rapid increase stage. The infrasound signals exhibit intermittent characteristics, with the primary frequency amplitude of these waves showing a strong correlation with temperature, evidenced by a correlation coefficient exceeding a certain threshold. (ii) The energy of infrasound waves is predominantly concentrated in the low-frequency range of 0–0.12 Hz during the process of CSC. The distribution of infrasound energy across different frequency bands varies significantly at different stages of CSC. As temperature increases, the proportion of energy within the 0.04–0.08 Hz frequency range progressively rises, while the energy proportion in the 0–0.04 Hz range diminishes. The observed increase in infrasound energy within the 0.04–0.08 Hz band serves as a critical precursor to the onset of CSC.

A comprehensive review on the application of complex acoustic wave techniques was conducted by Guo et al. [66]. All studies included and discussed in this review were conducted in laboratory-controlled environments. The majority of the studies converge in several key findings: (i) Combustion-generated sounds and externally applied acoustic signals can interact to create a dual-source composite acoustic wave. However, the effects of superposition and mutual interference between these source waves on the waveform and characteristics of the composite wave remain poorly understood. The application of dual-source acoustic thermometry is constrained by challenges in separating the composite wave spectrum and comprehending the influences of coal properties and gas composition on these interactions. (ii) The acoustic emissions generated by the spontaneous combustion of loose coal in mining areas are significantly influenced by several factors, including the degree of coal metamorphism, coal temperature, void characteristics, and surrounding gas conditions. Developing a relationship model that correlates these factors with acoustic signals will likely be a key advancement in the field of acoustic temperature measurement of coal.

3.2.3. CSC Detection Based on Magnetic Effects

Magnetometry surveys are effective in detecting underground anomalies induced by coal spontaneous combustion (CSC) due to changes in magnetic susceptibility resulting from thermal alteration of the medium. In a magnetometry survey conducted by Ide et al. [129] in the San Juan Basin, USA, the magnetic susceptibility of thermally altered overburden samples at the NCF was measured using a kappa bridge system. The study revealed that these altered samples exhibited magnetic susceptibilities significantly higher than the natural, unaltered values, by one to two orders of magnitude. Diurnal variations in the Earth's magnetic field and anomalies caused by metal objects were mitigated through filtering techniques. The results demonstrated high repeatability and enabled monitoring of the advancement of the combustion front. However, a notable limitation of this method is that geological structures affected by subsurface fires often contain substantial concentrations of magnetite, which can influence the accuracy of the measurements. Shao et al. [130] conducted a comprehensive field study on the Heshituoluogai coal fire in China, which encompasses 12 sub-fire areas across 57 km^2, totaling 1,444,220 m^2. Magnetic and self-potential anomaly data were collected using a magnetometer and an intensification-polarization instrument. Laboratory experiments revealed that rocks overlying a subsurface fire, when heated beyond the Curie temperature, exhibit significant changes in magnetic properties, such as magnetic susceptibility and thermoremanent magnetization. Data processing techniques, including diurnal fluctuation rectification, reduction to pole, and upward continuation, facilitated clearer interpretation of the results. The fire locations identified through magnetic and self-potential methods aligned with ground survey findings, demonstrating the effectiveness of these techniques for detecting coal fires. Additionally, the study found that the redox and Thomson potentials, resulting from coal combustion at high temperatures, induce self-potential anomalies: positive anomalies indicate shallow fires, while negative anomalies suggest deeper fires (beyond approximately 30 m).

3.2.4. Summary and Future Perspectives

Electromagnetic/electric, magnetic, and acoustic measurement techniques offer several advantages over gas-based detection methods. These techniques provide extensive spatial coverage beyond the immediate vicinity of the sensor, involve relatively low installation and maintenance costs, and are adaptable for detecting fires in underground coal seams, stockpiles, and coal silos. Despite a significant body of research on electromagnetic/electric-based detection of spontaneous coal combustion (CSC), several limitations affect their applicability. Challenges include the complexity of electrical phenomena, the variability in the electromagnetic properties of coal and geological structures, and the inconsistency of site conditions. While some studies (e.g., Kong et al. [127]) have successfully characterized subsurface fires using electromagnetic effects, further research is needed to generalize these findings across different sites, coal ranks, and geological morphologies. The mechanical–electrical and thermoelectrical coupling models depend on various mechanical, electrical, and thermophysical parameters that differ to a significant extent among coal ranks and geological formations.

Magnetometry has two principal limitations: (i) magnetic anomalies detectable by this method occur only at high temperatures, indicating that CSC is already fully developed, and (ii) the geological structures above the coal seams must contain a sufficient amount of magnetite. Furthermore, magnetic anomalies are highly sensitive to the magnetic concentration, which varies from site to site.

Acoustic techniques also present potential for CSC detection, with advanced systems capable of accurately locating the combustion front. However, these technologies are still under development and face significant challenges. Issues include the complex and heterogeneous nature of the propagation medium—comprising coal matrices with varying morphologies and gas- or air-filled voids—which affects acoustic wave propagation in ways that are not yet fully understood or consistently reproducible. Additionally, the technology has not yet achieved the maturity required for standardization and widespread use.

4. Surface and Combined Techniques

Surface techniques are mainly based on the temperature measurements, in general, by IR thermography. Surface techniques have several major advantages in comparison to underground techniques: (i) no physical contact is required; (ii) real-time heatmaps can be produced; (iii) with little effort, large surface areas can be monitored. However, it has to be noted that the surface heatmap does not always offer an accurate image of the CSC progression developing underground, and a set of hard-to-control factors can interfere with the quantitative measurements of surface techniques: subsurface geological structures morphology and dimensions; environmental factors (ambient temperature and humidity, wind speed and direction, solar radiation intensity), and radiative properties of the surface (emissivity and albedo).

Field studies attempting to characterize the CSC by means of IR thermography were reported by Pandey et al. [131], Misz-Kennan and Tabor [132], and Roy et al. [133]. Hu et al. [71] conducted an empirical investigation at a decommissioned coal waste dump in Changping, China. The experimental site was configured as a conical heap with a base diameter of 2.5 m and a height of 1.8 m, totaling approximately 2767 kg in mass. The coal waste comprised subbituminous coal, sandstone, clay, quartzite, sulfide mineralization, and various solid residues. Field surveys identified potential high-temperature zones on the surface of the dump and noted a significant absence of vegetation. The methodology developed in this study encompassed four key phases: field investigation, data pre-processing, data integration, and 3D visualization. The resulting 3D temperature distribution model enabled the classification of observed zones into three distinct categories based on varying temperature levels. Gao et al. [67] employed a ground penetrating radar operating at 900 MHz in a laboratory-controlled experiment to create underground images, which were further processed by means of deep learning to identify the position and size of an underground burning area. Based on the similarity theory, a scale model was built for a coal fire developing in a coal seam with the thickness of 4.3 m at the depth of 26.9 m.

The evolution ofCSC and the corresponding ground-penetrating radar (GPR) images are depicted in Figure 6. Due to the significant difference in dielectric constants between air and rock strata, where air exhibits a considerably lower dielectric constant, the incident electromagnetic waves generate a strong reflection at the air–rock boundary. This phenomenon is characterized by a distinct hyperbolic reflection pattern in the radar images. In contrast, coal has a dielectric constant slightly lower than that of the surrounding rock, and the medium demonstrates a relatively uniform composition. This results in a clear, linear reflection interface between the coal and the adjacent rock strata. During the early stages of combustion, the amount of ash produced is minimal, resulting in a continuous signal between the combustion cavity and the coal seam in the radar image, as illustrated in Figure 6a. By comparing the radar images obtained during testing with those produced by GPR, it becomes possible to identify and interpret waveform signals indicative of coal fires. Key spatial features associated with areas of coal combustion include the combustion cavity, the active combustion front, and the subsurface collapse zones—all of which serve as primary detection targets for GPR surveys. The combustion cavity is identified by a hyperbolic reflection response accompanied by fractures that form between the hyperbolic waveforms and linear interfaces. Meanwhile, the subsurface collapse surface due to combustion is represented by a continuous, curved waveform, characterized by irregular reflection signals emanating from below the collapsed zone.

Wang et al. (2020) [134] introduced an integrated approach for delineating coal fire zones, utilizing a multi-index data fusion method that combines temperature, gas, and radon concentration measurements. The methodology incorporates non-dimensional normalization to standardize different data types and applies weight analysis using the Analytic Hierarchy Process (AHP) to assess the relative importance of each parameter. The approach employs radon gas measurements, borehole temperature data, and gas concentration measurements to create a comprehensive index of fire zone delineation (IFZD). This IFZD, derived from the fusion of multi-dimensional data, provides a robust framework for accurately identifying and characterizing the extent of underground coal fire zones. Using fractal theory, the threshold for anomalies in the Index of Fire Zone Delineation (IFZD) was established through a piecewise linear regression in double-logarithmic coordinates. This process resulted in the development of a comprehensive multi-dimensional data fusion method for detecting underground fires, which was subsequently applied to a shallow coal seam fire in a typical integrated mining site. The analysis revealed that the weight coefficients assigned to various detection indices, namely carbon monoxide (CO), carbon dioxide (CO_2), sulfur dioxide (SO_2), temperature, and radon, were 0.12, 0.03, 0.05, 0.12, and 0.68, respectively. The lower boundary for IFZD anomalies was identified as 0.29. Utilizing the IFZD values and the corresponding anomaly thresholds, a detailed composite contour map was created, along with a plan view and a three-dimensional model of the anomalous zones. This mapping identified eight distinct areas, covering a total surface area of 27,061 square meters, with the anomalous regions extending from the vicinity of a decommissioned shaft and the main well toward the southwest and northeast. Notably, the densest contours and the highest values were found around the locations of the abandoned air shaft and main shaft, indicating severe coal oxidation in these areas. These findings underscore the need for intensified monitoring, particularly focusing on the inspection and sealing of ground fractures.

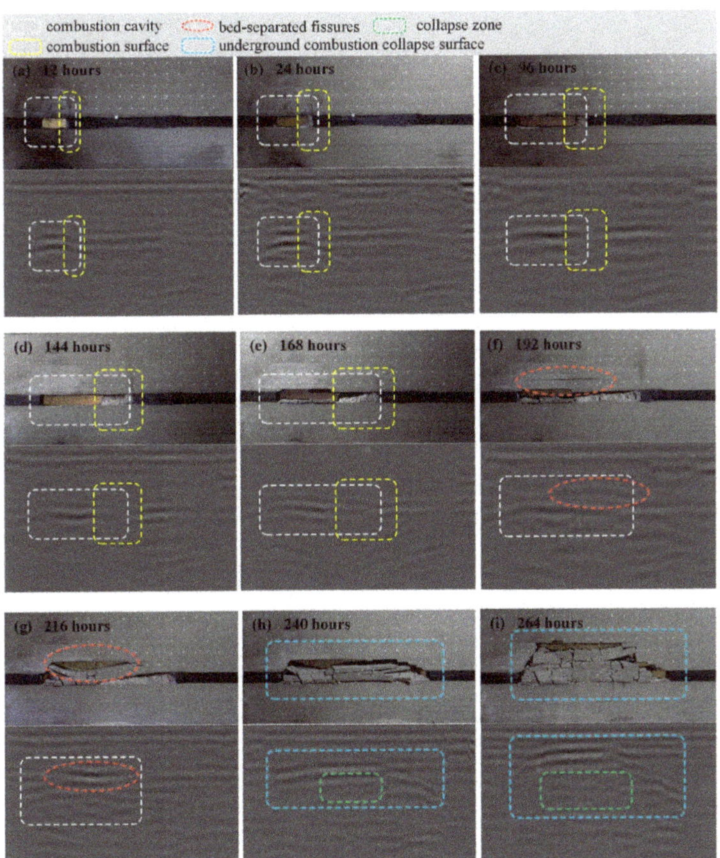

Figure 6. The advance of the simulated CSC and the corresponding radar images: (**a**) the initial stages revealed the combustion cavity in the form of a hyperbola in the radar image; (**b–d**) misalignment and fractures between the hyperbolic reflection and the linear interface occur due to ash formation; (**e**,**f**) thermal stress and gravity caused the roof collapse; (**g**) fissures in the adjacent strata; (**h**,**i**) collapse of the adjacent strata. Reproduced from [67] with permission from Elsevier.

A novel data-driven approach was proposed by Wang et al. (2024) [135] and applied to a coalfield in Xianjiang (China) in four characteristic regions, as depicted in Figure 7. By utilizing the atypical properties of shallow soil and surface strata as diagnostic indicators, a comprehensive evaluation framework is introduced to detect the incipient stages of combustion in coal fire areas through a data-centric methodology. A hierarchical assessment model is established for the early detection of combustion, integrating both terrestrial and atmospheric parameters. The combustion intensity at each survey point is determined by leveraging survey datasets to assign weighting factors to the assessment indicators. Geographic Information System (GIS) technology is applied to delineate the spatial distribution and boundary extents of varying combustion intensities within the surveyed region. The proposed assessment framework is validated using radon concentration measurements and infrared thermographic imaging techniques.

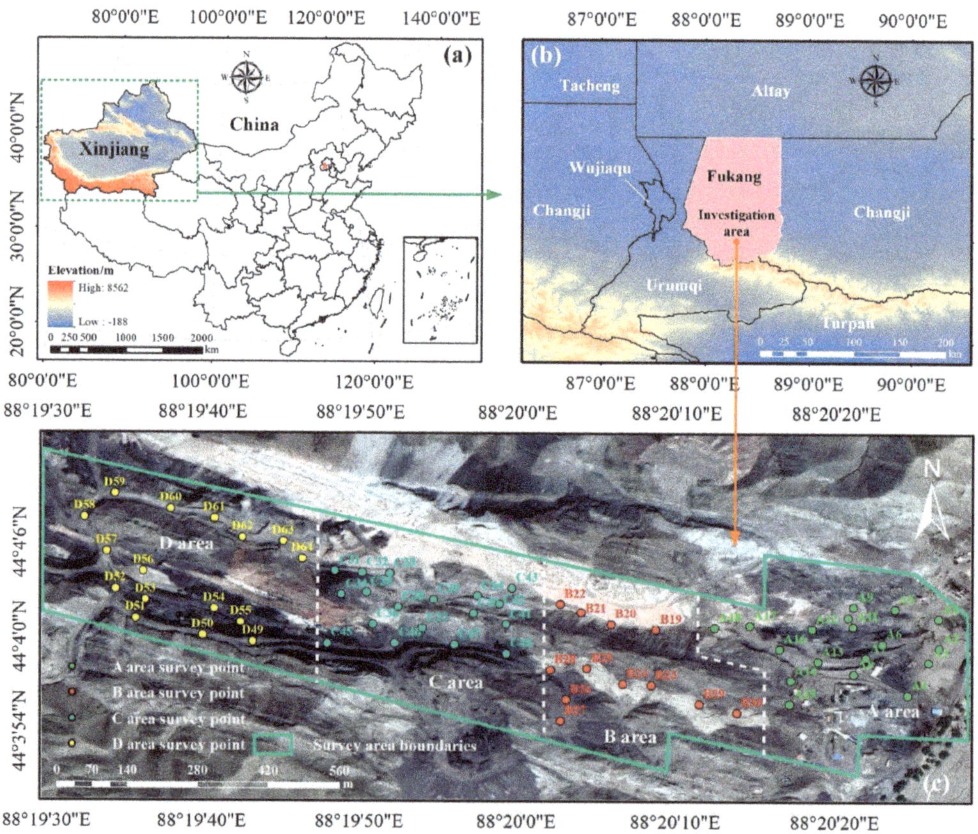

Figure 7. The coalfield where the data–driven approach presented in [135] was applied. (**a**) Map of the region Xinjiang; (**b**) Location of the survey area; (**c**) Survey boundaries and distribution points. Reproduced from [135] with permission from Elsevier.

To verify the precision of the assessment outcomes, field validation was carried out using radon gas detection and geological surveying techniques, as illustrated in Figure 8. Infrared thermal imaging was employed to examine the geological characteristics associated with risk assessment level IV, as shown in Figure 8a–d. In regions A, B, and C, where the risk level was classified as IV, surface temperatures were found to exceed 100 °C, revealing significant thermal anomalies. In contrast, no such high-temperature anomalies were detected in region D. This suggests that areas categorized under risk assessment level

IV exhibit more pronounced coal fire activity, consistent with the thermal irregularities typical of active combustion phases. The four regions (A, B, C, and D) display distinct spatial distributions of assessed risk levels. In regions A, B, and C, (locations evaluated at risk level IV) geological features such as fractures, subsidence pits, vents, and radon gas anomalies were reported. Notably, the maximum radon gas concentrations in these areas were 2672 Bq·m^{-3}, 2890 Bq·m^{-3}, and 3000 Bq·m^{-3}, respectively. This indicates that sites identified with assessment risk level IV correspond to periods of intensified coal fire combustion.

Rúa et al. [136] conducted field measurements on 23 parameters (coal seam specific as well as environmental) in two open-pit mines in Columbia with a history of spontaneous combustion events compiling a dataset with 21,000 entries. The logistic regression model revealed that the primary variables influencing the initiation of spontaneous combustion include seam temperature, atmospheric pressure, wind velocity, oxygen concentration, methane levels, altitude, ash content, volatile matter percentage, calorific value, vitrinite, liptinite, and the Hardgrove grindability index. Controlling certain variables can prevent the formation of conditions conducive to ignition. The key controllable factors are seam temperature, wind velocity, and oxygen concentration. Potential mitigation strategies involve sealing coal seams with materials that possess high thermal insulation and impermeability, as well as employing controlled blasting techniques such as precutting and buffer blasting to minimize heat transfer and restrict oxygen ingress, thereby inhibiting the oxidation reactions that promote combustion.

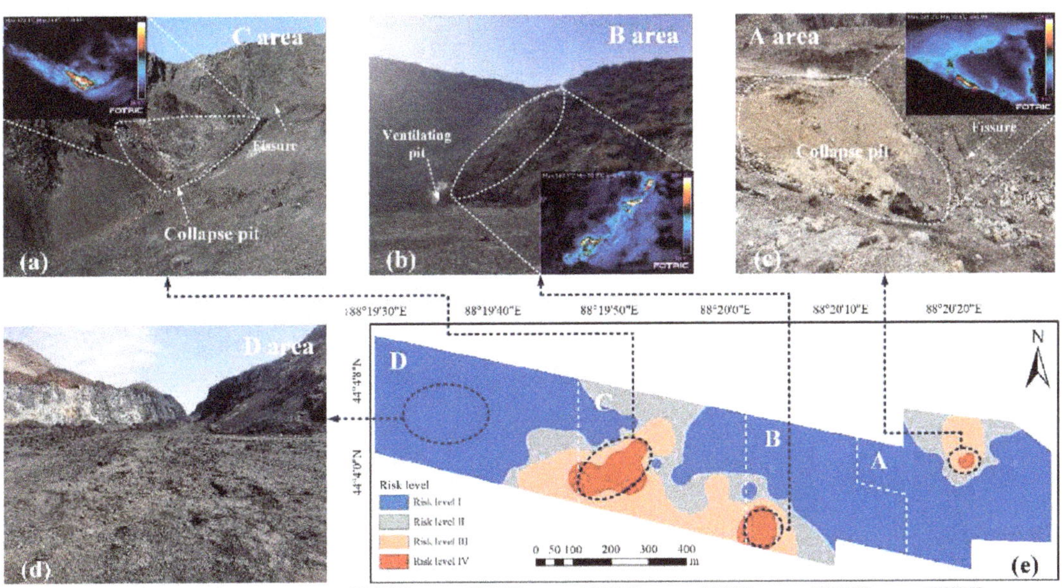

Figure 8. The results of the risk level assessment in four areas of the test region. (**a**) Field characteristics of area C; (**b**) field characteristics of area B; (**c**) field characteristics of area A; (**d**) field characteristics of area D; (**e**) risk level in the four areas considered. Reproduced from [135] with permission from Elsevier.

Abramovicz et al. [137] introduced and validated an innovative method for detecting subsurface smoldering fires by examining vegetation and soil condition changes in coal-waste dump areas affected by self-heating and spontaneous combustion. The study focused on the composition of plant species, life forms, and ecological groupings in relation to the heterogeneous particle size distribution of the deposited materials and fluctuating soil temperatures. These factors significantly influenced the dynamics of vegetation, leading to the development of a distinct plant community. Notably, hemicryptophytes and apophytes were predominant in regions experiencing active thermal processes. The distribution of vegetation was monitored over three distinct periods along a transect that encompassed three surface categories with varying thermal characteristics and subsurface smoldering directions. Subsurface temperatures at a depth of 20 cm ranged from 9.9 °C to 139 °C, while surface temperatures varied from 3.1 °C to 69.0 °C. The total organic carbon (TOC) content across all samples was between 1.7% and 7.6%, with concentrations from 3.1% to 4.5% in areas of active fire. Total nitrogen levels were measured between 0.023% and 0.29%, and soil pH ranged from 5.8 to 8.0 when analyzed in water. Temporal and spatial variations in vegetation patterns were indicative of the progression of fire spots. The findings underscore the critical influence of subsurface temperatures on the spatial organization and species diversity of plant communities in coal-waste disposal sites affected by spontaneous combustion.

A similar approach was applied in an experimental study conducted by Ren et al. [138]. A spatial analysis method was proposed to achieve early warning spontaneous combustion of coal waste dump after reclamation by integrating unmanned aerial vehicle (UAV) and vegetation (Medicago sativa/alfalfa) growth status. The experiment was carried out in two slope areas of a coal waste dump after reclamation in Shanxi province, China, which were known for spontaneous combustion propensity. Three alfalfa growth parameters, aboveground biomass (AGB), plant water content (PWC), and plant height (PH) of the study area, were estimated from UAV imagery features and used to assess the spontaneous combustion risk. Then, soil deep temperature points (25 cm depth) distributed evenly in the study area were collected to determine the underground temperature situation. It was reported that the UAV-derived rededge Chlorophyll index (CIrededge), canopy temperature depression (CTD), and canopy height model (CHM) achieved a better estimation of alfalfa AGB (R^2 = 0.81, RMSE = 99.2 g/m^2, and MAE = 74.9 g/m^2), PWC (R^2 = 0.68, RMSE = 3.9%, and MAE = 3.2%), and PH (R^2 = 0.77, RMSE = 9.79 cm, and MAE = 7.68 cm) of the study area, respectively. Another key observation was that three alfalfa parameters were highly correlated with the soil deep temperature, but differed in degree (R^2 = 0.46–0.81). They were consistent with the soil deep temperature in spatial distribution and could reveal the change direction of underground temperature, which could be helpful to detect potential spontaneous combustion areas. These results indicated that vegetation is a prior indicator to the changes in underground temperature of coal waste dump.

Wang et al. [139] introduced a novel technique for evaluating the three-dimensional spatial distribution and oxidation status of coal gangue dump fire zones by correlating surface temperature data with gas and radon concentration measurements and analyzing their interdependencies. This technique enables the precise determination of the location, depth, spread, degree of oxidation, and potential evolution of combustion areas within coal gangue dumps. A specific methodology for assessing the depth and oxidation levels of the fires was formulated, involving the calculation of a depth index derived from the analysis of thermal fluctuations and radon dispersion patterns. The oxidation degree was quantified by identifying anomalies in radon emissions and surface temperature deviations and establishing diagnostic threshold values. The results obtained from the depth index and oxidation degree assessments were in agreement with the three-dimensional distribution analysis, providing a comprehensive approach for assessing the spontaneous combustion risk in coal gangue dump fire areas. This integrated framework enhances the understanding of the underlying processes that drive such fires, enabling more effective surveillance and mitigation strategies.

Surface and Integrated Techniques—Essential Conclusions and Research Needs

Surface and integrated techniques rely on the manifestations of subsurface combustion at the ground level. The primary advantage of these methods is that they do not require the deployment of equipment underground, allowing for the monitoring of extensive areas with minimal instrumentation. Generally, these techniques are based on temperature measurements, typically obtained through infrared (IR) thermography and, less frequently, through direct contact sensors. For fires occurring below the surface, a notable change in surface temperature becomes detectable only after the fire has reached an advanced stage. However, surface measurement techniques offer other kind of qualitative information, which is not available in the case of other techniques. Propagation direction and speed can be easily assessed by means of surface temperature measurement techniques as well as a rough estimation of the fire size and trend, Song et al. [8]. During the early stages of coal seam combustion—specifically, low-temperature oxidation—the exothermic reactions do not produce sufficient thermal energy to cause a significant increase in surface temperature. Therefore, a combined approach, such as incorporating data from gas sensors, can provide early indications of the CSC state. Combined methodologies that integrate two or more independent detection techniques effectively mitigate the limitations of each individual method, resulting in a more robust and precise detection system. Data-driven methodologies, which utilize extensive datasets, offer a high level of generalizability. However, the effectiveness of the statistical algorithms used in these approaches is highly dependent on the quality of the data. While several studies have reported on data-driven methods, few have addressed issues related to data quality. Data quality encompasses aspects such as the data collection process, detection and removal of outliers, handling of missing values, bias control, stratification, and representativeness. Non-traditional techniques, such as soil and vegetation analysis, present some potential for the detection of CSC events. However, their effectiveness for timely and accurate identification remains limited, as they primarily assess the indirect effects of heat transfer from subsurface fires. Consequently, these techniques do not respond properly to the imperative of fast and accurate CSC detection in early stages.

5. Airborne and Spaceborne Techniques

5.1. Remote Sensing in Detection of CSC

Remote sensing is a pivotal technology in the field of Earth observation, employing advanced sensors mounted on satellite or aerial platforms to collect real-time data about the Earth's surface and atmosphere. Its applications span a wide range of disciplines, including environmental monitoring, land use mapping, and natural hazard assessment. In the context of detecting coal spontaneous combustion, remote sensing offers a unique advantage due to its ability to cover extensive areas, obtain data in real time, and penetrate through smoke or other obstructions using various spectral bands. Advanced techniques such as multispectral and hyperspectral imaging enable the identification of thermal anomalies and subtle changes in surface temperature, which are indicative of spontaneous combustion events in coal seams or stockpiles. Thermal infrared sensors, for example, can detect heat signatures that may not be apparent to the naked eye or standard optical instruments, allowing for early detection and mitigation of these hazardous events. Additionally, remote sensing can employ synthetic aperture radar to identify ground deformation associated with underground fires. By integrating remote sensing data with Geographic Information Systems and machine learning algorithms, it is possible to enhance the precision of combustion detection, forecast risk areas, and formulate effective intervention strategies. This multidisciplinary approach not only aids in preventing coal fire-related disasters but also contributes to minimizing associated environmental and economic impacts.

A comprehensive review by Syed et al. [140] focused on the remote sensing of coal fires in India, incorporating numerous studies related to fires in various regions across the country. The large volume of studies is attributed to the substantial losses caused by CSC events. Approximately 1.48 billion tons of coal remain unextracted due to around 70 surface and subsurface fires in the Jharia coalfield, a region that has been burning for over a century and spans an area of roughly 9 km^2. It is estimated that more than 37.6 million tons of coal have been consumed by fires in the Jharia coalfield alone. Based on the coal volume lost up to 2003 (37.6 million tons), the economic loss due to these fires, calculated using the 2018 import prices of coking and non-coking coal, is estimated to be around Rs. 280.74 billion. Several major research trends were identified as follows:

A. Thermal anomaly detection:
   A.1. Thermal IR sensors (TIR)
   A.2. Short-wave IR sensor (SWIR)
   A.3. Combined TIR and SWIR
B. Land subsidence monitoring
C. Geo-environmental indicators

This review critically analyzed the studies included and focused on limitations and shortcomings. Directions to improve the quality of observations and detection were identified and discussed, such as weighing factors—as spatial resolution, precision, detection potential, temporal constraints, and financial implications to determine the most suitable monitoring strategy. Integrating multiple techniques may provide a more effective solution by mitigating the drawbacks of each individual method. A discussion was included on how remote sensing could be employed to evaluate the CSC risk, develop fire mitigation strategies, and assessments of environmental impacts to comprehensively address the challenges posed by coal fire hazards.

Aerial monitoring of land for identification CSC is mainly based on IR thermal imaging. Thermal infrared (TIR) imaging is a remote sensing technique that identifies variations in the radiant energy emitted from the Earth's surface. Planck's law states that any object with a temperature above absolute zero (0 K) radiates energy in the thermal infrared range of the electromagnetic spectrum, making the primary challenge the accurate measurement of these variations. Fluctuations in radiant flux from the ground are primarily attributed to differences in the thermal properties of surface materials, such as rocks and soils, which affect their capacity to absorb solar radiation, retain it as heat (thermal capacity), and subsequently release it back into the atmosphere within the TIR spectrum. Several environmental and surface factors can influence radiant flux, thereby altering the tonal quality of TIR imagery. These include soil moisture content, the presence of bodies of water such as streams or wetlands, vertical stratification of vegetation (such as variations in tree canopy height), and active transpiration from dense vegetation areas, such as forests. Topographical features also play a significant role; for example, deep valleys may appear as warmer regions in TIR imagery due to reduced wind exposure or enhanced transpiration from vegetation. Additionally, large fauna, such as cattle or buffalo, may manifest as thermal anomalies or "hot spots" in TIR images, potentially leading to false positives in the interpretation of thermal data. Several recent studies on the CSC detection based on TIR imaging, discussing the hardware used and the processing algorithms, are presented in Table 6.

Color infrared (CIR) imaging employs modified color photographic film with altered spectral sensitivity, allowing it to capture reflected energy in the 0.7–0.9 μm range of the electromagnetic spectrum. A key advantage of CIR imaging is its capability to reveal the physiological condition of vegetation, often identifying signs of stress before they become apparent to an observer on the ground. Healthy vegetation, particularly in broadleaf forests, typically appears in various shades of red to magenta on CIR imagery, whereas stressed vegetation is represented by hues ranging from pink to blue, even when the stress is not yet visible to the naked eye. Vegetation in broadleaf forests can undergo stress due to a range of environmental factors, such as drought, disease, insect infestations, or other conditions that disrupt water uptake in the leaves. The capacity of CIR imaging to detect these early signs of stress has prompted its application in identifying the location and extent of coal fires. It has been hypothesized that the heat generated by a coal fire reduces the amount of water retained in the soil pores by surface tension. As a result, plants in the vicinity of a coal fire may progress more rapidly toward the wilting stage compared to those in unaffected areas. This differential stress response is detectable with CIR imaging, making it a valuable tool for early detection of coal fire impact zones.

Table 6. Hardware and processing approaches for CSC detection by means of TIR imaging.

Reference, Year	Platform	Sensor	Processing
Yuan et al. [141] 2021	UAV DJI M210 V2 drone with a Zenmuse XT 2 dual-light thermal infrared lens	TIR Resolution 640 × 512 TIR FOV: 32° × 26° RGB resolution: 4000 × 3000 RGB FOV: 57.12° × 42.44° TIR gain mode set to High, with the upper limit of the temperature range 150 °C	Two visual orthophotos (resolution of 2.54 cm) and two infrared images (resolution of 8.65 cm). The local variance and Shannon entropy are employed to explore the optimal observation scale and optimal coal fire area extraction scale for LST anomalies in coal fire areas.
Shao et al. [142] 2023	UAV DJI Matrice 210 RTK (Real-Time Kinematic) V2 quadrotor drone with a DJI Zenmuse XT2 dual-lens camera	12-megapixel RGB lens (FL: 8 mm, resolution: 4000 × 3000, pixel size 1.85 μm, FOV: 57.12° × 42.44°). FLIR Tau 2 IR lens (FL: 25 mm, TIR resolution 640 × 512, pixel size 17 μm, FOV: 25° × 20°). Temperature range: −25–135 °C (High gain); −40–550 °C (Low gain), thermal accuracy ±5 °C	The Rainbow High Contrast palette and temperature linear distribution mode were used to ensure that each image has a high contrast, color, and temperature consistency. Delineation and accurate identification of 3D temperature field were achieved by feature point matching and texture mapping.
Hu and Xia [71] 2017	TH9100MV/WV thermographic camera. RGB camera: Canon EOS 5D Mark digital camera for close range photogrammetry	TIR camera: 320 × 240 pixels and operates in the 8–14 μm range. The device has adjustable temperature measuring ranges from 0 °C to +500 °C with an accuracy of ±0.06 °C. RGB camera FL> 28 mm	Setting control points for the close-range photogrammetry. Setting the temperature-marked points (TMPs) for infrared thermograph. Bilinear interpolation algorithm provided the spatial coordinates and temperature values for each point in the grid.
He et al. [143] 2020	UAV DJI M210 Zenmuse XT2 cameras equipped with gimbal system	TIR camera: 640 × 512, 30 fps uncooled vanadium oxide (VOx) microbolometer for longwave radiation (7.5~13.5 μm); temperature range of −20 to 135 °C (high gain); FL 25 mm lens; File format raw 8-bit digital numbers; Acquisition rate: less than 9 Hz. RGB camera: 4K video, 12 megapixel photos	Maximum and minimum grey values of the TIR image were determined; based on the temperature range temperature anomaly regions were determined; coal fires and the locations of fire area were determined, with a ground resolution of 40 cm, corresponding to the pixel size.

Coal fires can induce alterations in the coloration of rock strata, degrade surrounding vegetation, release gases and smoke, and result in the accumulation of novel materials on the surface, Syed et al. [140]. Teodoro et al. [144] proposed a set of methodologies to monitor a decommissioned coal mine waste pile (São Pedro da Cova, Portugal) and its adjacent regions. Analysis of thermal and topographic data indicated a direct correlation between the temperature of the pile and ambient air temperature. However, in a localized area of approximately 200 m^2 near the base of the waste pile, spontaneous combustion was detected, with surface temperatures consistently exceeding 50 °C, irrespective of seasonal variations or fluctuations in air temperature. Land use and land cover (LULC) assessments, combined with the Normalized Difference Vegetation Index (NDVI) analysis, revealed that vegetation growth is confined to zones with comparatively lower surface temperatures. The presence of vegetation is critically important, as it plays a key role in mitigating soil erosion on the waste pile, thereby contributing to the stabilization of the structure. Additionally, a downward trend in altimetric measurements was observed, supporting the hypothesis that ongoing self-combustion processes are leading to a reduction in the volume of material within the waste pile. These findings underscore the complex interplay between thermal dynamics, vegetation cover, and topographical changes in areas affected by coal waste combustion. The persistent thermal anomalies and the lack of vegetation in high-temperature zones highlight the challenges in achieving ecological restoration and physical stabilization of coal mine waste piles. The ongoing material depletion, as evidenced by the decreasing altimetric values, suggests that self-combustion continues to actively alter the waste pile's composition and structure, potentially posing risks to both the environment and surrounding communities.

Spaceborne remote sensing represents a powerful tool for the detection and monitoring of coal spontaneous combustion (CSC), offering several distinct advantages over traditional ground-based methods. Utilizing satellite-based platforms, this approach enables continuous, large-scale observation of the Earth's surface, providing consistent and comprehensive data coverage across vast and often inaccessible areas. The high temporal frequency and extensive spatial resolution afforded by spaceborne sensors facilitate the early identification of thermal anomalies associated with CSC, even in remote or hazardous environments. Specific advantages include the ability to capture multispectral and thermal infrared imagery, which is critical for detecting subtle changes in surface temperature and heat flux that indicate ongoing combustion processes. Furthermore, the integration of various spectral bands—from visible to thermal infrared—allows for the assessment of associated changes in surface characteristics, such as vegetation stress and ground deformation, which are indicative of subsurface fires. The ability to gather data across different wavelengths, combined with advanced data processing techniques, enhances the accuracy and reliability of CSC detection, making spaceborne remote sensing an indispensable component of modern environmental monitoring and risk management strategies.

Landsat 8 is a highly advanced Earth observation satellite launched in 2013 as part of the Landsat program, a long-standing initiative for remote sensing led by NASA and the United States Geological Survey (USGS). It carries two primary sensors: (i) the Operational Land Imager (push-broom sensor with a four-mirror telescope and 12-bit quantization, collecting data for visible, near infrared, and short wave infrared spectral bands as well as a panchromatic band). Such data are crucial for a range of applications including land use mapping, vegetation monitoring, and water resource management. (ii) Thermal Infrared Sensor (collecting data for two more narrow spectral bands in the thermal region formerly covered by one wide spectral band on Landsats 4–7). The TIRS provides data in two thermal infrared bands, enabling accurate surface temperature measurements essential for detecting heat anomalies like those caused by coal spontaneous combustion (CSC). Landsat 8's enhanced radiometric resolution (12-bit), moderate spatial resolution (30 m across most spectral bands), and regular revisit interval (every 16 days) establish it as a critical tool for remote sensing applications, providing consistent and high-quality data for monitoring environmental changes over extended periods. The open accessibility of Landsat 8 data further facilitates its extensive utilization in scientific research, environmental monitoring, and disaster response, including the identification of thermal anomalies linked to subsurface coal fires and other combustion events.

Song et al. [145] conducted a comprehensive in situ investigations of coal fires in order to assess the fire suppression efforts carried out in the Wuda syncline in May 2014, following an initial site visit to the Wuda coalfield in 2012. These field observations were combined with thermal anomaly data derived from Landsat-8 and Landsat-7 satellite imagery, using an automated moving-window thermal anomaly detection tool, known as the regional anomaly extractor. This integrated approach enabled a detailed analysis of coal fire behavior, considering the effects of fire-fighting interventions and the current state of combustion activity within the Wuda syncline.

When using data from IR satellite-based remote sensing systems, one of the problems faced is the correct classification of pixels in "fire" and "no-fire" classes. Roy et al. [133] conducted a study based on ASTER Level 1B thermal data (summer time) and Landsat-8 TIRS (winter time) with the aim of generating a coal fire map by selecting an optimal cut-off temperature using a statistically driven, data-centric approach. The cut-off temperature was identified through the examination of radiant temperature values from uniformly distributed pixel clusters covering the entirety of the coal-bearing area. By employing this methodology, it was guaranteed that the selected cut-off temperature adequately addressed inherent variability and associated local complexities. Determining an appropriate cut-off temperature within thermal datasets to classify pixels into "fire" and "no fire" zones presents several challenges: the temperature of underlying terrain and background elements can exhibit significant seasonal fluctuations, and variations in the relative abundance of these elements across different sections of the coalfield may lead to inconsistencies in the cut-off temperature between sites. Additionally, the thermal contrast between active fire zones and surrounding areas, as captured in the remote sensing data, is highly dependent on factors such as the relative distribution of various rock types (e.g., sandstone, shale, and coal seams), land texture characteristics, the configuration of surface exposure (whether the fire is occurring at or beneath the surface), as well as the fire intensity and its spatial spread.

The algorithms used to process data from various satellite platforms are diverse but most of them employ the generalized single channel method, Jiménez-Muñoz and Sobrino [146] to determine the land surface temperature. The influence of atmosphere water content poses significant problems in processing the data. A selection of studies using data from Landsat, Sentinel, Envisat and ALOS PALSAR platforms and the data processing sequence is presented in Table 7. Techniques detecting landslides were also included in Table 7. These are even more complex than algorithms for land surface temperature determination.

Table 7. Typical algorithms for mapping the fire zones. LST and landslide detection studies.

Reference, Year	Remote Sensing Platform	Methodology	Objective Conclusions
Jiang et al. [147], 2017	Data before 2013: Landsat 5 Data after 2013: Landsat 7	Generalized single channel method (Jiménez-Muñoz and Sobrino [146] Planck's blackbody radiance law is divided by Taylor's Formula to obtain the approximate solution of land surface temperature). Natural breaks (Jenks [148]) clustering method was used to define four temperature classes. Eight scenes were selected including Landsat 5 and Landsat 7 images in the spring and fall from 2000 to 2015. (1) Radiance was calibrated using NASA calibration coefficients. (2) Geometric registration. (3) The FLAASH model was used to correct the atmosphere and obtained surface reflectance and moisture content. (4) Images were clipped for further processing.	Map the fire zone and assess the effectiveness of the CFSP (Coal Fire Suppression Project) over a period ranging from 2000 to 2015. CFSP successfully extinguished a significant percentage of surface area affected by coal fire while damaging the local landscape.
Wang et al. [149], 2022	Landsat 8	(1) Radiance correction and atmospheric correction were performed for multispectral data; the radiant brightness and brightness temperature of TIRS data are corrected to obtain the thermal radiation intensity value and the brightness temperature; (2) Terrain correction; (3) Calculation of surface emissivity; (4) Calculation of Atmospheric Water Vapor (AWV) based on NASA data on atmospheric transmissivity; (5) Generalized single channel method (Jiménez-Muñoz and Sobrino [146]) was used to determine the LST. Landsat-8 band 10 data were used for temperature inversion.	The average detection accuracy rate of the proposed method reaches 83.3%. The proposed coal fire detection approach can achieve good coal fire detection results regardless of using summer, winter, or annual data, with the annual data producing the best detection result.

Table 7. *Cont.*

Reference, Year	Remote Sensing Platform	Methodology	Objective Conclusions
Yu et al. [150], 2022	45 Landsat 8 images and 61 Sentinel-1 SAR images covering the same time span	(1) Utilizing the Radiative Transfer Equation (RTE) to retrieve land surface temperatures (LSTs) from multitemporal thermal infrared remote sensing data. (2) Normalizing the LSTs and applying an equidistant classification method to delineate thermal anomaly areas from remote sensing images obtained at various time intervals. (3) Quantifying the thermal anomaly frequencies (TAFs) across different regions. (4) Employing the Small Baseline Subset (SBAS) InSAR technique to derive temporal and spatial subsidence data. (5) Combining the mean subsidence values with the standard deviation to identify regions exhibiting significant subsidence within the study area. (6) Defining the threshold for thermal anomaly frequency (TAFT) by using land subsidence data as a constraint and designating an area as a coal fire zone if the observed frequency surpasses the TAFT. (7) Conducting a visual inspection of the results to eliminate any potential artificial heat sources. sources (such as industrial production areas) to further improve the identification accuracy.	To detect coal fire zones and evaluate their intensity and temporal evolution by integrating land surface temperature (LST) data, thermal anomaly metrics, and deformation information across a time series.
Jiang et al. [151], 2010	ENVISAT satellite (currently decommissioned), InSAR technique	A two-dimensional linear regression analysis of the differential interferometric phase was conducted on a set of 49 multi-master interferograms. The preliminary identification of Permanent Scatterers (PS) points was based on two selection criteria: (i) minimal temporal variability in the backscatter values, derived from co-registered Single Look Complex (SLC) intensity stacks, and (ii) the spectral characteristics of each individual SLC dataset.	The results obtained from InSAR analysis were cross-validated against GPS measurements acquired during two field campaigns in 2006 and 2008. A strong correlation was observed between the line-of-sight (LOS) displacement velocities derived from both methodologies, with an average absolute discrepancy of 5.4 mm/year and a standard deviation of 4.1 mm/year.
Mishra et al. [152], 2011	Landsat-7 ETM+ data of 29th March 2006 Ground data from TIR camera Jahria (India) of March, 2009	(1) Converting the image from digital number (DN) values to spectral radiance. (2) Transforming spectral radiance into radiant temperature. (3) Converting radiant temperature into surface temperature. The coal mine fire map derived from satellite imagery was categorized into three classifications: (i) Temperatures exceeding 52 °C were classified as high intensity, representing surface fires. (ii) Temperatures between 32 °C and 52 °C were identified as medium intensity, indicating subsurface fires. (iii) Temperatures at or below 32 °C were considered low intensity, corresponding to background or water bodies temperatures.	The discrepancy between surface temperature measurements and satellite-derived temperatures can be attributed to various factors, including wind direction and velocity, the angle and altitude of the thermal infrared (TIR) camera, atmospheric humidity, and the elevation of the terrain.

Table 7. *Cont.*

Reference, Year	Remote Sensing Platform	Methodology	Objective Conclusions
Biswal and Gorai [69], 2021	Landsat-5 (band 6) TM and Landsat-8 (band 10) Operational Land Imager/Thermal Infrared Sensor) satellite data of from 1989 to 2019.	(i) Landsat-8 sensor data were used for estimating the LST (Jiménez-Muñoz, J.C., Sobrino [146]) • Determination of atmospheric functions • Estimation of radiance • Estimation of LSE (ii) Determination of threshold LST for identifying the fire affected pixels (iii) Validation with ground temperature data	The fire-affected area measured 2.026, 3.009, 3.159, 3.991, 4.664, 8.656, and 9.957 km^2 for the years 1989, 1994, 1999, 2004, 2009, 2014, and 2019, respectively. The expansion of mining operations in the coalfield exposed additional coal seams to the atmosphere, leading to spontaneous heating and the formation of new coal fire pockets. Some fires were extinguished, either due to complete coal combustion or a cessation of the combustion process. The time-series analysis from 1989 to 2019 indicates both the formation of new fire pockets and the extinguishing of existing ones.
Zhou et al. [153], 2013	ALOS Phased Array type L-band Synthetic Aperture Radar	(i) Select pixels with coherence threshold (ii) Phase unwrapping for interferograms with short normal baselines (iii) Minimum Cost Flow method phase unwrapping (iv) Accumulated subsidence estimation (v) Average subsidence velocity calculation	Mean deformation rates of −8.52 cm/yr, −3.92 cm/yr, −4.6 cm/yr, −8.93 cm/yr and −6.07 cm/yr, were determined. Other areas outside the coal fire zones with similar subsidence signals were identified. Further investigation demonstrated that intense mining activities caused land subsidence in these areas.

5.2. Remote Sensing—Concluding Discussion

Remote sensing employs various sensors to acquire detailed information about objects and environments. It serves as an alternative to human-operated detection in inaccessible or hazardous areas, offering the advantage of rapidly collecting data with high spatial, spectral, and temporal resolution. Thermal infrared (TIR) data are particularly useful for detecting coal gangue pile fires, as it directly reflects surface temperature distribution. Remote sensing can be categorized into aerial and satellite-based methods, depending on the platform used.

Aerial remote sensing primarily involves color infrared or multispectral imaging, or the use of unmanned aerial vehicles (UAVs) equipped with TIR or hyperspectral sensors. Despite being influenced by factors such as surface albedo, terrain, and additional heat sources, these techniques provide high accuracy. Aerial methods detect surface features like thermal anomalies, heat-induced alterations in rock surfaces, and chemical byproducts of thermal processes. They are effective for large-area monitoring, such as coalfields, and hyperspectral imaging offers an additional advantage by correcting high-temperature saturation and sub-pixel issues. However, aerial surveillance is costly and impractical for continuous monitoring. In contrast, spaceborne remote sensing is more cost-effective and suitable for continuous, large-scale monitoring, particularly for tracking extensive coal fires and identifying fire propagation, Raju et al. [154], He et al. [155], Wang et al. [70], Singh et al. [156].

Spaceborne techniques employ three types of sensors: (i) multispectral sensors, such as those on the Landsat, MODIS, SPOT, and ASTER platforms, (ii) high spatial resolution sensors on platforms like Ikonos, Quickbird, and Worldview, and (iii) radar sensors. These sensors detect different features: multispectral sensors identify thermal anomalies, high spatial resolution sensors detect surface cracks and fissures, and radar sensors detect land subsidence and collapse. Radar methods, however, tend to have lower accuracy than multispectral and high-resolution techniques.

Very few studies exist though, attempting to integrate two or more remote sensing techniques, such as Elick [157] (thermal infrared imaging combined with CO/CO_2 measured along fire fronts to assess the progress and the fire depth). Karanam et al. [158] employed the methodology depicted in Figure 9 integrating thermal data from Landsat 8 and persistent Scatterer Interferometry analysis (60 Sentinel-1 C-band images) to monitor and map the coal fires in Jharia (India).

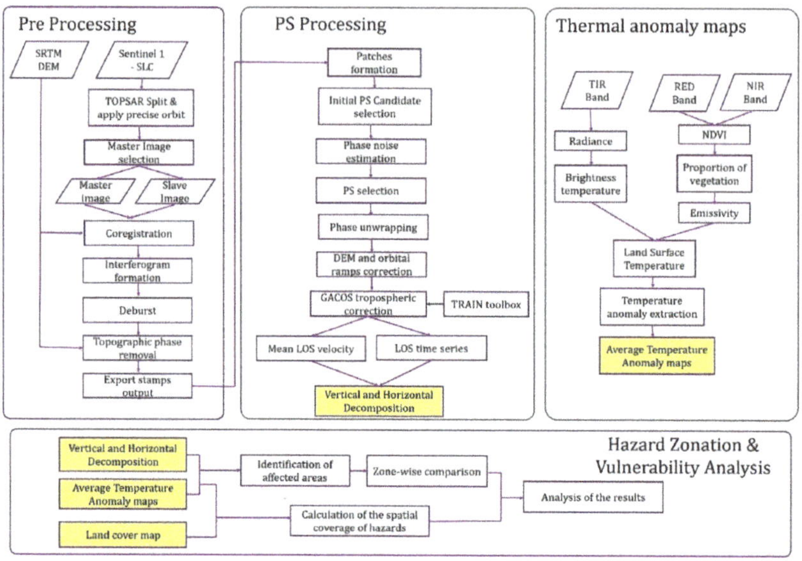

Figure 9. The methodology proposed by Karanam et al. [158] to combine TIR data with land subsidence information. Reproduced from [158] under CC BY 4.0.

6. Conclusions

The detection and monitoring of coal fires are critical for mapping their occurrence, modeling fire behavior, assessing environmental and public health impacts, and guiding fire-fighting engineering efforts. For instance, key model parameters such as fire depth, surface and subsurface temperatures, and fire propagation rates are fundamental inputs for computational simulations or serve as validation benchmarks for predictive models. These parameters also provide essential data for designing and implementing fire-fighting strategies. Additionally, monitoring emissions of greenhouse gases, toxic gases, and trace elements generated by coal fires is necessary for evaluating the environmental consequences and health risks posed by these fires.

The techniques discussed in this review can be broadly categorized into two main approaches: (i) the detection of coal spontaneous combustion (CSC) events during their initial stages, and (ii) the monitoring of fully developed fires—whether surface fires with visible flames or smoldering subsurface fires—to map their spatial extent, assess their progression, and evaluate the effectiveness of fire suppression interventions. From this review, several key observations and research directions can be identified to refine and narrow down CSC studies:

A. Site-Specificity of Detection Methods:

Many detection and monitoring methods are specifically designed or calibrated for particular coalfield sites, taking into account factors such as coal rank, composition, geological morphology, and climate. The literature lacks replication studies and discussions regarding the generalizability of these methods. This issue is particularly significant for early-stage CSC detection techniques, such as those targeting low-temperature oxidation. Experimental studies, whether conducted in the field or under controlled laboratory conditions, typically focus on specific types of coal and external conditions. Consequently, the accuracy and applicability of detection techniques designed for one site may not translate effectively to different locations with varying coal types and environmental factors.

B. Limitations of Laboratory-Controlled Experiments:

While laboratory-controlled experiments provide valuable insights, they cannot fully replicate the complex and dynamic conditions encountered at industrial coal sites. Real-world environments present numerous challenges, including variable gas emissions and absorption rates, fluctuating moisture levels, chemical spills, dust, noise, and other interfering factors. For instance, techniques such as thermal infrared imaging, though highly sensitive, may be overwhelmed by these environmental variables, limiting their practicality in industrial settings. There is a clear need for further research to bridge the gap between controlled laboratory findings and the realities of field conditions, ensuring that detection and monitoring techniques can be reliably applied in practical scenarios.

C. Development of Combined Techniques:

Although some progress has been made in developing ensemble or hybrid techniques that integrate multiple detection and monitoring approaches, these methods are still in their early stages. Combined techniques hold the potential to enhance robustness and accuracy by leveraging the strengths of individual methods while mitigating their weaknesses. For example, integrating thermal infrared imaging with gas detection sensors could improve the reliability of monitoring systems in variable environmental conditions. However, substantial research and development efforts are still required to advance these combined systems and validate their performance under real-world conditions.

In conclusion, while significant advancements have been made in the detection and monitoring of coal fires, several challenges remain. The site-specific nature of many techniques, the limitations of laboratory studies, and the early development of combined approaches all point to the need for further research to improve the accuracy, generalizability, and applicability of CSC detection and monitoring methods across diverse environments.

Author Contributions: Conceptualization, L..A and B.M.D.; methodology, L.A. and B.M.D.; formal analysis, B.M.D.; investigation, LA; resources, L.A.; data curation, L.A.; writing—original draft preparation, B.M.D.; writing—review and editing, B.M.D.; supervision, B.M.D.; All authors have read and agreed to the published version of the manuscript.

Funding: This research received no external funding.

Conflicts of Interest: The authors declare no conflict of interest.

References

1. Heffern, E.L.; Coates, D.A. Geologic history of natural coal-bed fires, Powder River basin, USA. *Int. J. Coal Geol.* **2004**, *59*, 25–47. [CrossRef]
2. Song, Z.; Zhu, H.; Jia, G.; He, C. Comprehensive evaluation on self-ignition risks of coal stockpiles using fuzzy AHP approaches. *J. Loss Prev. Process. Ind.* **2014**, *32*, 78–94. [CrossRef]
3. Liu, H.; Hong, R.; Xiang, C.; Lv, C.; Li, H. Visualization and analysis of mapping knowledge domains for spontaneous combustion studies. *Fuel* **2019**, *262*, 116598. [CrossRef]
4. Yang, F.; Qiu, D. Exploring coal spontaneous combustion by bibliometric analysis. *Process. Saf. Environ. Prot.* **2019**, *132*, 1–10. [CrossRef]
5. Huang, Z.; Ma, Z.; Song, S.; Yang, R.; Gao, Y.; Zhang, Y. Study on the influence of periodic weighting on the spontaneous combustion "three-zone" in a gob. *J. Loss Prev. Process. Ind.* **2018**, *55*, 480–491. [CrossRef]
6. Onifade, M.; Genc, B. A review of research on spontaneous combustion of coal. *Int. J. Min. Sci. Technol.* **2020**, *30*, 303–311. [CrossRef]
7. Zhang, J.; Ren, T.; Liang, Y.; Wang, Z. A review on numerical solutions to self-heating of coal stockpile: Mechanism, theoretical basis, and variable study. *Fuel* **2016**, *182*, 80–109. [CrossRef]
8. Song, Z.; Kuenzer, C. Coal fires in China over the last decade: A comprehensive review. *Int. J. Coal Geol.* **2014**, *133*, 72–99. [CrossRef]
9. Lu, X.; Deng, J.; Xiao, Y.; Zhai, X.; Wang, C.; Yi, X. Recent progress and perspective on thermal-kinetic, heat and mass transportation of coal spontaneous combustion hazard. *Fuel* **2022**, *308*, 121234. [CrossRef]
10. Li, Q.-W.; Xiao, Y.; Zhong, K.-Q.; Shu, C.-M.; Lü, H.-F.; Deng, J.; Wu, S. Overview of commonly used materials for coal spontaneous combustion prevention. *Fuel* **2020**, *275*, 117981. [CrossRef]
11. Onifade, M. Countermeasures against coal spontaneous combustion: A review. *Int. J. Coal Prep. Util.* **2021**, *42*, 2953–2975. [CrossRef]
12. Kong, B.; Li, Z.; Yang, Y.; Liu, Z.; Yan, D. A review on the mechanism, risk evaluation, and prevention of coal spontaneous combustion in China. *Environ. Sci. Pollut. Res.* **2017**, *24*, 23453–23470. [CrossRef] [PubMed]

13. Han, D.; Niu, G.; Zhu, H.; Chang, T.; Liu, B.; Ren, Y.; Wang, Y.; Song, B. Exploration and Frontier of Coal Spontaneous Combustion Fire Prevention Materials. *Processes* **2024**, *12*, 1155. [CrossRef]
14. Lu, W.; Cao, H.; Sun, X.; Hu, X.; Li, J.; Li, J.; Kong, B. A Review on the Types, Performance and Environmental Protection of Filling & Plugging Materials for Prevention and Control of Coal Spontaneous Combustion in China. *Combust. Sci. Technol.* **2022**, *196*, 2299–2335. [CrossRef]
15. Dai, J.; Tian, Z.-J.; Shi, X.-Y.; Lu, Y.; Chi, W.-L.; Zhang, Y. Research progress on gravity heat pipe technology to prevent spontaneous combustion in coal storage piles. *MRS Commun.* **2024**, *14*, 480–488. [CrossRef]
16. Kuenzer, C.; Zhang, J.; Sun, Y.; Jia, Y.; Dech, S. Coal fires revisited: The Wuda coal field in the aftermath of extensive coal fire research and accelerating extinguishing activities. *Int. J. Coal Geol.* **2012**, *102*, 75–86. [CrossRef]
17. Saini, V.; Gupta, R.P.; Arora, M.K. Environmental impact studies in coalfields in India: A case study from Jharia coal-field. *Renew. Sustain. Energy Rev.* **2015**, *53*, 1222–1239. [CrossRef]
18. Kuenzer, C.; Stracher, G.B. Geomorphology of coal seam fires. *Geomorphology* **2012**, *138*, 209–222. [CrossRef]
19. Kus, J. Impact of underground coal fire on coal petrographic properties of high volatile bituminous coals: A case study from coal fire zone No. 3.2 in the Wuda Coalfield, Inner Mongolia Autonomous Region, North China. *Int. J. Coal Geol.* **2017**, *171*, 185–211. [CrossRef]
20. Xu, Y.; Uhl, D.; Zhang, N.; Zhao, C.; Qin, S.; Liang, H.; Sun, Y. Evidence of widespread wildfires in coal seams from the Middle Jurassic of Northwest China and its impact on paleoclimate. *Palaeogeogr. Palaeoclim. Palaeoecol.* **2020**, *559*, 109819. [CrossRef]
21. Li, C.; Sun, J.; Shi, J.; Liang, H.; Cao, Q.; Li, Z.; Gao, Y. Mercury sources in a subterranean spontaneous combustion area. *Ecotoxicol. Environ. Saf.* **2020**, *201*, 110863. [CrossRef] [PubMed]
22. Carroll, M.; Gao, C.X.; Campbell, T.C.; Smith, C.L.; Dimitriadis, C.; Berger, E.; Maybery, D.; Ikin, J.; Abramson, M.J.; Sim, M.R.; et al. Impacts of coal mine fire-related PM2.5 on the utilisation of ambulance and hospital services for mental health conditions. *Atmos. Pollut. Res.* **2022**, *13*, 101415. [CrossRef]
23. Oliveira, M.L.; Pinto, D.; Tutikian, B.F.; da Boit, K.; Saikia, B.K.; Silva, L.F. Pollution from uncontrolled coal fires: Continuous gaseous emissions and nanoparticles from coal mines. *J. Clean. Prod.* **2019**, *215*, 1140–1148. [CrossRef]
24. Deng, J.; Ge, S.; Qi, H.; Zhou, F.; Shi, B. Underground coal fire emission of spontaneous combustion, Sandaoba coalfield in Xinjiang, China: Investigation and analysis. *Sci. Total. Environ.* **2021**, *777*, 146080. [CrossRef] [PubMed]
25. Xia, T.; Zhou, F.; Liu, J.; Kang, J.; Gao, F. A fully coupled hydro-thermo-mechanical model for the spontaneous combustion of underground coal seams. *Fuel* **2014**, *125*, 106–115. [CrossRef]
26. Rúa, M.O.B.; Aragón, A.J.D.; Baena, P.B. A study of fire propagation in coal seam with numerical simulation of heat transfer and chemical reaction rate in mining field. *Int. J. Min. Sci. Technol.* **2019**, *29*, 873–879. [CrossRef]
27. Xi, Z.; Suo, L.; Xia, T. Characterization of elementary reactions inducing coal spontaneous combustion. *Fuel* **2024**, *358*, 130138. [CrossRef]
28. Xiao, Y.; Ren, S.-J.; Deng, J.; Shu, C.-M. Comparative analysis of thermokinetic behavior and gaseous products between first and second coal spontaneous combustion. *Fuel* **2018**, *227*, 325–333. [CrossRef]
29. Yuan, H.; Restuccia, F.; Richter, F.; Rein, G. A computational model to simulate self-heating ignition across scales, configurations, and coal origins. *Fuel* **2019**, *236*, 1100–1109. [CrossRef]
30. Taraba, B.; Michalec, Z.; Michalcová, V.; Blejchař, T.; Bojko, M.; Kozubková, M. CFD simulations of the effect of wind on the spontaneous heating of coal stockpiles. *Fuel* **2014**, *118*, 107–112. [CrossRef]
31. Rong-Kun, P.; Chang, L.; Ke, Y.; Ming-Gao, Y. Distribution Regularity and Numerical Simulation Study on the Coal Spontaneous Combustion "three zones" Under the Ventilation Type of Ventilation Type of Ч + J. *Procedia Eng.* **2011**, *26*, 704–711. [CrossRef]
32. Zhuo, H.; Qin, B.; Qin, Q.; Su, Z. Modeling and simulation of coal spontaneous combustion in a gob of shallow buried coal seams. *Process. Saf. Environ. Prot.* **2019**, *131*, 246–254. [CrossRef]
33. Xiaomeng, L.; Minbo, Z.; Xiaoyan, Z.; Bo, T.; Saiyi, G. Numerical-simulation study on the influence of wind speed and segregation effect on spontaneous combustion of coal bunker. *Case Stud. Therm. Eng.* **2023**, *52*, 103678. [CrossRef]
34. Zhang, J.; Choi, W.; Ito, T.; Takahashi, K.; Fujita, M. Modelling and parametric investigations on spontaneous heating in coal pile. *Fuel* **2016**, *176*, 181–189. [CrossRef]
35. Yu, M.; Yang, N.; Li, H.; Wang, L.; Wu, M.; Wang, F.; Chu, T.; Wang, K. Numerical investigation on the effects of axial-stress loads on the temperature-programmed oxidation characteristics of loose broken coal. *Energy* **2024**, *289*, 129974. [CrossRef]
36. Plakunov, M.; Yavuzturk, C.; Chiasson, A. On the effects of temperature-dependent diffusion of carbon dioxide from underground coal fires. *Geothermics* **2020**, *85*, 101768. [CrossRef]
37. Qiao, L.; Deng, C.; Lu, B.; Wang, Y.; Wang, X.; Deng, H.; Zhang, X. Study on calcium catalyzes coal spontaneous combustion. *Fuel* **2022**, *307*, 121884. [CrossRef]
38. Deng, B.; Qiao, L.; Wang, Y.; Mu, X.; Deng, C.; Jin, Z. Study on the effect of inorganic and organic sodium on coal spontaneous combustion. *Fuel* **2023**, *353*, 129256. [CrossRef]
39. Zhang, Y.; Luo, Y.; Amini, S.H. Systematic investigation on the effect of particle size on low-rank coal spontaneous combustion under various extrinsic conditions. *Fuel* **2023**, *334*, 126844. [CrossRef]
40. Zhang, J.; Liang, Y.; Ren, T.; Wang, Z.; Wang, G. Transient CFD modelling of low-temperature spontaneous heating behaviour in multiple coal stockpiles with wind forced convection. *Fuel Process. Technol.* **2016**, *149*, 55–74. [CrossRef]

41. Li, D.-J.; Xiao, Y.; Lü, H.-F.; Xu, F.; Liu, K.-H.; Shu, C.-M. Effects of 1-butyl-3-methylimidazolium tetrafluoroborate on the exothermic and heat transfer characteristics of coal during low-temperature oxidation. *Fuel* **2020**, *273*, 117589. [CrossRef]
42. Said, K.O.; Onifade, M.; Lawal, A.I.; Githiria, J.M. An Artificial Intelligence-based Model for the Prediction of Spontaneous Combustion Liability of Coal Based on Its Proximate Analysis. *Combust. Sci. Technol.* **2020**, *193*, 2350–2367. [CrossRef]
43. Lei, C.; Deng, J.; Cao, K.; Xiao, Y.; Ma, L.; Wang, W.; Ma, T.; Shu, C. A comparison of random forest and support vector machine approaches to predict coal spontaneous combustion in gob. *Fuel* **2019**, *239*, 297–311. [CrossRef]
44. Sahu, H.; Mahapatra, S.; Sirikasemsuk, K.; Panigrahi, D. A discrete particle swarm optimization approach for classification of Indian coal seams with respect to their spontaneous combustion susceptibility. *Fuel Process. Technol.* **2011**, *92*, 479–485. [CrossRef]
45. Xie, Z.; Zhang, Y.; Jin, C. Prediction of Coal Spontaneous Combustion in Goaf Based on the BP Neural Network. *Procedia Eng.* **2012**, *43*, 88–92. [CrossRef]
46. Li, S.; Ma, X.; Yang, C. Prediction of spontaneous combustion in the coal stockpile based on an improved metabolic grey model. *Process Saf. Environ. Prot.* **2018**, *116*, 564–577. [CrossRef]
47. Sahu, H.; Padhee, S.; Mahapatra, S. Prediction of spontaneous heating susceptibility of Indian coals using fuzzy logic and artificial neural network models. *Expert Syst. Appl.* **2010**, *38*, 2271–2282. [CrossRef]
48. Wang, K.; Li, K.; Du, F.; Zhang, X.; Wang, Y.; Sun, J. Research on prediction model of coal spontaneous combustion temperature based on SSA-CNN. *Energy* **2024**, *290*, 130158. [CrossRef]
49. Li, J.; Fu, P.; Zhu, Q.; Mao, Y.; Yang, C. A lab-scale experiment on low-temperature coal oxidation in context of underground coal fires. *Appl. Therm. Eng.* **2018**, *141*, 333–338. [CrossRef]
50. Onifade, M.; Genc, B.; Carpede, A. A new apparatus to establish the spontaneous combustion propensity of coals and coal-shales. *Int. J. Min. Sci. Technol.* **2018**, *28*, 649–655. [CrossRef]
51. Wen, H.; Wang, H.; Liu, W.; Cheng, X. Comparative study of experimental testing methods for characterization parameters of coal spontaneous combustion. *Fuel* **2020**, *275*, 117880. [CrossRef]
52. Zhang, H.; Wang, Y.; Zhang, X.; Sasaki, K.; Sugai, Y.; Han, F.; Dong, W.; Thanh, H.V. Experimental study of moisture effects on spontaneous combustion of Baiyinhua lignite from individual particles to stockpile. *Fuel* **2023**, *334*, 126774. [CrossRef]
53. Yan, H.; Nie, B.; Liu, P.; Chen, Z.; Yin, F.; Gong, J.; Lin, S.; Wang, X.; Kong, F.; Hou, Y. Experimental investigation and evaluation of influence of oxygen concentration on characteristic parameters of coal spontaneous combustion. *Thermochim. Acta* **2022**, *717*, 179345. [CrossRef]
54. Zhao, J.; Yang, X.; Song, J.; Zhang, Y.; Shu, C.-M. Experimental study of coal spontaneous combustion high-temperature region spreading characteristics. *Process. Saf. Environ. Prot.* **2023**, *180*, 136–147. [CrossRef]
55. Fan, X.-L.; Ma, L.; Sheng, Y.-J.; Liu, X.-X.; Wei, G.-M.; Liu, S.-M. Experimental investigation on the characteristics of XG/GG/HPAM gel foam and prevention of coal spontaneous combustion. *Energy* **2023**, *284*, 128710. [CrossRef]
56. Zhang, Y.; Xu, J.; Wang, D. Experimental Study on the Inhibition Effects of Nitrogen and Carbon Dioxide on Coal Spontaneous Combustion. *Energies* **2020**, *13*, 5256. [CrossRef]
57. Zhang, X.; Gui, Y.; Lu, B.; Bai, G.; Qiao, L.; Li, C.; Zhang, X.; Gui, Y.; Lu, B.; Bai, G.; et al. Experimental study on the effect of inhibitor on coal spontaneous combustion under low temperature freezing. *Fuel* **2024**, *365*, 131225. [CrossRef]
58. Cheng, J.; Ma, Y.; Wang, C.; Wang, W.; Zhang, L.; Hu, X.; Zhang, M.; Ma, Z. A novel cement-based flexible spray coating for flame retardant. *Process. Saf. Environ. Prot.* **2023**, *177*, 366–379. [CrossRef]
59. Liu, Y.; Wen, H.; Guo, J.; Jin, Y.; Wei, G.; Yang, Z. Coal spontaneous combustion and N2 suppression in triple goafs: A numerical simulation and experimental study. *Fuel* **2020**, *271*, 117625. [CrossRef]
60. Tang, Y.; Hou, F.; Zhong, X.; Huang, A.; Jia, X.; Peng, B. Combination of heat energy extraction and fire control in underground high-temperature zones of coal fire areas. *Energy* **2023**, *278*, 127801. [CrossRef]
61. Hu, X.; Yang, S.; Zhou, X.; Yu, Z.; Hu, C. Coal spontaneous combustion prediction in gob using chaos analysis on gas indicators from upper tunnel. *J. Nat. Gas Sci. Eng.* **2015**, *26*, 461–469. [CrossRef]
62. Du, B.; Liang, Y.; Tian, F. Detecting concealed fire sources in coalfield fires: An application study. *Fire Saf. J.* **2021**, *121*, 103298. [CrossRef]
63. Wei, D.; Du, C.; Lei, B.; Lin, Y. Prediction and prevention of spontaneous combustion of coal from goafs in workface: A case study. *Case Stud. Therm. Eng.* **2020**, *21*, 100668. [CrossRef]
64. Ma, L.; Zou, L.; Ren, L.-F.; Chung, Y.-H.; Zhang, P.-Y.; Shu, C.-M. Prediction indices and limiting parameters of coal spontaneous combustion in the Huainan mining area in China. *Fuel* **2020**, *264*, 116883. [CrossRef]
65. Guo, J.; Wen, H.; Zheng, X.; Liu, Y.; Cheng, X. A method for evaluating the spontaneous combustion of coal by monitoring various gases. *Process. Saf. Environ. Prot.* **2019**, *126*, 223–231. [CrossRef]
66. Guo, J.; Shang, H.; Cai, G.; Jin, Y.; Wang, K.; Li, S. Early Detection of Coal Spontaneous Combustion by Complex Acoustic Waves in a Concealed Fire Source. *ACS Omega* **2023**, *8*, 16519–16531. [CrossRef]
67. Gao, R.; Zhu, H.; Liao, Q.; Qu, B.; Hu, L.; Wang, H. Detection of coal fire by deep learning using ground penetrating radar. *Measurement* **2022**, *201*, 111585. [CrossRef]
68. Zhou, B.; Wu, J.; Wang, J.; Wu, Y. Surface-based radon detection to identify spontaneous combustion areas in small abandoned coal mine gobs: Case study of a small coal mine in China. *Process. Saf. Environ. Prot.* **2018**, *119*, 223–232. [CrossRef]
69. Biswal, S.S.; Gorai, A.K. Studying the coal fire dynamics in Jharia coalfield, India using time-series analysis of satellite data. *Remote. Sens. Appl. Soc. Environ.* **2021**, *23*, 100591. [CrossRef]

70. Wang, Y.-J.; Tian, F.; Huang, Y.; Wang, J.; Wei, C.-J. Monitoring coal fires in Datong coalfield using multi-source remote sensing data. *Trans. Nonferrous Met. Soc. China* **2015**, *25*, 3421–3428. [CrossRef]
71. Hu, Z.; Xia, Q. An integrated methodology for monitoring spontaneous combustion of coal waste dumps based on surface temperature detection. *Appl. Therm. Eng.* **2017**, *122*, 27–38. [CrossRef]
72. Kong, B.; Wang, E.; Lu, W.; Li, Z. Application of electromagnetic radiation detection in high-temperature anomalous areas experiencing coalfield fires. *Energy* **2019**, *189*, 116144. [CrossRef]
73. Liu, W.; Qin, Y.; Yang, X.; Wang, W.; Chen, Y. Early extinguishment of spontaneous combustion of coal underground by using dry-ice's rapid sublimation: A case study of application. *Fuel* **2018**, *217*, 544–552. [CrossRef]
74. Shi, X.; Zhang, C.; Jin, P.; Zhang, Y.; Jiao, F.; Xu, S.; Cao, W. Eliminating scale effect: Development and attenuation of coal spontaneous combustion. *Fuel* **2024**, *357*, 130073. [CrossRef]
75. Zhai, X.; Wu, S.; Wang, K.; Drebenstedt, C.; Zhao, J. Environment influences and extinguish technology of spontaneous combustion of coal gangue heap of Baijigou coal mine in China. *Energy Procedia* **2017**, *136*, 66–72. [CrossRef]
76. Cheng, W.; Hu, X.; Xie, J.; Zhao, Y. An intelligent gel designed to control the spontaneous combustion of coal: Fire prevention and extinguishing properties. *Fuel* **2017**, *210*, 826–835. [CrossRef]
77. Cheng, J.; Wu, Y.; Dong, Z.; Zhang, R.; Wang, W.; Wei, G.; Chu, T.; Yu, Z.; Qin, Y.; Liu, G.; et al. A novel composite inorganic retarding gel for preventing coal spontaneous combustion. *Case Stud. Therm. Eng.* **2021**, *28*, 101648. [CrossRef]
78. Fan, Y.-J.; Zhao, Y.-Y.; Hu, X.-M.; Wu, M.-Y.; Xue, D. A novel fire prevention and control plastogel to inhibit spontaneous combustion of coal: Its characteristics and engineering applications. *Fuel* **2020**, *263*, 116693. [CrossRef]
79. Li, Y.; Xiao, G.; Li, F.; Guo, Y.; Chen, C.; Chen, C.; Li, R.; Yang, Z. A novel H-TiO_2/gel co-stabilized three-dimensional network synergistic fire-retardant foam gel for coal-pile. *Colloids Surf. A Physicochem. Eng. Asp.* **2022**, *650*, 129642. [CrossRef]
80. Shi, B.; Su, H.; Li, J.; Qi, H.; Zhou, F.; Torero, J.L.; Chen, Z. Clean Power Generation from the Intractable Natural Coalfield Fires: Turn Harm into Benefit. *Sci. Rep.* **2017**, *7*, 5302. [CrossRef]
81. Su, H.; Zhou, F.; Qi, H.; Li, J. Design for thermoelectric power generation using subsurface coal fires. *Energy* **2017**, *140*, 929–940. [CrossRef]
82. Xiao, Y.; Yin, L.; Tian, Y.; Li, S.-G.; Zhai, X.-W.; Shu, C.-M.; Ren, S.-J. Sustainable utilisation and transformation of the thermal energy from coalfield fires: A comprehensive review. *Appl. Therm. Eng.* **2023**, *233*, 121164. [CrossRef]
83. Xiao, Y.; Gao, L.-H.; Meng, X.; Liu, J.-W.; Lu, X.; Tian, Y.; Shu, C.-M. Thermal removal from coal fires by a two-phase closed thermosyphon with a CuO nanofluid. *Case Stud. Therm. Eng.* **2023**, *42*, 102745. [CrossRef]
84. Liang, Y.; Zhang, J.; Wang, L.; Luo, H.; Ren, T. Forecasting spontaneous combustion of coal in underground coal mines by index gases: A review. *J. Loss Prev. Process. Ind.* **2018**, *57*, 208–222. [CrossRef]
85. Dong, X.; Wen, Z.; Wang, F.; Meng, Y. Law of gas production during coal heating oxidation. *Int. J. Min. Sci. Technol.* **2019**, *29*, 617–620. [CrossRef]
86. Xueqiu, H.; Xianfeng, L.; Baisheng, N.; Dazhao, S. FTIR and Raman spectroscopy characterization of functional groups in various rank coals. *Fuel* **2017**, *206*, 555–563. [CrossRef]
87. Zhao, J.; Deng, J.; Wang, T.; Song, J.; Zhang, Y.; Shu, C.-M.; Zeng, Q. Assessing the effectiveness of a high-temperature-programmed experimental system for simulating the spontaneous combustion properties of bituminous coal through thermokinetic analysis of four oxidation stages. *Energy* **2018**, *169*, 587–596. [CrossRef]
88. Cliff, D.; Davis, R.; Bennet, T.; Galvin, G.; Clarksosn, F. Large Scale Laboratory Testing of the Spontaneous Combustibility of Australian Coals. Queensland Mining Industry Health and Safety Conference Proceedings. 1998. Available online: https://www.osti.gov/etdeweb/biblio/296307 (accessed on 23 August 2024).
89. Xie, J.; Xue, S.; Cheng, W.; Wang, G. Early detection of spontaneous combustion of coal in underground coal mines with development of an ethylene enriching system. *Int. J. Coal Geol.* **2010**, *85*, 123–127. [CrossRef]
90. Qing, G.; Wanxing, R.; Wei, L. Risk evaluation of coal spontaneous combustion from the statistical characteristics of index gases. *Thermochim. Acta* **2022**, *715*, 179287. [CrossRef]
91. Ma, T.; Zhai, X.-W.; Xiao, Y.; Bai, Y.-E.; Shen, K.; Song, B.-B.; Hao, L.; Ren, L.-F.; Chen, X.-K. Study on the influence of key active groups on gas products in spontaneous combustion of coal. *Fuel* **2023**, *344*, 128020. [CrossRef]
92. Yang, S.; Hu, X.; Liu, W.V.; Cai, J.; Zhou, X. Spontaneous combustion influenced by surface methane drainage and its prediction by rescaled range analysis. *Int. J. Min. Sci. Technol.* **2018**, *28*, 215–221. [CrossRef]
93. MDG 1006 Spontaneous Combustion Management—Technical Reference. Available online: https://www.resourcesregulator.nsw.gov.au/ (accessed on 21 August 2024).
94. Liu, H.; Li, Z.; Li, J.; Yang, Y.; Zhang, Y. CO-based early warning threshold for coal spontaneous combustion. *Mater. Chem. Phys.* **2024**, *313*, 128730. [CrossRef]
95. Xu, Y.-L.; Liu, Z.-J.; Wang, L.-Y.; Lv, Z.-G.; Wu, J.-D.; Li, M.-J. Hysteresis characteristics of oxidation-thermodynamic for residual coal in goaf under uniaxial stress. *Fuel* **2021**, *306*, 121750. [CrossRef]
96. Danish, E.; Onder, M. Application of Fuzzy Logic for Predicting of Mine Fire in Underground Coal Mine. *Saf. Health Work.* **2020**, *11*, 322–334. [CrossRef] [PubMed]
97. Singh, A.K.; Singh, R.; Singh, M.P.; Chandra, H.; Shukla, N. Mine fire gas indices and their application to Indian underground coal mine fires. *Int. J. Coal Geol.* **2006**, *69*, 192–204. [CrossRef]

98. Yan, H.; Nie, B.; Liu, P.; Chen, Z.; Yin, F.; Gong, J.; Lin, S.; Wang, X.; Kong, F.; Hou, Y. Experimental assessment of multi-parameter index gas correlation and prediction system for coal spontaneous combustion. *Combust. Flame* **2023**, *247*, 112485. [CrossRef]
99. Lu, H.; Li, J.; Lu, W.; Xu, Z.; Li, J.; He, Q. Variation laws of CO2/CO and influence of key active groups on it during low-temperature oxidation of coal. *Fuel* **2023**, *339*, 127415. [CrossRef]
100. Wang, C.-P.; Deng, Y.; Zhang, Y.-T.; Xiao, Y.; Deng, J.; Shu, C.-M. Coal oxidation characteristics and index gases of spontaneous combustion during the heating and cooling processes. *Fuel* **2022**, *307*, 121806. [CrossRef]
101. Li, L.; Ren, T.; Zhong, X.; Wang, J. Study of ambient temperature oxidation in low metamorphic coal and the oxidation mechanism. *Energy* **2022**, *252*, 124039. [CrossRef]
102. Liu, H.; Wang, F.; Ren, T.; Qiao, M.; Yan, J. Influence of methane on the prediction index gases of coal spontaneous combustion: A case study in Xishan coalfield, China. *Fuel* **2020**, *289*, 119852. [CrossRef]
103. Zhang, D.; Cen, X.; Wang, W.; Deng, J.; Wen, H.; Xiao, Y.; Shu, C.-M. The graded warning method of coal spontaneous combustion in Tangjiahui Mine. *Fuel* **2020**, *288*, 119635. [CrossRef]
104. Cheng, J.; Ma, Y.; Lu, W.; Liu, G.; Cai, F. Using inverting CO critical value to predict coal spontaneous combustion severity in mine gobs with considering air leakages—A case study. *Process. Saf. Environ. Prot.* **2022**, *167*, 45–55. [CrossRef]
105. Chang, G.; Chang, H. Underground abnormal sensor condition detection based on gas monitoring data and deep learning image feature engineering. *Heliyon* **2023**, *9*, e22026. [CrossRef] [PubMed]
106. Marwan, N.; Carmenromano, M.; Thiel, M.; Kurths, J. Recurrence plots for the analysis of complex systems. *Phys. Rep.* **2007**, *438*, 237–329. [CrossRef]
107. Lu, Z.; Zhou, B.; Wang, J.; Chan, Z.; Yang, Q.; Dong, Z.; Dong, K. Study on the formation mechanism of a radon source during coal spontaneous combustion in a goaf. *Fuel* **2023**, *336*, 127135. [CrossRef]
108. Zhou, B.; Deng, C.; Hao, J.; An, B.; Wu, R. Experimental study on the mechanism of radon exhalation during coal spontaneous combustion in goaf. *Tunn. Undergr. Space Technol.* **2021**, *113*, 103776. [CrossRef]
109. Wen, H.; Xiaojiao, C.; Yanhui, X.; Li, M.; Jun, G.; Kiyin, J.; Baoqi, W.; Junchang, J. Law of radon precipitation and migration in loose coal during spontaneous combustion process. *J. China Coal Soc.* **2019**, *44*, 9. [CrossRef]
110. Chan, Z.; Zhou, B.; Wang, J.; Lu, Z.; Yang, Q.; Dong, Z.; Dong, K. Long-distance migration law of radon in overburden of abandoned goaf during coal spontaneous combustion. *J. Environ. Radioact.* **2023**, *270*, 107284. [CrossRef]
111. Wen, H.; Cheng, X.; Fan, S.; Xu, Y.; Ren, S.; Guo, J. A method for detecting hidden fire source in deep mine goafs based on radon measurement and its experimental verification. *Appl. Geochem.* **2020**, *117*, 104603. [CrossRef]
112. Ding, W.-C.; He, L.; Huang, H.-Q.; Xu, W.-F.; Li, S.-B.; He, T. Study on alpha cup technique for monitoring of soil radon levels. *J. Radioanal. Nucl. Chem.* **2017**, *314*, 1635–1641. [CrossRef]
113. Hu, X.; Sun, Q.; Shi, Q.; Wang, N.; Geng, J.; Xue, S. Radon exhalation characteristics after pyrolysis of long flame coal. *Sci. Total. Environ.* **2023**, *904*, 167228. [CrossRef]
114. Gbadamosi, A.; Onifade, M.; Genc, B.; Rupprecht, S. Analysis of spontaneous combustion liability indices and coal recording standards/basis. *Int. J. Min. Sci. Technol.* **2020**, *30*, 723–736. [CrossRef]
115. de Boer, C.B.; Dekkers, M.J.; van Hoof, T.A. Rock-magnetic properties of TRM carrying baked and molten rocks straddling burnt coal seams. *Phys. Earth Planet. Inter.* **2001**, *126*, 93–108. [CrossRef]
116. Kong, B.; Li, Z.; Wang, E.; Lu, W.; Chen, L.; Qi, G. An experimental study for characterization the process of coal oxidation and spontaneous combustion by electromagnetic radiation technique. *Process. Saf. Environ. Prot.* **2018**, *119*, 285–294. [CrossRef]
117. Wang, E.; He, X.; Liu, X.; Li, Z.; Wang, C.; Xiao, D. A non-contact mine pressure evaluation method by electromagnetic radiation. *J. Appl. Geophys.* **2011**, *75*, 338–344. [CrossRef]
118. Kong, G.; Zhou, L.; Peng, H.; Gu, H. Reduction rate of dragload and downdrag of piles by taper angles. *Trans. Tianjin Univ.* **2016**, *22*, 434–440. [CrossRef]
119. Kong, B.; Wang, E.; Li, Z.; Niu, Y. Time-varying characteristics of electromagnetic radiation during the coal-heating process. *Int. J. Heat Mass Transf.* **2017**, *108*, 434–442. [CrossRef]
120. Kong, B.; Wang, E.; Li, Z. Regularity and coupling correlation between acoustic emission and electromagnetic radiation during rock heating process. *Geomech. Eng.* **2018**, *15*, 1125–1133. [CrossRef]
121. Zhu, H.; Qu, B.; Liao, Q.; Xie, L.; Wang, J.; Hu, L.; Wang, H.; Gao, R. Evolution and mechanism for the terahertz dielectric spectrum of coal during oxidation. *Infrared Phys. Technol.* **2022**, *127*, 104412. [CrossRef]
122. Zhu, H.; Wang, W.; Wang, H.; Zhao, H.; Xin, M. Study on electrical properties of coal at spontaneous combustion characteristic temperature. *J. Appl. Geophys.* **2018**, *159*, 707–714. [CrossRef]
123. Karaoulis, M.; Revil, A.; Mao, D. Localization of a coal seam fire using combined self-potential and resistivity data. *Int. J. Coal Geol.* **2014**, *128-129*, 109–118. [CrossRef]
124. Shi, X.; Wu, K. Research and Application of Comprehensive Electromagnetic Detection Technique in Spontaneous Combustion Area of Coalfields. *Procedia Earth Planet. Sci.* **2011**, *3*, 195–202. [CrossRef]
125. Ren, S.-J.; Zhang, Y.-N.; Song, Z.-Y.; Xiao, Y.; Deng, J.; Shu, C.-M. Initial exploration on potential fire hazards detection from coal spontaneous combustion applied by acoustic wave. *Sci. Total. Environ.* **2023**, *897*, 165475. [CrossRef] [PubMed]
126. Deng, J.; Qu, G.; Ren, S.; Wang, C.; Xiao, Y.; Wang, J.; Sa, B.; Duan, X.; Yang, N.; Zhao, X. Investigation on high temperature point detection of spontaneous combustion of loose coal based on optimal acoustic signal. *Process. Saf. Environ. Prot.* **2024**, *185*, 423–434. [CrossRef]

127. Kong, B.; Zhong, J.; Wei, J.; Lu, W.; Sun, X.; Yang, G.; Zhao, X.; Ma, L. Study on sound wave kinematic characteristics and temperature sensing mechanism during the warming process of loose coals. *Energy* **2024**, *307*, 132753. [CrossRef]
128. Liu, J.; Ma, Y.; Kong, B.; Bing, Y.; Yang, T.; Zhao, X.; Ma, L. Study on the precursor characteristics of coal energy spontaneous combustion process using infrasound wave monitoring and warning. *Energy* **2024**, *292*, 130406. [CrossRef]
129. Ide, T.S.; Crook, N.; Orr, F. Magnetometer measurements to characterize a subsurface coal fire. *Int. J. Coal Geol.* **2011**, *87*, 190–196. [CrossRef]
130. Shao, Z.; Wang, D.; Wang, Y.; Zhong, X. Theory and application of magnetic and self-potential methods in the detection of the Heshituoluogai coal fire, China. *J. Appl. Geophys.* **2014**, *104*, 64–74. [CrossRef]
131. Pandey, J.; Kumar, D.; Mishra, R.K.; Mohalik, N.K.; Khalkho, A.; Singh, V.K. Application of Thermography Technique for Assessment and Monitoring of Coal Mine Fire: A Special Reference to Jharia Coal Field, Jharkhand, India. *Int. J. Adv. Remote Sens.* **2013**, *2*, 138–147.
132. Misz-Kennan, M.; Tabor, A. The Thermal History of Select Coal-Waste Dumps in the Upper Silesian Coal Basin, Poland. In *Coal and Peat Fires: A Global Perspective*; Stracher, G.B., Prakash, A., Sokol, E.V., Eds.; Elsevier Inc.: Amsterdam, The Netherlands, 2015; Volume 3, pp. 432–462. [CrossRef]
133. Roy, P.; Guha, A.; Kumar, K.V. An approach of surface coal fire detection from ASTER and Landsat-8 thermal data: Jharia coal field, India. *Int. J. Appl. Earth Obs. Geoinf.* **2015**, *39*, 120–127. [CrossRef]
134. Wang, H.; Fang, X.; Li, Y.; Zheng, Z.; Shen, J. Research and application of the underground fire detection technology based on multi-dimensional data fusion. *Tunn. Undergr. Space Technol.* **2020**, *109*, 103753. [CrossRef]
135. Wang, H.; Fan, C.; Chen, L.; Chen, X.; Zhang, J.; Zhong, H. Research on early identification of burning status in a fire area in Xinjiang based on data-driven. *Case Stud. Therm. Eng.* **2024**, *60*, 104685. [CrossRef]
136. Rúa, M.B.; Aragón, A.D.; Baena, P.B.; Botero, J.O. Statistical analysis to establish an ignition scenario based on extrinsic and intrinsic variables of coal seams that affect spontaneous combustion. *Int. J. Min. Sci. Technol.* **2019**, *29*, 731–737. [CrossRef]
137. Abramowicz, A.; Rahmonov, O.; Chybiorz, R.; Ciesielczuk, J. Vegetation as an indicator of underground smoldering fire on coal-waste dumps. *Fire Saf. J.* **2021**, *121*, 103287. [CrossRef]
138. Ren, H.; Zhao, Y.; Xiao, W.; Zhang, J.; Chen, C.; Ding, B.; Yang, X. Vegetation growth status as an early warning indicator for the spontaneous combustion disaster of coal waste dump after reclamation: An unmanned aerial vehicle remote sensing approach. *J. Environ. Manag.* **2022**, *317*, 115502. [CrossRef]
139. Wang, H.; Fang, X.; Du, F.; Tan, B.; Zhang, L.; Li, Y.; Xu, C. Three-dimensional distribution and oxidation degree analysis of coal gangue dump fire area: A case study. *Sci. Total. Environ.* **2021**, *772*, 145606. [CrossRef]
140. Syed, T.H.; Riyas, M.J.; Kuenzer, C. Remote sensing of coal fires in India: A review. *Earth-Sci. Rev.* **2018**, *187*, 338–355. [CrossRef]
141. Yuan, G.; Wang, Y.; Zhao, F.; Wang, T.; Zhang, L.; Hao, M.; Yan, S.; Dang, L.; Peng, B. Accuracy assessment and scale effect investigation of UAV thermography for underground coal fire surface temperature monitoring. *Int. J. Appl. Earth Obs. Geoinf.* **2021**, *102*, 102426. [CrossRef]
142. Shao, Z.; Deng, R.; Zhang, G.; Li, Y.; Tang, X.; Zhang, W. 3D thermal mapping of smoldering coal gangue pile fires using airborne thermal infrared data. *Case Stud. Therm. Eng.* **2023**, *48*, 103146. [CrossRef]
143. He, X.; Yang, X.; Luo, Z.; Guan, T. Application of unmanned aerial vehicle (UAV) thermal infrared remote sensing to identify coal fires in the Huojitu coal mine in Shenmu city, China. *Sci. Rep.* **2020**, *10*, 13895. [CrossRef]
144. Teodoro, A.; Santos, P.; Marques, J.E.; Ribeiro, J.; Mansilha, C.; Melo, A.; Duarte, L.; de Almeida, C.R.; Flores, D. An Integrated Multi-Approach to Environmental Monitoring of a Self-Burning Coal Waste Pile: The São Pedro da Cova Mine (Porto, Portugal) Study Case. *Environments* **2021**, *8*, 48. [CrossRef]
145. Song, Z.; Kuenzer, C.; Zhu, H.; Zhang, Z.; Jia, Y.; Sun, Y.; Zhang, J. Analysis of coal fire dynamics in the Wuda syncline impacted by fire-fighting activities based on in-situ observations and Landsat-8 remote sensing data. *Int. J. Coal Geol.* **2015**, *141–142*, 91–102. [CrossRef]
146. Jiménez-Muñoz, J.C.; Sobrino, J.A. A generalized single-channel method for retrieving land surface temperature from remote sensing data. *J. Geophys. Res. Atmos.* **2003**, *108*, D22. [CrossRef]
147. Jiang, W.; Jia, K.; Chen, Z.; Deng, Y.; Rao, P. Using spatiotemporal remote sensing data to assess the status and effectiveness of the underground coal fire suppression efforts during 2000–2015 in Wuda, China. *J. Clean. Prod.* **2017**, *142*, 565–577. [CrossRef]
148. Jenks, G.F. The data model concept in statistical mapping. *Int. Yearb. Cartogr.* **1967**, *7*, 186–190.
149. Wang, T.; Wang, Y.; Zhao, F.; Feng, H.; Liu, J.; Zhang, L.; Zhang, N.; Yuan, G.; Wang, D. A spatio-temporal temperature-based thresholding algorithm for underground coal fire detection with satellite thermal infrared and radar remote sensing. *Int. J. Appl. Earth Obs. Geoinf.* **2022**, *110*, 102805. [CrossRef]
150. Yu, B.; She, J.; Liu, G.; Ma, D.; Zhang, R.; Zhou, Z.; Zhang, B. Coal fire identification and state assessment by integrating multitemporal thermal infrared and InSAR remote sensing data: A case study of Midong District, Urumqi, China. *ISPRS J. Photogramm. Remote. Sens.* **2022**, *190*, 144–164. [CrossRef]
151. Jiang, L.; Lin, H.; Ma, J.; Kong, B.; Wang, Y. Potential of small-baseline SAR interferometry for monitoring land subsidence related to underground coal fires: Wuda (Northern China) case study. *Remote. Sens. Environ.* **2010**, *115*, 257–268. [CrossRef]
152. Mishra, R.; Bahuguna, P.; Singh, V. Detection of coal mine fire in Jharia Coal Field using Landsat-7 ETM+ data. *Int. J. Coal Geol.* **2011**, *86*, 73–78. [CrossRef]

153. Zhou, L.; Zhang, D.; Wang, J.; Huang, Z.; Pan, D. Mapping Land Subsidence Related to Underground Coal Fires in the Wuda Coalfield (Northern China) Using a Small Stack of ALOS PALSAR Differential Interferograms. *Remote. Sens.* **2013**, *5*, 1152–1176. [CrossRef]
154. Raju, A.; Singh, A.; Chandniha, S.K. A synergetic approach for quantification and analysis of coal fires in Jharia Coalfield, India. *Phys. Chem. Earth Parts A/B/C* **2023**, *131*, 103441. [CrossRef]
155. He, D.; Le, B.T.; Xiao, D.; Mao, Y.; Shan, F.; Ha, T.T.L. Coal mine area monitoring method by machine learning and multispectral remote sensing images. *Infrared Phys. Technol.* **2019**, *103*, 103070. [CrossRef]
156. Singh, N.; Chatterjee, R.; Kumar, D.; Panigrahi, D. Spatio-temporal variation and propagation direction of coal fire in Jharia Coalfield, India by satellite-based multi-temporal night-time land surface temperature imaging. *Int. J. Min. Sci. Technol.* **2021**, *31*, 765–778. [CrossRef]
157. Elick, J.M. Mapping the coal fire at Centralia, Pa using thermal infrared imagery. *Int. J. Coal Geol.* **2011**, *87*, 197–203. [CrossRef]
158. Karanam, V.; Motagh, M.; Garg, S.; Jain, K. Multi-sensor remote sensing analysis of coal fire induced land subsidence in Jharia Coalfields, Jharkhand, India. *Int. J. Appl. Earth Obs. Geoinf.* **2021**, *102*, 102439. [CrossRef]

fire

MDPI

Article

Experimental Study of the Thermoelectric Conversion Characteristics of a Device Combining a TPCT and TGs

Haining Qi [1] and Hetao Su [2,*]

[1] School of Safety Engineering, China University of Mining and Technology, Xuzhou 221116, China; tb17120013b2@cumt.edu.cn
[2] School of Engineering and Technology, China University of Geosciences (Beijing), Beijing 100083, China
* Correspondence: h.su@cugb.edu.cn

Abstract: In this paper, the thermoelectric conversion characteristics of a device combining a TPCT and TGs are studied. The experimental devices consist of four parts: TPCT heat transfer module, cooling and heat dissipation module, thermoelectric power generation module, and data collection module. The effects of different heating powers (100 W, 200 W, 400 W, and 600 W) and different liquid filling rates of the TPCT (10%, 25%, 35%, and 45%) on the heat transfer performance and the power generation performance of the device are studied. The research indicates that the impact of the liquid filling rate on heat transfer and power generation performance is less significant than that of heating power. As the heating power increases, both the heat transfer and power generation performance of the device will improve and is finally in a relatively stable state. The thermal resistance at the liquid filling rate of 10% is the smallest, roughly around 0.11 °C/W. At a heating power of 200 W, the TPCT at the liquid filling rate of 10% has the largest heat transfer efficiency, which is 83.36%. The maximum values of power generation efficiency and net power generation efficiency are 2.27% and 3.10%, respectively.

Keywords: coalfield fires; two-phase closed thermosiphon; thermoelectric generators; thermoelectric conversion; liquid filling rate

1. Introduction

Coalfield fires, which are the burning or overcast combustion of underground coal seams, are a widespread geological challenge worldwide. Coalfield fires cause great waste of coal resources and serious damage to the ecological environment. It is also a serious threat to human health and safety and hinders social and economic development [1]. Traditional means of coal fire management mainly include stripping, drilling, water injection, grouting, and covering loess, which are prone to ecological damage and have limited cooling effect, and the coalfield fire is not easily extinguished completely [2]. A large amount of water resources is wasted in the process of fire extinguishing, which easily causes groundwater pollution. About 1000 GW of waste heat energy from coalfield fires worldwide every year is wasted and not effectively utilized, which is equivalent to 2.5 times of the total power generation of 500 nuclear power plants worldwide and more than the sum of global water power generation [3]. If the waste heat resources can be recovered and effectively used, then the coalfield fire treatment process can be accelerated and water resources can be saved to avoid ecological damage and resource waste, which will produce huge ecological and environmental value and social and economic benefits.

In order to recover waste heat resources and convert them into renewable resources, a good solution is to apply two-phase closed thermosiphons (TPCTs). The TPCT has a simple structure, low cost, and high heat transfer efficiency. The heat can only be transferred from the bottom to the top of the TCPT in one direction, due to the effect of gravity [4]. TPCTs are used in many fields and have the advantages of excellent heat transfer performance, environmental adaptability, low economic cost, energy saving, and environmental protection [5]. In the field of permafrost ground temperature control and road snow melting, the excellent thermal conductivity of the TPCT, the absence of external power source and maintenance, and the good applicability to different engineering environments make it become an important method [6]. Studies have shown that TPCTs can reasonably solve the air–TPCT–soil coupled heat transfer problem of TPCT embankments [7]. It effectively attenuates the thermal effect of sunward slope, cools the subsurface permafrost, and ensures the stability of the permafrost foundation [8,9]. TPCTs are used to transfer the heat stored in the subsurface soil or groundwater to the road surface to melt snow and ice. It can overcome the drawbacks of traditional snow melting techniques, substantially increase the temperature of concrete pavements, and prove that the system can operate for a long time [10–12]. In industrial production, TPCTs are also widely used. In the nuclear industry, passive heat dissipation from TPCTs is used to remove decay heat from spent fuel pools in the event of a nuclear power plant blackout [13–15]. Adsorption chillers with a separated two-phase closed thermosiphon heating process allow the system to have fewer moving parts and reduced building size, while improving the performance factor and cooling power ratio of the cooling system [16]. The use of TPCTs can improve the fluid temperature distribution in the oilfield wellbore and can effectively transfer the heat contained in the high-temperature fluid at the bottom of the wellbore to the lower-temperature fluid in the upper part of the wellbore, increasing the oil production rate [17,18]. Energy-efficient technology for phosphate production was developed using two-phase closed thermosiphon [19]. A cryogenic TPCT was made using copper tubes, flexible bellows hoses with nitrogen fluid, and its heat transfer properties were used to design a thermally stable system for RED100 detectors that can provide heat transfer rates of up to 100 W in the temperature range of 100–80 K [20]. The application of TPCTs to the field of geothermal extraction has improved the efficiency of heat extraction [21]. The heat transfer performance of solar collectors was enhanced using TPCTs [5,22]. In the field of coalfield fires, researchers have also used TPCTs to remove heat sources and extract heat energy. Li et al. [23] established a coal pile–TPCT heat transfer model and used a TPCT to control the coal pile temperature below 80 °C, which had a significant effect on coal pile heat removal. Zhong et al. [24] used a TPCT to suppress heat production during coal spontaneous combustion, and the use of CuO nanofluidic material could improve the heat transfer performance of the TPCT by 27.9%. However, the study only considered the transfer of heat energy from coalfield fires, and further research is needed on heat utilization and heat conversion.

Therefore, after waste heat from coalfield fires is extracted and recovered, the next issue to be considered is to solve the problem of how waste heat from coalfield fires is utilized and converted, and thermoelectric power generation is a very suitable solution. Thermoelectric power generation uses the Seebeck effect of thermoelectric materials to directly convert thermal energy into electrical energy, which is a new green, environmentally friendly, and pollution-free power generation technology [25]. Currently, commercial thermoelectric power generation modules are generally composed of multiple PN galvanic arms connected in series, and the hot and cold ends are covered using ceramic sheets to form a temperature differential generator with multiple PN junctions [26]. Temperature differential generators (from AltaRock Energy, Inc., Sausalito, CA, USA) can convert low-taste heat into clean electrical energy and have the advantages of flexible installation, noiseless operation, environmental friendliness, high applicability, low maintenance cost, and long service life [27]. Therefore, thermoelectric power generation is also used in many fields. The combination of TGs with biomass burners can power small-scale electrical devices [28–30]. It is combined with solar energy for the co-production of solar thermal and electrical energy [31,32]. In the field of low-temperature geothermal utilization, it converts heat flow from the upper soil layer into electrical energy [33,34]. It is combined with small nuclear reactors to supply electrical energy to space probes [35,36]. In the automotive field, it recycles the heat energy from the exhaust of the car [37–39]. Using the function of TGs energized and cooled to dissipate heat for processor chips and electronic devices [40–42]. In the field of coalfield fire utilization, researchers have also applied thermoelectric power generation. Su et al. [1] concluded that thermoelectric power generation has the advantages of good heat source adaptation, easy assembly, and no location dependence by comparing a series of heat recovery modes of KC (Carina cycle) and ORC (Rankine cycle). It was proved to be practical in the field of coalfield fire utilization. Shi et al. [3] used the temperature difference between the heat transfer medium and the cooling medium to provide energy for thermoelectric power generation, and achieved a single-hole power generation power of 174.6 W in the Daquan Lake fire area in Xinjiang. Su et al. [2] carried out thermoelectric power generation model experiments, analyzed the main performance indexes of thermoelectric power generation, and designed a distributed thermoelectric power generation device, which can achieve a maximum output power of 700 W at a temperature difference of 80 °C. However, the study only considers the thermoelectric power generation technology for heat energy, and there is less research on heat transfer.

The above studies on the recovery and utilization of waste heat resources in coalfield fires have considered only TPCT heat transfer or thermoelectric power generation alone. Recently, the system combining TPCT and thermoelectric power generation technology has been studied and applied. Su et al. [43] studied the temperature gradient of the TPCT and the characteristics of temperature difference power generation. Deng et al. [44] established a thermal energy recovery system combining TPCT and thermoelectric power generation in the Sandaoba fire area in Xinjiang and evaluated the recovered heat. However, these studies do not consider the influence of TPCT heat transfer and temperature difference power generation together. Studies that consider how the heat transfer from the heat source and the TPCT parameters together affect the thermoelectric conversion performance of the system are lacking.

Coalfield fires are a global challenge, causing huge waste of resources and environmental problems. The vast heat resources from coalfield fires have great economic and ecological value [1]. The efficient heat transfer performance of two-phase closed thermosiphons (from Chenyi Heat pipe Company, Harbin, China) (TPCTs) and the direct power generation performance of thermoelectric generators (from Seebeck Company, Changsha, China) (TGs) are well suited for the recovery and utilization of waste heat resources in coalfield fire areas. It is necessary to study the recovery and utilization of waste heat resources from coalfield fires by studying TPCT heat transfer and thermoelectric power generation. Analyzing how different heat source heat transfers and TPCT parameters affect the system heat transfer performance and power generation performance, which is helpful to select the appropriate TPCT parameters for temperature level of heat sources for coalfield fires, will ensure that the thermoelectric conversion performance is optimal. This paper preliminarily investigates the thermoelectric conversion characteristics based on a two-phase closed thermosiphon and a temperature difference generator. By building an experimental setup, the effects of heating power (100 W, 200 W, 400 W, and 600 W) and liquid filling rate of the TPCT (10%, 25%, 35%, and 45%) on the heat transfer performance and power generation performance of the device are experimentally investigated. Combining a two-phase closed thermosiphon with thermoelectric power generation can bring out the advantages of each.

It will have broad application prospects in the field of coalfield fire waste heat resource recovery and utilization.

2. Materials and Methods

2.1. Experimental Devices

Figure 1 shows the experimental devices of thermoelectric conversion based on a TPCT and TGs. The effect of the liquid filling rate of the TPCT and heating power on the heat recovery and utilization of the experimental device is studied and analyzed in terms of both heat transfer and power generation. The experimental devices consist of four parts: TPCT heat transfer module, cooling and heat dissipation module, thermoelectric power generation module, and data collection module.

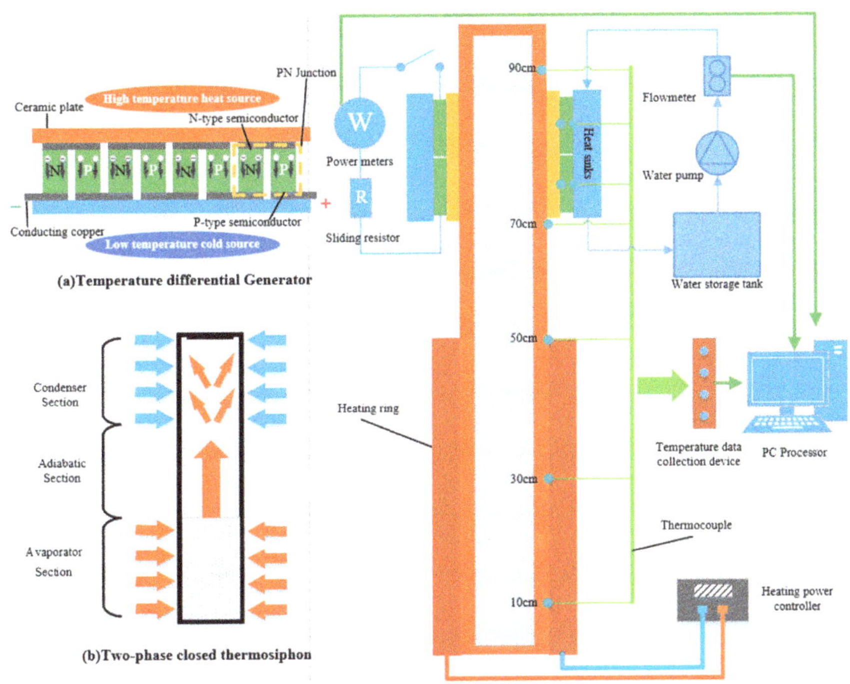

Figure 1. Experimental devices for thermoelectric conversion based on a TCPT and TGs. (**a**) Temperature differential Generator. (**b**) Two-phase closed thermosiphon.

2.1.1. TPCT Heat Transfer Module

The TPCT heat transfer module consists of a heating ring, a heating power controller and a TPCT. The heating power of the heater ring (from Safety Valley, Xuzhou, China) is in the range of 0–600 W and can be adjusted. The TPCT evaporation section is heated by heat conduction. The heating power controller can change the heating power of the heating ring by adjusting the voltage and current, and the adjustment range is 0–600 W. The parameters of TCPT are as follows: the length is 1000 mm; evaporation section is 500 mm; condensation section is 500 mm; outer diameter is 38 mm; thickness of TPCT wall is 3 mm; liquid filling ratio is 10%, 25%, 35%, and 45%; liquid filling medium is distilled water. The evaporation section of the TPCT is entirely located in the heating ring. The liquid in the evaporation section of the TPCT is heated to evaporate and liquefies in the condensation section when it is cold, releasing heat. Then, the condensing section transfers the heat to the thermoelectric power generation module by heat conduction.

2.1.2. Cooling and Heat Dissipation Module

The cooling and heat dissipation module consists of a water storage tank, a water pump and a heat sink. The water pump can increase the flow of cold water in the radiator, thus reducing the temperature of the cold end of the thermoelectric power generation module. There are channels inside the radiator that can circulate cold water, it has cooling fins arranged on the outer surface, and there are two cooling methods: water cooling and air cooling, which can enhance the cooling effect.

2.1.3. Thermoelectric Power Generation Module

The thermoelectric power generation module uses four TGs (from Seebeck Company, Changsha, China). The model number is TEG1–241-1.4–1.2. The two TGs on each side of the TPCT are connected in series and then in parallel. The thermoelectric power generation module is connected to the sliding resistor and then connected to the power meter. When the water in the cooling heat sink module flows through the heat sink, a temperature difference is created on both sides of the thermoelectric power generation module. Adjusting the radiator water flow can change the temperature of the condensing section of the TPCT, which can change the temperature difference of the temperature differential power module. The average of the temperatures measured by the two K-type thermocouples at the hot end of the differential temperature generation module is defined as the hot end temperature. The average of the temperatures measured by the two K-type thermocouples at the cold end is defined as the cold end temperature.

2.1.4. Data Acquisition Module

The data acquisition module consists of a PC processor, a temperature data acquisition device, nine K-type thermocouples, a power meter, and a flow meter. K-type thermocouple model is HH-K-24-SLE (from Zetian Company, Xuzhou, China), measuring temperature range is −73−704 °C, and accuracy is 0.5 °C. There are 13 thermocouples laid out, distributed on the surface of the TPCT wall and the hot and cold ends of the thermoelectric power generation assembly. The power meter is used to measure the open-circuit voltage, load voltage, and load power of the temperature difference generation module. The flow meter is used to measure the flow of cold water in the cooling heat sink module, its model is vortex flow meter MJ-A68-1 (from Zetian Company, Xuzhou, China), and its applicable flow range 0.2–6 L/min. Temperature, voltage, power, flow rate, and other data are uniformly processed by the PC processor.

2.1.5. Experimental Methods

To study and analyze the effect of the liquid filling rate of the TCPT and heating power on the heat recovery and utilization of the experimental devices, the heat transfer performance and power generation performance of TPCTs with four liquid filling rates (10%, 25%, 35%, and 45%) at different heating powers (100 W, 200 W, 400 W, and 600 W) are studied. The TPCT is heated for 50 min at four specific heating powers until the temperatures were steady. Then, the cooling module is turned on to observe the parameters of TCPTs and TGs under different temperature conditions. During the monitoring experiment, the temperature at different distances along the TCPT (10 cm, 30 cm, 50 cm, 70 cm, and 90 cm), cold and hot end temperature of TGs, ambient temperature, cooling water flow, open-circuit voltage of TGs, load current of TGs, and load power of TGs are monitored.

2.2. Principle and Calculation

During the two-phase closed thermosiphon heat transfer and thermoelectric conversion experiment, the main destination of the heat released from the heating source is two parts. One part of the heat is transferred to the condensing section of the TPCT through the evaporation section of the TPCT by heat absorption. The other part of the heat is released to the ambient atmosphere in the form of radiation and convection heat transfer at the wall of the evaporation section of the TPCT. Then, the heat that is transferred to the condensing section of the TPCT goes to three main parts. One part is converted into electricity by the temperature difference generator. One part is carried away by the radiator in the form of hot water. One part is released by the exposed condensing section of the TPCT in the form of radiation and convection heat transfer.

Therefore, in order to accurately calculate parameters such as TPCT heat transfer efficiency and thermoelectric efficiency, a series of calculations are required for the heat transfer process of the device. The heat transfer losses in the experimental heating process of the TPCT are radiation heat transfer losses and convective heat transfer losses. Because the convective heat transfer coefficient between the TPCT material and the air is very low, it is known that the main heat transfer loss is the radiation heat transfer loss.

2.2.1. Heat Transfer Loss

Radiation Heat Transfer Loss

TPCT radiation heat transfer losses include radiation heat transfer losses in the evaporation section and radiation heat transfer losses in the condensation section. The radiative heat transfer from the closed cavity consisting of two blackbody surfaces is:

$$\Phi_{1,2} = A_1 E_{b1} X_{1,2} - A_2 E_{b2} X_{2,1} = A_1 X_{1,2}(E_{b1} - E_{b2}) = A_2 X_{2,1}(E_{b1} - E_{b2}) \quad (1)$$

$\Phi_{1,2}$ is the radiation heat transfer. A_1 is the area of object 1. A_2 is the area of object 2. $X_{1,2}$ and $X_{1,2}$ are the angular coefficients of radiative heat transfer. E_{b1} is the radiation force of object 1, $E_{b1} = \sigma {T_1}^4$. E_{b2} is the radiation force of object 2, $E_{b2} = \sigma {T_2}^4$.

$$\begin{aligned} X_{1,2} = \frac{2}{\pi xy}\{\ln\left[\frac{(1+x^2)(1+y^2)}{1+x^2+y^2}\right]^{\frac{1}{2}} - x\arctan x + x\sqrt{1+y^2}\arctan\frac{x}{\sqrt{1+y^2}} \\ -y\arctan y + y\sqrt{1+x^2}\arctan\frac{y}{\sqrt{1+x^2}}\} \end{aligned} \quad (2)$$

σ is the Stephan–Boltzmann constant, which is the blackbody radiation constant. 5.67×10^{-8} W/(m^2·K^4).

TPCT radiation heat transfer is consistent with radiation heat transfer from a closed cavity consisting of two diffuse ash surfaces. The radiative heat transfer in the closed cavity consisting of two diffuse ash surfaces is:

$$\Phi_{1,2} = \frac{E_{b1} - E_{b2}}{\frac{1-\varepsilon_1}{\varepsilon_1 A_1} + \frac{1}{A_1 X_{1,2}} + \frac{1-\varepsilon_2}{\varepsilon_2 A_2}} \quad (3)$$

Using A_1 as the calculated area, the above equation can be rewritten as:

$$\Phi_{1,2} = \frac{A_1(E_{b1} - E_{b2})}{\left(\frac{1}{\varepsilon_1} - 1\right) + \frac{1}{X_{1,2}} + \frac{A_1}{A_2}\left(\frac{1}{\varepsilon_2} - 1\right)} = \varepsilon_s A_1 X_{1,2}(E_{b1} - E_{b2}) \quad (4)$$

In Equation (5):

$$\varepsilon_s = \frac{1}{1 + X_{1,2}\left(\frac{1}{\varepsilon_1} - 1\right) + X_{2,1}\left(\frac{1}{\varepsilon_2} - 1\right)} \quad (5)$$

ε_1 is the emissivity of object 1. ε_2 is the emissivity of object 2. ε_s is the emissivity of the system. When the radiating surface of object 1 is a flat or convex surface, $X_{1,2} = 1$, then:

$$\Phi_{1,2} = \varepsilon_s A_1 (E_{b1} - E_{b2}) \tag{6}$$

The system emissivity is:

$$\varepsilon_s = \frac{1}{\frac{1}{\varepsilon_1} + \frac{A_1}{A_2}\left(\frac{1}{\varepsilon_2} - 1\right)} \tag{7}$$

When the surface area A_2 is much larger than A_1, $A_1/A_2 \to 0$. Surface 1 is a radiant heat transfer system with a non-concave surface. Then, Equation (4) is:

$$\Phi_{1,2} = \varepsilon_1 A_1 (E_{b1} - E_{b2}) \tag{8}$$

In this experiment, the TPCT radiation heat transfer is in accordance with the description in Equation (8). Therefore, the TPCT radiation heat transfer loss is:

$$Q_2 = \varepsilon_1 A_1 \times 5.67\ \mathrm{W/m^2{\cdot}}K^4 \left[(T_1/100)^4 - (T_2/100)^4 \right] \tag{9}$$

ε_1 is the TPCT wall emissivity, taken as 0.8.

The above can be obtained from the TPCT convection heat transfer loss being much smaller than the TPCT radiation heat transfer loss. Therefore, the TPCT heat transfer loss is taken as the TPCT radiation heat transfer loss value. TPCT heat transfer losses are:

$$Q_s = Q_2 \tag{10}$$

TPCT heat transfer power Q_c is the difference between heating power Q_j and TPCT heat transfer loss Q_s:

$$Q_c = Q_j - Q_s \tag{11}$$

Convective Heat Transfer Loss

TPCT convective heat transfer losses include convective heat transfer losses in the evaporation section and convective heat transfer losses in the condensation section. The convection between the TPCT and the air exists as natural convection and forced convection. From the equation ($Gr/Re^2 \geq 10$) of judgment, we know that the effect of forced convection is negligible for natural convection. So, the TPCT convection heat transfer for the large space is natural convection heat transfer. The convective heat transfer between the air and the evaporative section of the TPCT conforms to the large space natural convective heat transfer with uniform wall temperature boundary conditions.

The experimental correlation equation for natural convection in large spaces is:

$$Nu_m = C(GrPr)^n_m \tag{12}$$

Nu_m is the number of Nu consisting of the average surface heat transfer coefficient. The subscript m indicates that the arithmetic average temperature of the boundary layer ($t_m = (t_\infty + t_w)/2$) is used for the qualitative temperature. Pr is the Prandtl number. Gr is the dimensionless Grashov number with H as the characteristic length: $Gr = \frac{g\alpha_V \Delta t H^3}{v^2}$. g is the acceleration of gravity. α_V is the coefficient of volume change. Ideal gas: α_V is the inverse of T. Δt is the temperature difference. $\Delta t = t_w - t_\infty$. t_w is the wall surface temperature. t_∞ is the ambient temperature. The TPCT conforms to the following inequality for the vertical cylinder.

$$\frac{d}{H} \geq \frac{35}{Gr_{\mathrm{H}}^{1/4}} \tag{13}$$

d is the diameter of the TPCT. H is the length of the evaporating section of the TPCT. The constant C is 0.11 and the coefficient n is 1/3 [45]. The Nu number is composed of the average surface heat transfer coefficient.

$$Nu_{\mathrm{m}} = C(GrPr)_{\mathrm{m}}^{\mathrm{n}} \tag{14}$$

Nu_{m} is also called the dimensionless number Nussle number with H as the characteristic length.

$$Nu_{\mathrm{m}} = \frac{h_{\mathrm{H}}H}{\lambda} \tag{15}$$

h_{H} is the surface heat transfer coefficient for convective heat transfer in the evaporation section of the TPCT. λ is the thermal conductivity of the air fluid. h_{H} can be obtained from above. Newton's formula for convective heat transfer in heat transfer:

$$Q_1 = A_{\mathrm{Z}} h_{\mathrm{H}} \Delta T_1 \tag{16}$$

Q_1 is the convective heat exchange between air and TPCT evaporation section. A_{Z} is the exterior area of evaporation section. h_{H} is the heat transfer coefficient of the surface of the TPCT evaporation section.

$$A_{\mathrm{Z}} = \pi d_{\mathrm{o}} L_1 \tag{17}$$

Δt is the temperature difference of convective heat exchange.

$$\Delta t = t_{\mathrm{w}} - t_\infty \tag{18}$$

d_{o} is the diameter of the outer surface of the TPCT evaporation section. L_1 is the length of the evaporation section of the TPCT. The above can be obtained from the TPCT convection heat transfer loss Q_1 is much smaller than the TPCT radiation heat transfer loss Q_2. So, the TPCT heat transfer loss is taken as the TPCT radiation heat transfer loss value.

2.2.2. Thermal Resistance of TPCT

A typical TPCT is divided into three sections: the evaporation section, the adiabatic section, and the condensation section. The latent heat of phase change of the working medium is utilized by the TPCT. This allows for rapid heat transfer from the evaporative section to the condensing section. A schematic diagram of the structure of a TPCT is shown in Figure 1b. The heat transfer process consists of six interrelated phase processes.

Heat is transferred from the heating source to the evaporation section liquid through the wall of the TPCT.

1. The liquid in the evaporation section is heated to evaporate and absorb heat.
2. The evaporated vapor is transferred along the TPCT to the condensing section of the TPCT.
3. Vapor in the condensing section of the wall of the tube is exothermic condensation.
4. Heat is transferred from the vapor–liquid through the wall of the TPCT to the cold source.
5. The liquid condensed in the condensing section wall flows back to the evaporating section due to gravity.

The thermal resistance per section of the TPCT in the heat transfer process is as follows Table 1.

Table 1. The thermal resistance per section of the TPCT in the heat transfer process.

Definition	Expressions	
The heat transfer thermal resistance R_0 from the heating source to the outer wall of the evaporation section of the TPCT	$R_0 = \frac{1}{\pi d_e l_e h_{o,e}}$	(19)
The thermal conductivity thermal resistance R_1 from the outer wall to the inner wall of the evaporation section of the TPCT	$R_1 = \frac{1}{2\pi\lambda l_e} ln \frac{d_e}{d_{ie}}$	(20)
The liquid-vapor evaporative heat transfer thermal resistance R_2 of the medium inside the evaporation section	$R_2 = \frac{1}{\pi d_{ie} l_e h_{i,e}}$	(21)
The thermal resistance R_3 caused by the pressure drop in the vapor flow from the evaporation section to the condensation section of the TPCT	$R_3 \approx 0$	(22)
The heat transfer resistance R_4 of the condensing section medium vapor–liquid condensation	$R_4 = \frac{1}{\pi d_{ic} l_c h_{i,c}}$	(23)
The thermal conductivity thermal resistance R_5 from the inner wall to the outer wall of the condensing section of the TPCT	$R_5 = \frac{1}{2\pi\lambda l_c} ln \frac{d_c}{d_{ic}}$	(24)
The heat transfer thermal resistance R_6 between the outer wall of the condensing section of the TPCT and the heat sink	$R_6 = \frac{1}{\pi d_c l_c h_{o,c}}$	(25)

$h_{o,e}$ is the total surface heat transfer coefficient between the wall of the evaporation section of the TPCT and the heating source. $h_{o,c}$ is the total surface heat transfer coefficient between the wall of the condensing section of the TPCT and the heat sink. d_e is the outside diameter of the evaporating section of the TPCT. d_c is the outside diameter of the condensing section of the TPCT. d_{ie} is the inner diameter of the TPCT evaporation section. d_{ic} is the inner diameter of the condensing section of the TPCT. λ is the thermal conductivity of the TPCT material. $h_{i,e}$ is the surface heat transfer coefficient of the TPCT evaporation heat transfer. $h_{i,c}$ is the surface heat transfer coefficient of the TPCT condensation heat transfer.

R_1, R_2, R_3, R_4, and R_5 are the thermal resistances inside the TPCT, and the total internal resistance is:

$$R = R_1 + R_2 + R_3 + R_4 + R_5 \\ = \frac{1}{2\pi\lambda l_e} ln \frac{d_e}{d_{ie}} + \frac{1}{\pi d_{ie} l_e h_{i,e}} + 0 + \frac{1}{\pi d_{ic} l_c h_{i,c}} + \frac{1}{2\pi\lambda l_c} ln \frac{d_c}{d_{ic}} \quad (26)$$

The heat transferred from the evaporation section to the surface of the condensation section per unit time is the TPCT heat transfer power Q_c. It is the difference between the heating power Q_j and the TPCT heat transfer loss Q_s.

$$Q_c = Q_j - Q_s \quad (27)$$

$$Q_c = \frac{\Delta t_{ec}}{R} \quad (28)$$

$$R = \frac{\Delta t_{ec}}{Q_j - Q_s} \quad (29)$$

$$\eta_c = \frac{Q_c}{Q_j} \times 100\% \quad (30)$$

Δt_{ec} is the temperature difference between the surface of the evaporation section and the surface of the condensation section of the TPCT. η is the heat transfer efficiency.

2.2.3. Thermoelectric Power Generation

As shown in Figure 1b, thermoelectric power generation uses the Seebeck effect of thermoelectric materials. It converts thermal energy directly into electrical energy, which is a new green, environmentally friendly, and pollution-free power generation technology. Copper, a good conductive metal, is connected to one end of the P-type thermoelectric material and the N-type thermoelectric material. The other end is connected with copper separately to form a PN junction, also known as a PN coupler arm. At this time, there is a high-temperature heat source at one end of the PN coupling arm to provide heat, forming the hot end. At the other end of the PN coupler arm, a low-temperature cold source dissipates energy, forming the cold end. This results in a temperature difference between the two ends of the PN junction. The holes (P-type thermoelectric material) and electrons (N-type thermoelectric material) at the high-temperature end of the PN coupler arm are driven by the temperature gradient and begin to diffuse toward the low-temperature end. This creates an electrical potential difference between the two ends of the PN coupler arm. Multiple PN coupler arms are connected to a load resistor, at which point a current is generated in the circuit. Due to the small electric potential that can be generated by a single PN coupler arm, multiple PN coupler arms are generally connected in series in commercial thermoelectric power generation modules today in order to obtain a larger output power. The hot and cold ends are covered with ceramic sheets to form a temperature differential generator with multiple PN junctions.

Thermoelectric materials are characterized by low resistance, low thermal conductivity, and high electrical conductivity. TGs composed of thermoelectric materials can generate DC voltage.

$$E = \alpha(T_h - T_c) = \Delta T_{hc} \quad (31)$$

$$E = \alpha(T_h - T_c) = \Delta T_{hc} \quad (32)$$

E is the open-circuit electric potential of TGs. α is the Seebeck coefficient. T_h is the temperature of the hot end of TGs. T_c is the temperature of the cold end of TGs. ΔT_{hc} is the temperature difference between the hot end of TGs and the end of TGs.

The thermoelectric conversion efficiency of a thermoelectric material depends mainly on the euphoria factor ZT.

$$ZT = \frac{\alpha^2\sigma}{\lambda}T \quad (33)$$

σ is the electrical conductivity of the thermoelectric material. λ is the thermal conductivity of the thermoelectric material. T is the absolute temperature. A larger value of the thermoelectric optimum ZT indicates a better thermoelectric performance of the thermoelectric material.

When the load resistance R_L is the same as the internal resistance R_m of TGs, there is the maximum output power P_{max}.

$$P_{\max} = \frac{E^2}{4R_{\mathrm{m}}} \tag{34}$$

$$\eta_{\mathrm{e1}} = \frac{P}{Q_{\mathrm{j}}} \times 100\% \tag{35}$$

$$\eta_{\mathrm{e2}} = \frac{P}{Q_{\mathrm{c}}} \times 100\% \tag{36}$$

P is the load power. Q_{j} is the heating power of heating source. Q_{c} is the heat transfer power of the evaporation section of the TPCT. η_{e1} is the power generation efficiency. η_{e2} is the net power generation efficiency.

Therefore, the open-circuit voltage E and load power P of TGs are important parameters to measure the thermoelectric conversion performance of the experimental device.

From Equation (29), it is known that the factors affecting the open-circuit voltage and load power of TGs are the Seebeck coefficient α and the temperature difference of TGs $T_{\mathrm{h}} - T_{\mathrm{c}}$. The Seebeck coefficient mainly depends on the nature of the thermoelectric material of TGs, while the main external factor is the temperature difference between the hot end and the cold end of TGs. Then, the temperature difference of TGs is mainly affected by the temperature of the evaporating section of the TPCT and the temperature of the condensing section of the TPCT. Therefore, it is necessary to study and analyze this.

3. Results and Discussion

3.1. Heat Transfer Performance

3.1.1. TPCT Heating and Isothermal Characteristics

Figure 2a–d shows the variation in temperature at different distances along the TCPT with time (increasing the heating power from 100 W to 600 W) at the liquid filling rates of 10%, 25%, 35%, and 45%, respectively. It is observed that the temperature at 10 cm, 30 cm, 50 cm, 70 cm, and 90 cm along the TCPT increases as the heating power increases. When the heating power is 100 W, the variation in temperature at 90 cm with time is relatively small, the heat transfer power of the TCPT is low and the heat cannot reach the top of TCPT. According to the heating power, the temperature rise process of TCPT can be divided into four stages. During each stage, the temperature of TCPT increases with time, and the rate of temperature increase gradually decreases. When the cooling and heat dissipation module is started, the temperature decreases quickly. When the heating power is changed, the temperature increases rapidly. The variations in temperature with different liquid filling rates are basically the same. With the increase in the heating power, the temperature of the TPCT increase. It shows that the heating power is the main factor affecting the temperature of the TPCT, and the liquid filling rate has a smaller effect on the temperature increase of the TPCT.

Figure 3a–d shows the variation in temperature with distance along the TCPT at different heating powers at the liquid filling rates of 10%, 25%, 35%, and 45%, respectively. There is a large temperature gradient in the evaporation section (from 0 cm to 50 cm), and there is a small temperature gradient in the condensation section (from 50 cm to 100 cm). As the wall of evaporation section absorbs heat and boils, there are a large number of bubbles, and the temperature distribution is more uneven. While the condensation section of the TPCT condenses and exerts heat, the temperature distribution is more uniform. It is observed that the temperature with distance along the TCPT increase as the heating power increase at all liquid filling rates. The temperature of the TCPT does not change obviously with the increase in the liquid filling rate. It can be concluded that the heating power is the main factor that affecting the temperature of TCPT and the liquid filling rate is a secondary factor. Finally, the temperature at each position of the condensation section tends to be consistent and stable at about 160 °C. It indicates that with the increase in heating power, the temperature consistency of evaporation section becomes weakening, and the temperature of the condensation section tends to be consistent and more average.

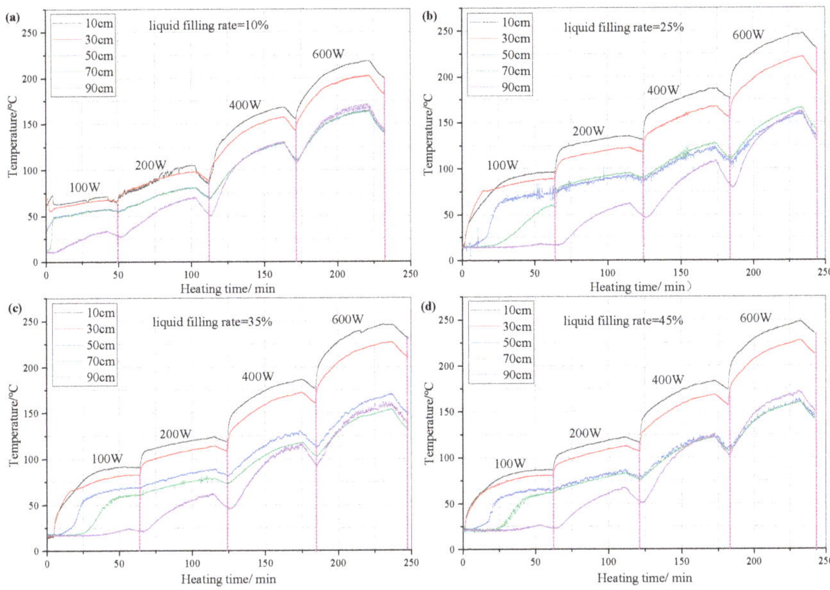

Figure 2. Variation in temperature with time at different distances along the TCPT at the liquid filling rates of (**a**) 10%, (**b**) 25%, (**c**) 35%, and (**d**) 45%.

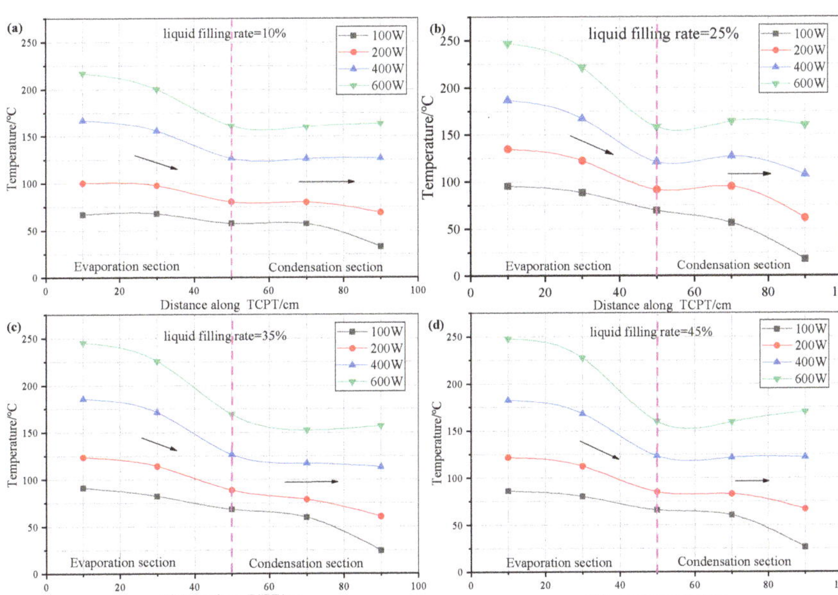

Figure 3. Variation in temperature with distance along the TCPT at different heating powers at liquid filling rates of (**a**) 10%, (**b**) 25%, (**c**) 35%, and (**d**) 45%.

3.1.2. TPCT Heat Transfer Characteristics Analysis

The variation in the thermal resistance of the TPCT with heating power at different liquid filling rates is shown in Figure 4. It is observed that the thermal resistance of the TPCT decreases continuously with the increase in heating power. When the heating power is 100 W and 200 W, the rate of the variation in the thermal resistance is larger. When the heating power is 400 W and 600 W, the rate of the variation in the thermal resistance becomes smaller, consistent. The thermal resistance at the liquid filling rates of 10%, 25%, 35%, and 45% less variable, with some approximate value of 0.10 °C/W, 0.14 °C/W, 0.19 °C/W, and 0.18 °C/W, respectively. At the heating powers of 100 W, 200 W, 400 W, and 600 W, the thermal resistance of the TPCT at the liquid filling rate of 10% is the smallest, and with the increase in heating power, the thermal resistance does not change much, roughly around 0.11 °C/W. At the heating powers of 200 W–600 W, the thermal resistance of the TPCT with the 35% liquid filling rate is the largest; at the heating power of 100 W, the thermal resistance of the TPCT with the 25% liquid filling rate is the largest.

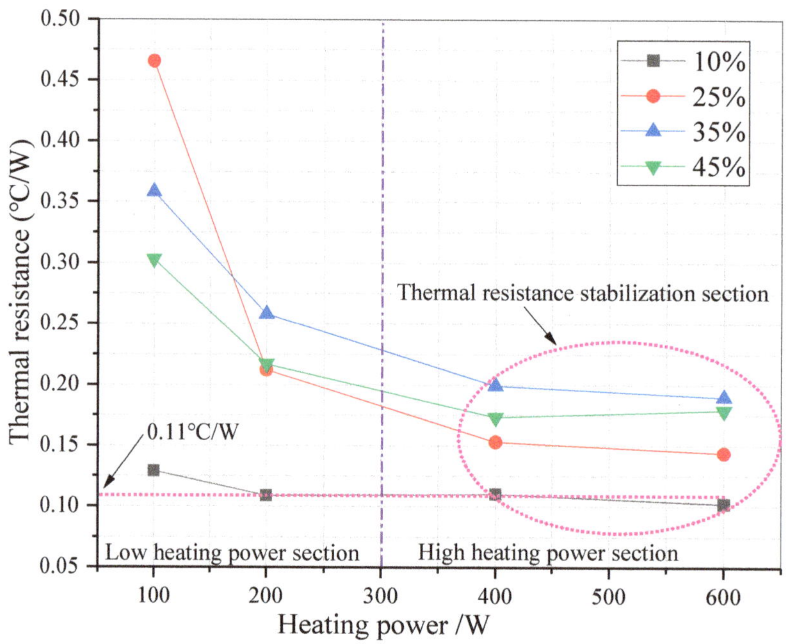

Figure 4. Variation in thermal resistance with heating power at the liquid filling rates of 10%, 25%, 35%, and 45%.

The variation in heat transfer loss of the TPCT with heating power at different liquid filling rates is shown in Figure 5. It is observed that the heat transfer loss has been increasing as the heating power increases. When the heating power is 100 W and 200 W, the difference in heat transfer loss between TPCTs with different liquid filling rates is small. It indicates that the liquid filling rate has less effect on the heat transfer loss at low heating power. But there is also a minimum point of heat transfer loss, the heat transfer loss of the TPCT with the 10% liquid filling rate is the smallest. When the heating power is 100 W and 200 W, the smallest heat transfer loss is 18.43 W and 33.28 W, respectively. When the heating power is 400 W and 600 W, the difference in heat transfer loss between TPCTs with different liquid filling rates is large. It indicates that the liquid filling rate has a more obvious effect on the heat transfer loss at high heating power. When the heating power is 400 W and 600 W, the smallest heat transfer loss is for the TPCT with the 25% liquid filling rate, and is 71.43 W and 119.48 W, respectively.

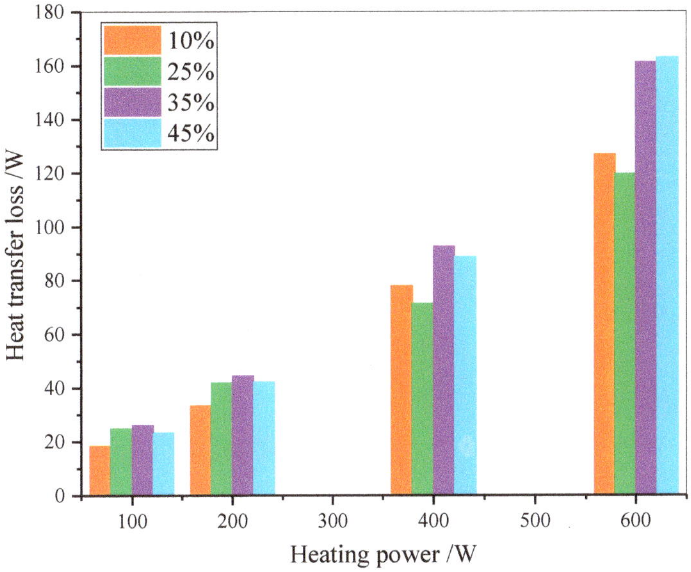

Figure 5. Variation in heat transfer loss with heating power at the liquid filling rates of 10%, 25%, 35%, and 45%.

The variation in the heat transfer power of the TPCT with heating power at different liquid filling rates is shown in Figure 6. It is observed that that as the heating power increases, the heat transfer power keeps increasing. And the heating power and heat transfer power are roughly linear. When the heating power is 100 W and 200 W, the heat transfer power varies smoothly with the liquid filling rate, and the heat transfer power is basically the same at different liquid filling rates. At the heating power of 100 W, the heat transfer power of the TPCT at different liquid filling rates is about 82 W or less; at the heating power of 200 W, the heat transfer power is about 162 W or less. It means that the liquid filling rate has less influence on the heat transfer power at low heating power. Moreover, the heat transfer power of the TPCT at the liquid filling rate of 35% is the smallest, and the heat transfer power of the TPCT at the liquid filling rate of 10% is the largest at low heating power. When the heating power is 400 W and 600 W, the variation in heat transfer power with liquid filling rate is larger. It indicates that the liquid filling rate has a more obvious effect on the heat transfer power at high heating power. At high heating power, the TPCT at the liquid filling rate of 25% has the largest heat transfer power. At the high heating power of 400 W and 600 W, the largest heat transfer powers of the TPCT are 480 W and 328 W, respectively.

The variation in the heat transfer efficiency of the TPCT with heating power at different liquid filling rates is shown in Figure 7. It is observed that the heat transfer efficiency of TPCTs with different liquid filling rates shows a trend of first increasing and then decreasing as the heating power increases. TPCTs with liquid filling rates of 10%, 35%, and 45% exhibit the highest heat transfer efficiency at a heating power of 200 W, with efficiencies of 83.36%, 78.84%, and 77.71%, respectively. The TPCT with a liquid filling rate of 25% achieves the greatest heat transfer efficiency of 82.14% at a heating power of 400 W. It is observed that that the TPCT with the liquid filling rate of 10% has the largest heat transfer efficiency at the low heating powers of 100 W and 200 W, which are 81.57% and 83.36%, respectively. At the high heating powers of 400 W and 600 W, the TPCT with the liquid filling rate of 25% has the largest heat transfer efficiency, which is 82.14% and 80.09%, respectively. Therefore, the heat transfer efficiency of the TPCT is larger at low liquid filling rates. At the heating powers of 100 W, 200 W and 400 W, the TPCT with a liquid filling rate of 35% has the smallest heat transfer efficiency. At a heating power of 600 W, the heat transfer efficiency of the TPCT is essentially the same for a liquid filling rate of 35% and a liquid filling rate of 45%. Therefore, it can be considered that the smallest heat transfer efficiency of the TPCT at heating power (from 100 W to 600 W) is found in the TPCT at the liquid filling rate of 35%.

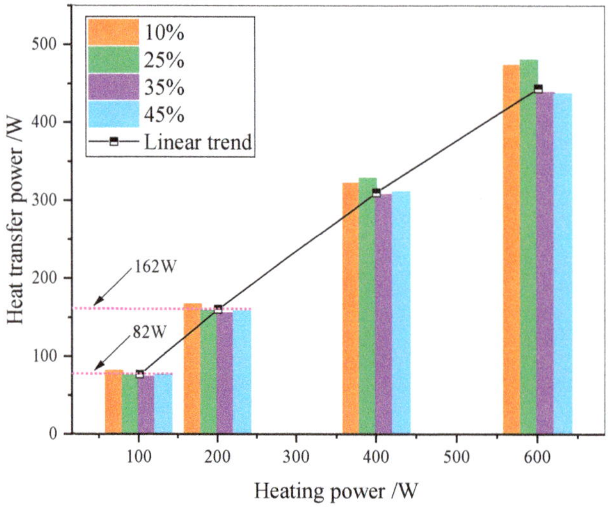

Figure 6. Variation in heat transfer power with heating power at the liquid filling rates of 10%, 25%, 35%, and 45%.

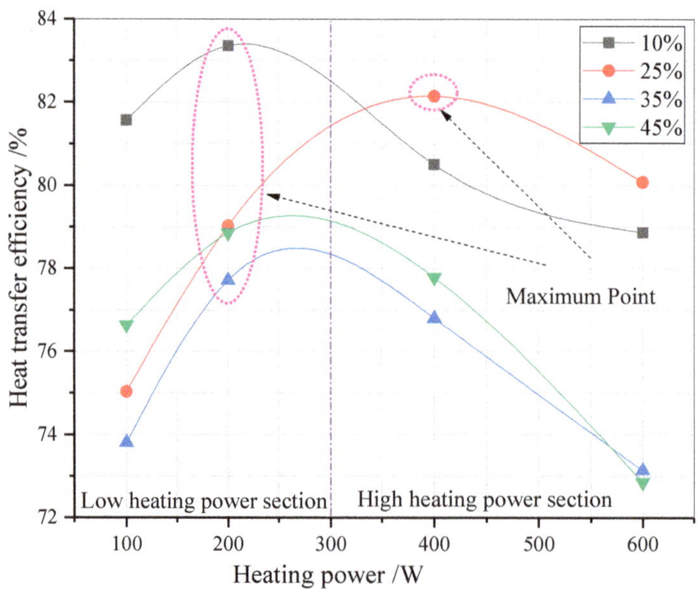

Figure 7. Variation in heat transfer efficiency with heating power at the liquid filling rates of 10%, 25%, 35%, and 45%.

3.2. Power Generation Performance Analysis

3.2.1. Temperature Parameters Analysis

The variation in ΔT_{hc} with liquid filling rate at the heating powers of 100 W, 200 W, 400 W, and 600 W is shown in Figure 8. It is observed that that ΔT_{hc} increases as the heating power increases. When the liquid filling rate of the TPCT is 10%, 25%, and 35%, the ΔT_{hc} change curve is closer. When the liquid filling rate is 45%, ΔT_{hc} increases significantly. At the heating powers of 200 W, 400 W, and 600 W, the maximum ΔT_{hc} are 40.62 °C, 68.30 °C, and 90.03 °C, respectively, for the TPCT at the liquid filling rate of 45%. It shows that the heating power has a greater effect on ΔT_{hc} at higher liquid fill rates of the TPCT. It is observed that ΔT_{hc} shows a trend of decreasing and then increasing with the increase in the liquid filling rate of the TPCT. When the heating power is 100 W and 200 W, the variation in the temperature difference of TGs is smaller. When the high heating power is 400 W and 600 W, the variation is larger. It shows that at low heating power, the liquid filling rate has a smaller effect on the temperature difference of the power generator. At high heating power, the effect of the liquid filling rate on ΔT_{hc} is larger.

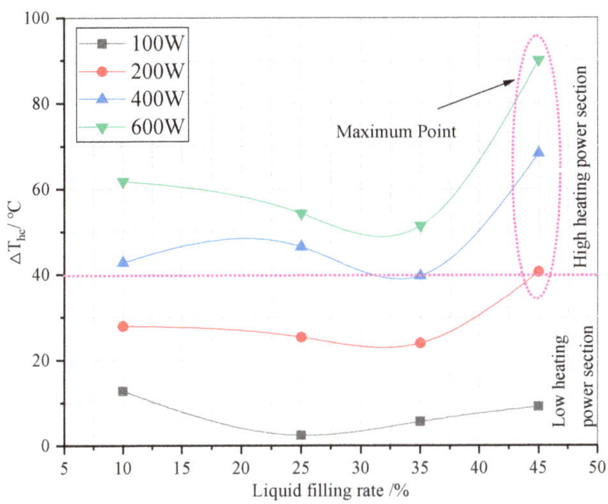

Figure 8. Variation in ΔT_{hc} with liquid filling rate at the heating powers of 100 W, 200 W, 400 W, and 600 W.

The variation in Δt_{ec} with liquid filling rate at the heating powers of 100 W, 200 W, 400 W, and 600 W is shown in Figure 9. It is observed that Δt_{ec} gradually increases as the heating power increases. As the liquid filling rate increases, Δt_{ec} first increases and then decreases. At the heating powers of 200 W, 400 W, and 600 W, Δt_{ec} at the liquid filling rate of 35% is the largest, which is 40.09 °C, 61.24 °C and 83.35 °C, respectively. At the heating power of 100 W, Δt_{ec} at the liquid filling rate of 25% is the largest, which is 34.93 °C.

The variation in the evaporation section temperature with liquid filling rate at four heating powers is shown in Figure 10. It is observed that the evaporation section temperature of the TPCT increases gradually with the increase in heating power. When the heating power is greater than or equal to 400 W and the liquid filling rate is greater than or equal to 25%, the variation in the evaporation section temperature is roughly the same. It means that the evaporation section temperature of the TPCT is only proportional to the heating power at this time. When the heating power is 100 W and 200 W, the evaporation section temperature of the TPCT with the liquid filling rate of 25% is larger; when the heating power is 400 W and 600 W, the evaporation section temperature of the TPCT with liquid filling rates of 25%, 35%, and 45% tends to be the same. So, it means that in all heating power (from 100 W to 600 W), the evaporation section of the TPCT at the liquid filling rate of 25% has a higher temperature and better heating performance.

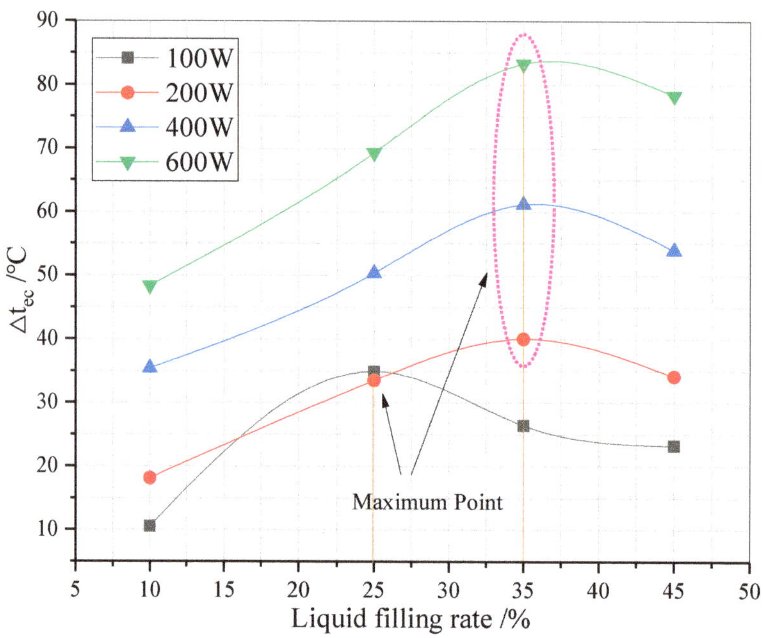

Figure 9. Variation in Δt_{ec} with liquid filling rate at the heating powers of 100 W, 200 W, 400 W, and 600 W.

3.2.2. Power Generation Performance Indicators Analysis

Figure 11a–c shows the variation in open-circuit voltage, load current, and load power with heating power at the liquid filling rates of 10%, 25%, 35%, and 45%. It is observed that as the increase in heating power, the open-circuit voltage, load current, and measured load power increase. When the heating power is 200 W, 400 W, and 600 W, the values of power generation parameters, such as open-circuit voltage, load current, and load power, do not differ much for different liquid filling rates. It means that the open-circuit voltage, load current, and load power are less affected by the liquid filling rate and more affected by the heating power. However, when the heating power is 100 W, the value of the power generation parameter is larger for a 10% liquid-filled TPCT. This is because at low heating power, the TPCT with the lower liquid filling rate is easier to start and has better heat transfer characteristics. It is observed that with the increase in the liquid filling rate, when the heating power is 100 W and 200 W, the open-circuit voltage, load current, and load power show a trend of first decreasing and then increasing, and the TPCT with a liquid filling rate of 10% have the maximum values. When the high heating power is 400 W and 600 W, the open-circuit voltage, load current, and load power show a trend of first increasing and then decreasing, and the TPCTs with liquid filling rates of 25% and 35% have the maximum values. Table 2 shows the maximum values of open-circuit voltage, load current, and load power at different heating powers and liquid filling rates.

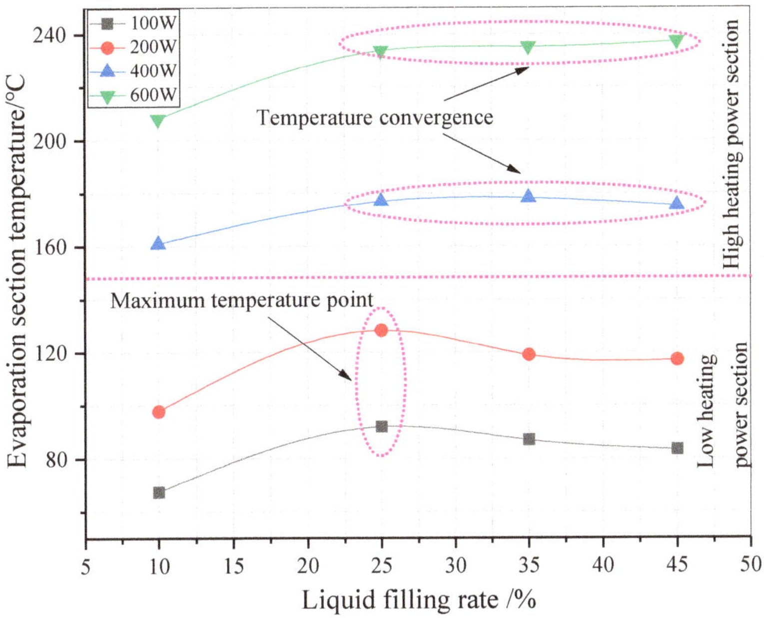

Figure 10. Variation in evaporation section temperature with liquid filling rate at the heating powers of 100 W, 200 W, 400 W, and 600 W.

Figure 12a,b shows the variation in power generation efficiency and net power generation efficiency with heating power at the liquid filling rates of 10%, 25%, 35%, and 45%. It is observed that the power generation efficiency and net power generation efficiency increase as the heating power increases. When the heating power reaches 400 W, the increases in both power generation efficiency and net power generation efficiency become slower and gradually level off. It indicates that the heating power has more influence on the power generation efficiency and net power generation efficiency at the heating powers of 100 W and 200 W; the heating power has a smaller effect on the power generation efficiency and net power generation efficiency at the high heating powers of 400 W and 600 W. Finally, the power generation efficiency and net power generation efficiency converge to 2.08% and 2.68%. It is observed that with the increase in liquid filling rate, at the heating powers of 100 W and 200 W, the power generation efficiency and net power generation efficiency both show a trend of first decreasing and then increasing, and they have maximum values when the liquid filling rate is 10%. At the high heating powers of 400 W and 600 W, the power generation efficiency and net power generation efficiency both show a trend of first increasing and then decreasing, and they have maximum values when the liquid filling rates are 25% and 35%, respectively. Table 3 shows the maximum values of power generation efficiency and net power generation efficiency at different heating powers and liquid filling rates.

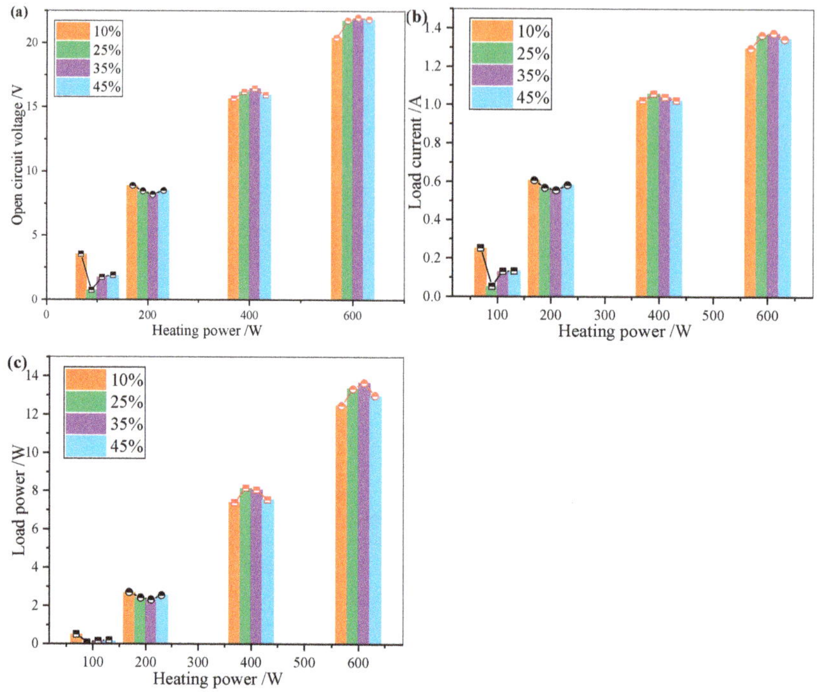

Figure 11. (**a**) Variation in (**b**) open-circuit voltage, (**c**) load current, and load power with heating power at the liquid filling rates of 10%, 25%, 35%, and 45%.

Table 2. Power generation parameters at different heating powers.

Heating Power	Maximum Open-Circuit Voltage/V	Maximum Load Current/A	Maximum Load Power/W
100 W	3.51 (10%)	0.25 (10%)	0.47 (10%)
200 W	8.89 (10%)	0.61 (10%)	2.68 (10%)
400 W	16.43 (35%)	1.06 (25%)	8.13 (25%)
600 W	21.93 (35%)	1.37 (35%)	13.64 (35%)

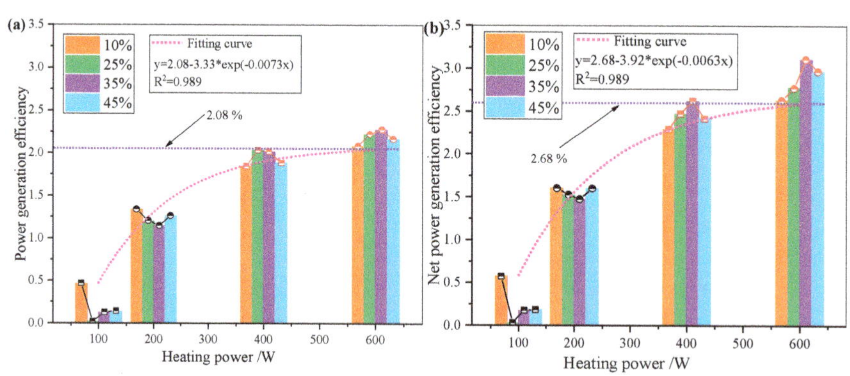

Figure 12. Variation in power generation efficiency and net power generation efficiency with heating power at the liquid filling rates of 10%, 25%, 35%, and 45%. (**a**) Power generation efficiency; (**b**) Net power generation efficiency.

Table 3. Maximum power generation efficiency and maximum net power generation efficiency at different heating powers.

Heating Power	Maximum Power Generation Efficiency/%	Maximum Net Power Generation Parameters Under/%
100 W	0.4657 (10%)	0.5709 (10%)
200 W	1.3375 (10%)	1.6046 (10%)
400 W	2.0313 (25%)	2.6217 (35%)
600 W	2.2736 (35%)	3.1086 (35%)

4. Conclusions

The effects of heating power and liquid filling rate on heat transfer performance and power generation performance are studied. The following conclusions are drawn:

(1) The thermal resistance decreases as heating power increases, and the heat transfer power increases as heating power increases. The important indexes of power generation performance, for example, open-circuit voltage, load current, load power, power generation efficiency, and net power generation efficiency, increase as heating power increases. This shows that as the heating power increases, heat transfer performance and power generation performance of the device will also increase. As the heating power increases, the heat transfer loss increases and the heat transfer efficiency increases and then decreases. The power generation efficiency and net power generation efficiency increase as the heating power increases, and converge to 2.18% and 2.68%. This indicates that as the heating power increases, the heat transfer performance and power generation performance become slow to improve and finally stabilize at a relatively steady state.

(2) Overall, the effect of the liquid filling rate is smaller than that of the heating power for both heat transfer and power generation performance. At the low heating powers of 100 W and 200 W, the device with the TPCT at the lower liquid filling rate of 10% has better heat transfer performance and power generation performance. At the high heating powers of 400 W and 600 W, the device with the TPCT at the higher liquid filling rates of 25%, and 35% has better heat transfer performance and power generation performance.

(3) The thermal resistance of the TPCT at the liquid filling rate of 10% is the smallest, and with the increase in heating power, the thermal resistance does not change much, roughly around 0.11 °C/W. At a heating power of 200 W, TPCTs at liquid filling rates of 10%, 35%, and 45% have the maximum heat transfer efficiency, which is 83.36%, 78.84%, and 77.71%, respectively. At a heating power of 400 W, the TPCT at the liquid filling rate of 25% has the largest heat transfer efficiency, which is 82.14%.

The maximum values of power generation efficiency and net power generation efficiency are 2.27% and 3.10%, respectively.

Author Contributions: Conceptualization, H.Q. and H.S.; methodology, H.Q. and H.S.; software, H.Q.; validation, H.Q. and H.S.; formal analysis, H.Q.; investigation, H.Q.; resources, H.Q.; data curation, H.Q.; writing—original draft preparation, H.Q.; writing—review and editing, H.Q.; visualization, H.Q.; supervision, H.S.; project administration, H.S.; funding acquisition, H.S. All authors have read and agreed to the published version of the manuscript.

Funding: This research was funded by the Fundamental Research Funds for the Central Universities, grant number [2652018098].

Institutional Review Board Statement: Not applicable.

Informed Consent Statement: Not applicable.

Data Availability Statement: The data that support this study cannot be publicly shared due to privacy reasons. The raw data from this study will be made available upon reasonable request to the corresponding author.

Conflicts of Interest: The authors declare no conflict of interest.

References

1. Su, H.T.; Zhou, F.B.; Shi, B.B.; Qi, H.N.; Deng, J.C. Causes and detection of coalfield fires, control techniques, and heat energy recovery: A review. *Int. J. Miner. Metall. Mater.* **2020**, *27*, 3–19. [CrossRef]
2. Su, H.; Zhou, F.; Qi, H.; Li, J. Design for thermoelectric power generation using subsurface coal fires. *Energy* **2017**, *140*, 929–940. [CrossRef]
3. Shi, B.; Su, H.; Li, J.; Qi, H.; Zhou, F.; Torero, J.L.; Chen, Z. Clean Power Generation from the Intractable Natural Coalfield Fires: Turn Harm into Benefit. *Sci. Rep.* **2017**, *7*, 5302. [CrossRef] [PubMed]
4. Faghri, A. *Heat Pipe Science & Technology*; Taylor & Francis Group: Abingdon, UK, 1995.
5. Esen, M.; Esen, H. Experimental investigation of a two-phase closed thermosyphon solar water heater. *Sol. Energy* **2005**, *79*, 459–468. [CrossRef]
6. Fontaine, P.-O.; Marcotte, D.; Pasquier, P.; Thibodeau, D. Modeling of horizontal geoexchange systems for building heating and permafrost stabilization. *Geothermics* **2011**, *40*, 211–220. [CrossRef]
7. Vasilyeva, M.; Stepanov, S.; Sirditov, I. Reduced dimension model for heat transfer of ground heat exchanger in permafrost. *J. Phys. Conf.* **2017**, *937*, 012056. [CrossRef]
8. Zhang, B.; Sheng, Y.; Chen, J.; Li, J. In-Situ Test Study on the Cooling Effect of Two-Phase Closed Thermosyphon in Marshy Permafrost Regions along the Chaidaer–Muli Railway, Qinghai Province, China. *Cold Reg. Sci. Technol.* **2011**, *65*, 456–464. [CrossRef]
9. Song, Y.; Jin, L.; Zhang, J. In-situ study on cooling characteristics of two-phase closed thermosyphon embankment of Qinghai-Tibet Highway in permafrost re gions. *Cold Reg. Sci. Technol.* **2013**, *93*, 12–19. [CrossRef]
10. Zorn, R.; Steger, H.; Kolbel, T. De-Icing and Snow Melting System with Innovative Heat Pipe Technology. In Proceedings of the World Geothermal Congress 2015, Melbourne, VIC, Australia, 19–25 April 2015.
11. Sakata, Y.; Chishimba, J.; Mochizuki, M.; Nagano, K. Evaluating Groundwater Flow Effects for Enhancement of Ground-Source Heat Pipes in the Case of the Toyohira River Alluvial Fan, Japan. *Hydrology* **2021**, *8*, 135. [CrossRef]
12. Chen, F.; Su, X.; Ye, Q.; Fu, J. Experimental Investigation of Concrete Runway Snow Melting Utilizing Heat Pipe Technology. *Sci. World J.* **2018**, *2018*, 4343167. [CrossRef]
13. Jeong, Y.S.; Kim, K.M.; Kim, I.G.; Bang, I.C. Hybrid heat pipe based passive in-core cooling system for advanced nuclear power plant. *Appl. Therm. Eng.* **2015**, *90*, 609–618. [CrossRef]
14. Shi, S.; Liu, Y.; Yilgor, I.; Sabharwall, P. A two-phase three-field modeling framework for heat pipe application in nuclear reactors. *Ann. Nucl. Energy* **2022**, *165*, 108770. [CrossRef]
15. Kirsch, M.; Castro, S.I.C. Experimental Investigation on the Long-term Operational Behaviour of Two-phase Closed Thermosyphon Bundles for Passive Heat Transfer from Spent Fuel Pools under Normal, Abnormal, and Accident Conditions. *Atw-Int. J. Nucl. Power* **2022**, *67*, 74–77.
16. Lu, Z.; Wang, R. Novel adsorption refrigerators with separate type two phase closed thermosyphon designs. *Int. J. Energy Res.* **2015**, *39*, 1681–1688. [CrossRef]
17. Ma, C.; Wu, X.; Gao, S. Analysis and applications of a two-phase closed thermosyphon for improving the fluid temperature distribution in wellbores. *Appl. Therm. Eng.* **2013**, *55*, 1–6. [CrossRef]
18. Zhang, Y.; Zhao, L.; Fan, Z.; Wu, X.; Fu, L.; Xu, B.; Kong, F.; Jiang, S. Temperature distribution of fluids in a two-section two-phase closed thermosyphon wellbore. *Pet. Sci.* **2014**, *11*, 287–292. [CrossRef]
19. Tyurin, M.P.; Borodina, Y.S.; Osmanov, Z.N. Investigation of Processes of Heat and Mass Exchange in a Closed Two-Phase Thermosiphon for the Development of Energy Conserving Technologies in the Production of Edible Phosphates. *Fibre Chem.* **2018**, *49*, 388–393. [CrossRef]
20. Bolozdynya, A.I.; Efremenko, Y.V.; Khromov, V.A.; Shafigullin, R.R.; Shakirov, A.V.; Sosnovtsev, V.V.; Tolstukhin, I.A. Thermostabilization system based on two-phase closed cryogenic thermosyphon for RED100 detector. *Phys. Procedia* **2015**, *74*, 431–434. [CrossRef]
21. Hassanpour, A.; Borji, M.; Ziapour, B.M.; Kazemi, A. Performance Analysis of a Cascade PCM Heat Exchanger and Two-Phase Closed Thermosiphon: A Case Study of Geothermal District Heating System. Sustain. *Energy Technol. Assess.* **2020**, *40*, 100755. [CrossRef]
22. Abreu, S.L.; Colle, S. An Experimental Study of Two-Phase Closed Thermosyphons for Compact Solar Domestic Hot-Water Systems. *Sol. Energy* **2004**, *76*, 141–145. [CrossRef]

23. Li, B.; Deng, J.; Xiao, Y.; Zhai, X.W.; Shu, C.M.; Gao, W. Heat transfer capacity of heat pipes: An application in coalfield wildfire in China. *Heat Mass Transf.* **2018**, *54*, 1755–1766. [CrossRef]
24. Zhong, K.Q.; Xiao, Y.; Lu, X.; Deng, J.; Yin, L.; Tian, Y.; Shu, C.-M. Nanofluidic two-phase closed thermosyphon-assisted thermoelectric generator for heat recovery from coal spontaneous combustion. *Appl. Therm. Eng.* **2021**, *197*, 117397. [CrossRef]
25. Janak, L.; Ancik, Z.; Vetiska, J.; Hadas, Z. Thermoelectric Generator Based on MEMS Module as an Electric Power Backup in Aerospace Applications. Mater. *Today Proc.* **2015**, *2*, 865–870.
26. Tohidi, F.; Holagh, S.G.; Chitsaz, A. Thermoelectric Generators: A comprehensive review of characteristics and applications. *Appl. Therm. Eng.* **2022**, *201*, 117793. [CrossRef]
27. Li, K.; Garrison, G.; Moore, M.; Zhu, Y.; Liu, C.; Horne, R.; Petty, S. An Expandable Thermoelectric Power Generator and the Experimental Studies on Power Output. *Int. J. Heat Mass Transf.* **2020**, *160*, 120205. [CrossRef]
28. Nuwayhid, R.; Shihadeh, A.; Ghaddar, N. Development and testing of a domestic woodstove thermoelectric generator with natural convection cooling. *Energy Conver. Manag.* **2004**, *46*, 1631–1643. [CrossRef]
29. O'Shaughnessy, S.; Deasy, M.; Doyle, J.; Robinson, A. Field trial testing of an electricity-producing portable biomass cooking stove in rural Malawi. *Energy Sustain. Dev.* **2014**, *20*, 1–10. [CrossRef]
30. Champier, D.; Bédécarrats, J.; Rivaletto, M.; Strub, F. Thermoelectric power generation from biomass cook stoves. *Energy* **2010**, *35*, 935–942. [CrossRef]
31. Singh, R.; Tundee, S.; Akbarzadeh, A. Electric power generation from solar pond using combined thermosyphon and thermoelectric modules. *Solar Energy* **2014**, *48*, 453–463. [CrossRef]
32. Zhang, M.; Miao, L.; Kang, Y.P.; Tanemura, S.; Fisher, C.A.J.; Xu, G.; Li, C.X.; Fan, G.Z. Efficient, low-cost solar thermoelectric cogenerators comprising evacuated tubular solar collectors and thermoelectric modules. *Appl. Energy* **2013**, *109*, 51–59. [CrossRef]
33. Whalen, S.A.; Dykhuizen, R.C. Thermoelectric energy harvesting from diurnal heat flow in the upper soil layer. *Energy Convers. Manag.* **2012**, *64*, 397–402. [CrossRef]
34. Suter, C.; Jovanovic, Z.R.; Steinfeld, A. A 1 kWe Thermoelectric Stack for Geothermal Power Generation—Modeling and Geometrical Optimization. *Appl. Energy* **2012**, *99*, 379–385. [CrossRef]
35. El-Genk, M.S.; Saber, H.H. Performance analysis of cascaded thermoelectric converters for advanced radioisotope power systems. *Energy Convers. Manag.* **2005**, *46*, 1083–1105. [CrossRef]
36. El-Genk, M.; Saber, H.; Caillat, T. Efficient segmented thermoelectric unicouples for space power applications. *Energy Convers. Manag.* **2003**, *44*, 1755–1772. [CrossRef]
37. Karri, M.; Thacher, E.; Helenbrook, B. Exhaust energy conversion by thermoelectric generator: Two case studies. *Energy Convers. Manag.* **2011**, *52*, 1596–1611. [CrossRef]
38. Ono, K.; Suzuki, R.O. Thermoelectric Power Generation: Converting Low-Grade Heat into Electricity. *Energy Resour.* **1998**, *50*, 49–51. [CrossRef]
39. Crane, D.T.; Jackson, G.S. Optimization of cross flow heat exchangers for thermoelectric waste heat recovery. *Energy Convers. Manag.* **2004**, *45*, 1565–1582. [CrossRef]
40. Hendricks, T.J. Thermal System Interactions in Optimizing Advanced Thermoelectric Energy Recovery Systems. *J. Energy Resour. Technol.* **2007**, *129*, 223–231. [CrossRef]
41. Gould, C.A.; Shammas, N.Y.A.; Grainger, S.; Taylor, I. Thermoelectric Cooling of Microelectronic Circuits and Waste Heat Electrical Power Generation in a Desktop Personal Computer. Mater. *Sci. Eng. B* **2011**, *176*, 316–325. [CrossRef]
42. Putra, N.; Yanuar; Iskandar, F.N. Application of nanofluids to a heat pipe liquid block and the thermoelectric cooling of electronic equipment. *Exp. Therm. Fluid Sci.* **2011**, *35*, 1274–1281. [CrossRef]
43. Su, H.; Qi, H.; Liu, P.; Li, J. Experimental investigation on heat extraction using a two-phase closed thermosyphon for thermoelectric power generation. Energy Sources Part A Recover. *Util. Environ. Eff.* **2018**, *40*, 1485–1490.
44. Deng, J.; Zhou, F.; Shi, B.; Torero, J.L.; Qi, H.; Liu, P.; Ge, S.; Wang, Z.; Chen, P.C. Waste heat recovery, utilization and evaluation of coalfield fire applying heat pipe combined thermoelectric generator in Xinjiang, China. *Energy* **2020**, *207*, 118303. [CrossRef]
45. Yang, S.M.; Zhang, Z.Z. An Experimental Study of Natural Convection Heat Transfer from a Horizontal Cylinder in High Rayleigh Number Laminar and Turbulent Regions. In *Institution of Chemical Engineers Symposium Series*; Hemisphere Publishing Corporation: London, UK, 1994; Volume 135, p. 185.

fire

Article

Research on Coal and Gas Outburst Prediction and Sensitivity Analysis Based on an Interpretable Ali Baba and the Forty Thieves–Transformer–Support Vector Machine Model

Yanping Wang, Zhixin Qin *, Zhenguo Yan, Jun Deng, Yuxin Huang, Longcheng Zhang, Yuqi Cao and Yiyang Wang

College of Safety Science and Engineering, Xi'an University of Science and Technology, Xi'an 710054, China; wang_yanping@xust.cn (Y.W.); yanzg@xust.edu.cn (Z.Y.); dengj518@xust.edu.cn (J.D.); 20120089017@stu.xust.edu.cn (Y.H.); 23220089040@stu.xust.edu.cn (L.Z.); 19403070424@stu.xust.edu.cn (Y.C.); 22320226003@stu.xust.edu.cn (Y.W.)

* Correspondence: 23220089060@stu.xust.edu.cn

Abstract: Coal and gas outbursts pose significant threats to underground personnel, making the development of accurate prediction models crucial for reducing casualties. By addressing the challenges of highly nonlinear relationships among predictive parameters, poor interpretability of models, and limited sample data in existing studies, this paper proposes an interpretable Ali Baba and the Forty Thieves–Transformer–Support Vector Machine (AFT-Transformer-SVM) model with high predictive accuracy. The Ali Baba and the Forty Thieves (AFT) algorithm is employed to optimise a Transformer-based feature extraction, thereby reducing the degree of nonlinearity among sample data. A Transformer-SVM model is constructed, wherein the Support Vector Machine (SVM) model provides negative feedback to refine the Transformer feature extraction, enhancing the prediction accuracy of coal and gas outbursts. Various classification assessment methods, such as TP, TN, FP, FN tables, and SHAP analysis, are utilised to improve the interpretability of the model. Additionally, the permutation feature importance (PFI) method is applied to conduct a sensitivity analysis, elucidating the relationship between the sample data and outburst risks. Through a comparative analysis with algorithms such as eXtreme gradient boosting (XGBoost), k-nearest neighbour (KNN), radial basis function networks (RBFNs), and Bayesian classifiers, the proposed method demonstrates superior accuracy and effectively predicts coal and gas outburst risks, achieving 100% accuracy in the sample dataset. The influence of parameters on the model is analysed, highlighting that the coal seam gas content is the primary factor driving the outburst risks. The proposed approach provides technical support for coal and gas outburst predictions across different mines, enhancing emergency response and prevention capabilities for underground mining operations.

Keywords: transformer-based feature extraction; classification assessment methods; interpretable model; SHAP analysis; permutation feature importance

1. Introduction

Coal remains a vital component of global energy systems, playing a significant role in ensuring energy security and driving industrial progress. However, with the depletion of shallow coal resources and increasing mining depths, coal and gas outburst accidents have become critical challenges that threaten the normal operation of mines [1,2]. As of

2019, over 878 coal and gas outburst incidents were reported across 22 underground coal mines in Australia [3]. On 12 January 2024, a coal and gas outburst accident in the Tianan Coal Mine, Henan Province, China, resulted in 16 fatalities and 5 injuries. Similarly, on 7 June 2024, a gas outburst occurred in the Pingdingshan Coal Mine, Yunnan Province, China, causing three fatalities and five injuries. Despite the rapid advancements in artificial intelligence and machine learning technologies in recent years, which have introduced numerous new methods and theories to address the challenges of coal and gas outbursts, such accidents continue to occur frequently. The prevention and control of coal and gas outbursts remain a significant challenge, and urgent measures are required to address these issues [4,5].

To tackle the problem of coal and gas outbursts, significant scientific research and engineering practices have been conducted, leading to notable achievements. Early international research widely recognised the comprehensive action theory, which posits that coal and gas outbursts are determined by three factors: ground stress, coal properties, and gas content [6,7]. J. Hanes et al. [8] conducted studies in Australia on geological conditions, coal properties, and gas parameters, concluding that coal and gas outbursts are jointly triggered by stress and gas content, with gas content being the dominant factor. Similarly, K. Sato et al. [9] utilised digital seismographs across entire coal mines in Japan to analyse the impact of geological variations on gas outbursts. J. Shepherd et al. [10] studied parameters such as ground stress and gas content in Germany, identifying them as the most critical factors influencing coal and gas outbursts.

With the rapid development of artificial intelligence and machine learning models, numerous new theories have been applied to the study of coal and gas outbursts in mines. Peng Ji et al. [11] developed a coal mine data model based on the HPO-BiLSTM algorithm, achieving a method for coal and gas outburst early warning. Junqi Zhu et al. [12] constructed a coal and gas outburst risk identification model using the RS-GA-BP hybrid model, which significantly improved the risk identification speed. Other researchers have employed mathematical theories and data mining techniques to establish coal and gas outburst risk evaluation and identification systems. For instance, Xie Xuecai et al. [13] utilised data mining and the Apriori algorithm to analyse the causes of coal and gas outbursts, developing a Bayesian network model to conduct a sensitivity analysis of accident occurrences. Wei Wang et al. [14] applied extension theory to construct a risk prediction and risk grading indicator system for coal and gas outbursts, successfully predicting risks in 12 high-gas mines. David R. Hanson et al. [15] analysed MSHA accident data and applied various algorithms to coal samples in Pennsylvania, examining the probability of outburst accidents based on stratigraphic chemistry and lithofacies data. Furthermore, some researchers have explored traditional algorithms such as decision tree models. Zheng Xiaoliang et al. [16] combined meta-heuristic algorithms with the XGBoost theory to achieve quantitative analyses of gas outburst predictions. Zhonghui Li et al. [17] investigated a risk assessment for coal and gas outbursts based on logistic regression models, constructing a non-contact EMR index and achieving a regression prediction accuracy of 94%. Finally, geophysical methods have also been employed to develop various evaluation approaches and equipment. For example, V. Frid et al. [18] monitored high-frequency electromagnetic waves emitted from rock fractures to detect outburst risks and compared these with laboratory results, bridging the gap between laboratory and field predictions. Similarly, Janathan P. Mathews et al. [19] utilised X-ray computed tomography to study the expansion and contraction of gas absorption and desorption in coal under confining pressure, providing novel methodologies for coal and gas outburst research.

In practical coal mining operations, predictive models for coal and gas outbursts must effectively address challenges such as limited sample data, significant nonlinear relationships between parameters, and poor model interpretability, thereby achieving high generalisation capability and prediction accuracy. To address these challenges, this study constructs an interpretable Ali Baba and the Forty Thieves–Transformer–Support Vector Machine (AFT-Transformer-SVM) model based on multidimensional data from actual mining faces and designs a series of computational experiments. By employing a SHAP analysis to interpret feature contributions and combining TP, TN, FP, and FN metrics for comprehensive performance evaluation, the model's interpretability is significantly enhanced. Additionally, the permutation feature importance (PFI) method is adopted to quantify the sensitivity of raw data features, enabling the development of a coal and gas outburst early warning model under real-world conditions. Through comparative validation with various classical algorithms and field data, the proposed model demonstrates superior prediction accuracy and efficiency, offering robust support for the development of high-performance, interpretable coal and gas outburst prediction models.

2. Materials and Methods

Figure 1 represents the primary model proposed in this study, which integrates the coal and gas outburst prediction model with the feature sensitivity analysis model. Initially, nine features were selected, encompassing three dominant factors: the physical and mechanical properties of coal, gas-related factors, and ground stress. These features were input into the AFT-Transformer-SVM model for analysis. The Ali Baba and the Forty Thieves (AFT) algorithm was employed to optimise the feature extraction capability of the Transformer module. During this process, the prediction accuracy of the coal and gas outburst model, derived from the SVM, was utilised as a convergence function to provide negative feedback, thereby adjusting the Transformer module's layer structure. This iterative optimisation ultimately facilitated accurate feature extraction by the Transformer.

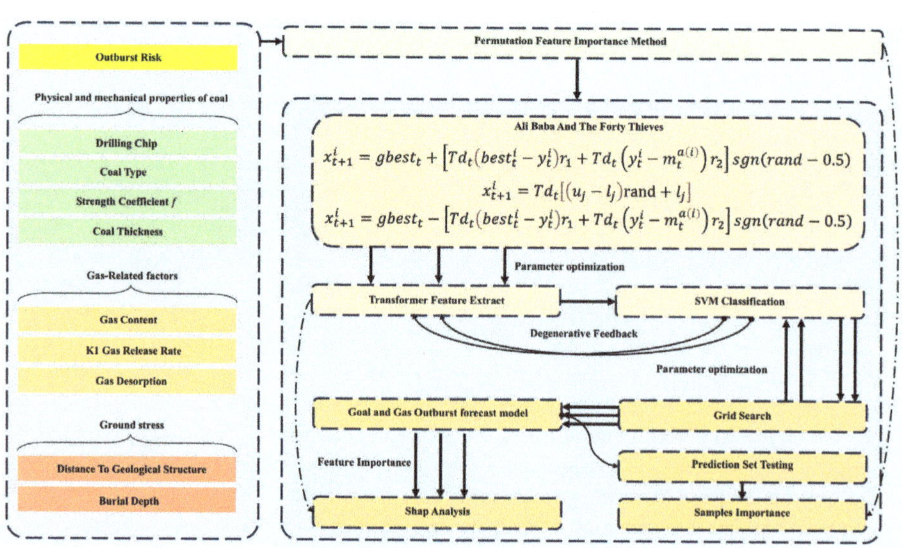

Figure 1. Schematic diagram of the overall model workflow.

Subsequently, a grid search algorithm was utilised to optimise the SVM layer structure, enabling coal and gas outburst testing on the test dataset. A SHAP analysis [20] was applied to examine the distribution of features extracted by the Transformer, offering interpretability for the current model. To quantify the contribution of the original sample data, the permutation feature importance (PFI) method [21] was employed to randomly shuffle the columns of the original feature data. The model was then retrained and predictions were performed. The prediction accuracy was used as a criterion to evaluate the contribution of the original data to the model.

2.1. Collection of Raw Data

To adapt to the conditions of on-site working environments, the primary factors influencing coal and gas outbursts were selected. These factors include gas-related parameters, the physical and mechanical properties of coal, and ground stress. The selected sample features consist of nine key parameters: coal failure type, initial velocity of gas desorption from coal, coal firmness coefficient, gas content in coal seams, K1 gas desorption volume, drilling debris volume, distance to geological structures, burial depth of the coal, and coal seam thickness. A total of 569 data samples were collected from an intelligent mine analysis platform for testing and validation purposes [22].

(1) Gas-Related Factors

Gas content: The gas content within coal seams is recognised as one of the critical factors contributing to gas-related disasters. Coal seams with high gas content, particularly those with loose structures and significant porosity, are prone to gas accumulation, thereby increasing the risk of outbursts. The gas storage capacity of coal seams is closely associated with the permeability of the surrounding rock. When the surrounding rock is dense and exhibits low permeability, gas release becomes restricted, further heightening the probability of outbursts.

K1 gas release rate: The K1 value represents the capacity of coal samples to desorb gas from the coal body, reflecting the gas storage conditions within the coal seam. A higher K1 value indicates a stronger gas release capability, which typically corresponds to a looser coal structure that facilitates gas release, thus elevating the risk of gas outburst accidents.

Gas desorption: The initial velocity of gas desorption reflects the rate at which gas is released from the coal body. This value is measured by applying a specific pressure to a coal sample and observing the pressure change within one minute. A higher gas desorption velocity indicates a faster release of gas, increasing the likelihood of gas outbursts.

(2) Physical and Mechanical Properties of Coal

Coal type: Under the influence of geological activities, including structural stress and ground stress, the original structure of coal may be damaged, resulting in the formation of numerous pore spaces. Coal failure types are classified into intact coal, fractured coal, highly fractured coal, pulverised coal, and completely pulverised coal, depending on the degree of damage. The extent of structural damage directly affects the mechanical properties and gas storage conditions of coal, thereby influencing the risk of gas outbursts.

Strength coefficient: The strength coefficient of coal is utilised to measure its compressive strength, hardness, and brittleness. A higher value indicates greater resistance to compressive forces and less susceptibility to damage. Conversely, a lower strength coefficient implies that the coal is more vulnerable to external forces, facilitating gas release under pressure and, thereby, increasing the probability of outbursts.

Drilling chip volume: Drilling chip volume refers to the amount of coal powder generated during the drilling process in coal seams. This parameter is closely related to the strength and brittleness of the coal. Weaker or more brittle coal seams produce larger volumes of drilling chips, indicating a higher propensity for structural damage, thereby elevating the risk of gas outbursts.

Coal thickness: The thickness of coal seams directly determines the volume of gas stored within the coal body. Thicker coal seams generally contain more gas. When the coal body exhibits low strength, the accumulated gas is more likely to breach the constraints of the coal structure and escape, resulting in gas outbursts.

(3) Ground Stress

Distance to geological structures: Coal seams located near geological structures, such as faults and folds, are subjected to significant structural stress, increasing the risk of gas outbursts. As mining operations progress, the redistribution of structural stress may lead to gas accumulation. This is particularly evident in stress-concentrated regions, where the likelihood of outbursts is significantly heightened.

Burial depth: With increasing mining depth, ground stress intensifies progressively. Deeply buried coal seams are subjected to higher ground stress, resulting in tighter coal structures that hinder the release of gas. When gas accumulation reaches a critical threshold, an outburst disaster may occur. Therefore, the burial depth of coal seams and the associated ground stress are critical factors influencing the risk of gas outbursts.

2.2. *AFT-Transformer-SVM Model for Coal and Gas Outburst Prediction*

Figure 2 illustrates the Transformer-SVM coal and gas outburst prediction model, designed for multidimensional, nonlinear, and small-sample prediction tasks. The diagram demonstrates the connection between the raw data, the Transformer model framework, and the SVM classification model. Initially, raw sample data are fed into the Transformer model for training. The Transformer model is primarily composed of multi-head attention and positional embedding. Positional embedding explicitly incorporates sequence position information, while multi-head attention models the correlation within the sequence by feeding the input data into multiple attention heads to learn different feature representations. These features are then concatenated to allow the model to extract more comprehensive and multidimensional feature information. Subsequently, the extracted features are passed to a Flatten Layer, which converts the sequence into a one-dimensional vector, flattening the multidimensional features into a single vector. The vector is then processed through a Dense Layer, where nonlinear mapping is applied to extract higher-level and abstract features. Following this, the SVM model adopts a degenerative feedback Transformer structure to enhance feature extraction and improve the prediction performance. In the diagram, the left section represents the raw data, consistent with Figure 1, and depicts the original sample. The middle section represents the Transformer feature extraction module, showcasing various layers of the Transformer structure. The right section illustrates the SVM classification module, where the extracted features are learnt and compared against the outburst risk parameter to evaluate the prediction accuracy.

2.2.1. Ali Baba and the Forty Thieves Optimisation Algorithm

The AFT algorithm is an intelligent optimisation method suitable for solving multidimensional objective functions to obtain optimal solutions [23]. The fundamental principles of the algorithm are described as follows:

(1) Acquiring Information and Pursuing "Alibaba":

The algorithm first gathers information and pursues the optimal solution, as described in Equation (1):

$$x_{t+1}^{i} = gbest_t + \left[Td_t\left(best_t^{i} - y_t^{i}\right)r_1 + Td_t\left(y_t^{i} - m_t^{a(i)}\right)r_2\right]sgn(rand - 0.5) \tag{1}$$

Here, x_{t+1}^i denotes the position of the ith individual in the $t+1$ iteration; $gbest$ represents the global best position, indicating the global optimal solution identified during the current search process; Td_t is the tracking distance, which controls the magnitude of position updates; y_t^i is the current position of individual i in the tth iteration; $best_t^i$ is the historical best position of the individual i; and $m_t^{a(i)}$ is the reference position of the individual i.

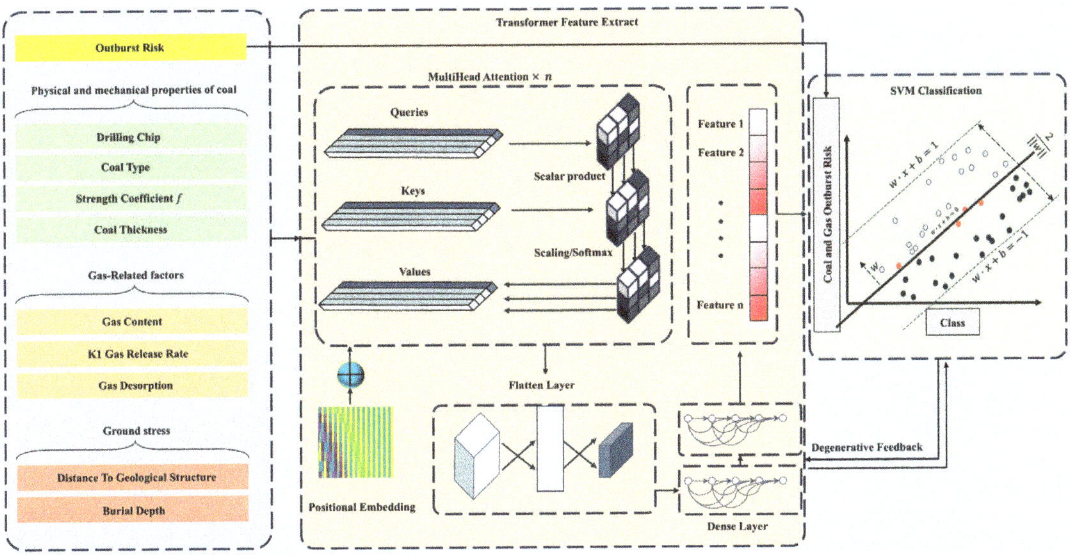

Figure 2. Schematic diagram of the Transformer-SVM model structure.

(2) Random Exploration and Deception:

Individuals are deceived and randomly explore the solution space, as described in Equation (2):

$$x_{t+1}^i = Td_t\left[(u_j - l_j)\text{rand} + l_j\right] \quad (2)$$

Here, u_j and l_j represent the upper and lower bounds, respectively, of the jth dimension in the search space.

(3) Balancing Global and Local Search:

The algorithm ensures a balance between global exploration and local exploitation to prevent premature convergence to local optima. This balance is achieved as described in Equation (3):

$$x_{t+1}^i = gbest_t - \left[Td_t\left(best_t^i - y_t^i\right)r_1 + Td_t\left(y_t^i - m_t^{a(i)}\right)r_2\right]sgn(rand - 0.5) \quad (3)$$

This optimisation approach enables efficient exploration of the solution space, ensuring a balance between exploitation and exploration for achieving the global optimal solution.

2.2.2. Transformer Feature Extraction

In the Transformer model, a positional encoding layer and self-attention mechanism are introduced to effectively process the sequential information of the input data. Initially, a positional embedding layer is incorporated, which utilises an embedding mechanism to assign a unique vector to each input, enabling the model to perceive positional information.

Through an additive mechanism, positional embeddings are combined with the original features, allowing the Transformer model to fully comprehend the spatial relationships within the sequential data.

Given the input sequence $X = [x_1, x_2, \ldots, x_n]$, where $x_i \in \mathbb{R}^d$ represents the feature vector at the ith position of the input, the positional encoding $P = [p_1, p_2, \ldots, p_n]$ generates the positional embedding vector $p_i \in \mathbb{R}^d$ for each position i. The positional encoding is then additively combined with the input features, as described in Equation (4):

$$X' = X + P \tag{4}$$

where X' represents the input data after the addition of positional encoding.

Next, a multi-head self-attention layer is introduced. By parallel computing multiple attention weight matrices, the model is enabled to capture critical information from the input data across different subspaces simultaneously. Through the stacking of two self-attention layers, the model effectively learns the dependencies among features at various levels, enhancing its capability to recognise complex patterns. The computation for each attention head is as follows:

(1) Calculation of Query (Q), Key (K), and Value (V):

For each attention head h, the Query Q_h, Key K_h , and Value V_h are computed as follows:

$$Q_h = X'W_Q^{(h)} \tag{5}$$

$$K_h = X'W_K^{(h)} \tag{6}$$

$$V_h = X'W_V^{(h)} \tag{7}$$

where $W_Q^{(h)}$, $W_K^{(h)}$, and $W_V^{(h)}$ are the learnt weight matrices.

(2) Attention Score:

The attention score is calculated using the Query and Key as follows:

$$Attention(Q_h, K_h) = \frac{Q_h K_h^T}{\sqrt{d_k}} \tag{8}$$

In the equation, d_k represents the dimensionality of the Key vector, and the normalisation factor $\sqrt{d_k}$ is used to prevent the values from becoming excessively large.

(3) Weighted Sum Calculation:

The attention scores are used to compute the weighted sum of the values, as shown in Equation (9).

$$Output_h = softmax\left(\frac{Q_h K_h^T}{\sqrt{d_k}}\right) V_h \tag{9}$$

(4) Multi-Head Attention Combination:

The outputs from all attention heads are concatenated and passed through a linear transformation to obtain the final attention output, as shown in Equation (10):

$$MultiHeadOutput = Concat(Output_1, \ldots, Output_h) W_O \tag{10}$$

Here, W_O is a learnt linear transformation matrix.

Finally, a fully connected layer and classification output are applied. Based on the self-attention mechanism, the extracted features are flattened through a flattening layer and further subjected to a series of fully connected (Dense) layers for nonlinear mapping. The output layer utilises the softmax activation function for multi-class classification, providing the final prediction for the coal and gas outburst discrimination task, as shown in Equations (11) and (12):

(5) Flattening and Fully Connected Layer:

The flattened output Z is mapped through the first fully connected layer:

$$F_1 = Relu(ZW_1 + b_1) \tag{11}$$

Here, W_1 and b_1 represent the weight matrix and bias term of the first fully connected layer, respectively, and Relu is the nonlinear activation function.

(6) Output Layer:

Multi-class classification is performed through the second fully connected layer and the softmax activation function:

$$\hat{y} = softmax(F_1 W_2 + b_2) \tag{12}$$

Here, W_2 and b_2 are the weight matrix and bias term of the second fully connected layer, respectively, and $\hat{y}$ represents the predicted probability distribution output by the model.

2.2.3. Grid-Optimised Support Vector Machine Algorithm

Grid search is an exhaustive search method used to optimise the hyperparameters of a Support Vector Machine (SVM). The performance of an SVM model relies heavily on the proper selection of hyperparameters, such as the kernel function parameters and the regularisation parameter. The primary objective of grid optimisation is to identify the optimal combination of these hyperparameters to maximise the model's performance on the validation set.

(1) Fundamental Formula of SVM

Since coal and gas outburst data are nonlinear, even after feature extraction using the Transformer model, a Gaussian radial basis function (RBF) kernel is required to map the data to a higher-dimensional space. The formula for the Gaussian RBF kernel is given in Equation (13):

$$K(\mathbf{x}_i, \mathbf{x}_j) = \exp\left(-\frac{\| \mathbf{x}_i - \mathbf{x}_j \|^2}{2\sigma^2}\right) \tag{13}$$

(2) Grid-Optimised SVM

Grid optimisation identifies the optimal hyperparameter combination that achieves the best model performance. It involves optimising the regularisation parameter C and the kernel parameter σ. The steps for grid optimisation are as follows:

Define the hyperparameter search space:

$$\mathcal{H} = \{(C, \gamma) \mid C \in \mathcal{C}, \gamma \in \Gamma\} \tag{14}$$

Here, $\mathcal{C}$ represents the set of candidate values for C, and Γ represents the set of candidate values for the RBF kernel parameter $\gamma = 1/(2\sigma^2)$.

Cross-validation evaluation: For each hyperparameter combination $(C, \gamma) \in \mathcal{H}$, k-fold cross-validation is performed to evaluate the model's performance:

$$\mathrm{CV}_{\mathrm{score}}(C, \gamma) = \frac{1}{k} \sum_{i=1}^{k} \mathrm{Score}_i(C, \gamma) \tag{15}$$

Here, $\text{Score}_i(C,\gamma)$ represents the model's accuracy on the ith fold of the data, and k denotes the number of folds in the cross-validation process.

Hyperparameter selection: All hyperparameter combinations are iterated, and the combination that maximises the cross-validation score is selected as follows:

$$(C^*,\gamma^*) = \arg \underset{(C,\gamma)\in\mathcal{H}}{max} \text{CV}_{\text{score}}(C,\gamma) \tag{16}$$

Model training with optimal hyperparameters: Finally, the SVM model is retrained on the training set using the optimal hyperparameters (C^*,γ^*):

$$\hat{f}(x) = \text{SVM}(x; C^*,\gamma^*) \tag{17}$$

2.3. Data Sensitivity Analysis

In the process of data analysis, a combination of SHAP analysis and permutation feature importance (PFI) methods is utilised to evaluate the importance of data features.

2.3.1. Permutation Feature Importance

The PFI method is applied within the Transformer-SVM model by randomly permuting the values of the specific feature j. The permuted data are then used for coal and gas outburst prediction with the Transformer-SVM model to assess the model's performance. The prediction accuracy is analysed to measure the impact of different features on the overall prediction performance. The fundamental formulas are as follows:

First, let $M(model, X)$ denote the prediction accuracy of the model on data X, and let $X_{perm}^{(j)}$ represent the new matrix obtained by randomly permuting the jth column of the feature matrix X. The model performance on the permuted data is expressed using Equation (18):

$$M_j^{(perm)} = M\left(AFT - Transformer - SVM, X_{perm}^{(j)}\right) \tag{18}$$

Based on the performance degradation, the reduction in model performance is defined by Equation (19):

$$PFI_j = M_{base} - M_j^{(perm)} \tag{19}$$

2.3.2. SHAP Analysis for Interpreting the Coal and Gas Outburst Model

The SHAP analysis is used to explain the influence of parameters extracted by the Transformer-SVM model on the model's predictions. The Shapley value from game theory is employed to determine the contribution of each participant (feature) to the final prediction of the coal and gas outburst model. The model is expressed as shown in Equation (20):

$$f(x) = SVM(T(x)) \tag{20}$$

Here, $T(x)$ represents the process in which the Transformer maps the original sample x to the new feature $p = T(x)$.

For a single sample $x \in \mathbb{R}^d$ and the model $f(\cdot)$, let the feature set $F = \{1,2,\ldots,d\}$. The SHAP value $\phi_j(x)$ for the jth feature is calculated as defined in Equation (21):

$$\phi_j(x) = \sum_{S\subseteq F\{j\}} \frac{|S|!(d-|S|-1)!}{d!}[f(S\cup\{j\} - f(S)] \tag{21}$$

where $|S|$ represents the size of the feature subset S; and $f(S)$ denotes the model's prediction for sample x using only the feature subset S.

In practical applications, to reduce the computational complexity of enumerating all subsets S, the kernel SHAP method is employed to approximate the SHAP values and minimise the computational burden.

3. Results and Discussion

This study utilised a dataset of 569 samples to construct the AFT-Transformer-SVM model using the TensorFlow library within Python 3.9. A series of computational comparison experiments were designed to assess and validate the coal and gas outburst risk based on real-world data, providing technical support for safe production in actual mining operations.

Table 1 presents a subset of the sample data used for the analysis, which is consistent with Figure 1 and the descriptions provided earlier. Nine variables were selected as sample features, with the outburst risk levels represented by drilling operations at the mining site. No specific measures taken were considered as no outburst risk, represented by 0.1. Conducting 12–15 sets of pressure relief boreholes indicated a moderate outburst risk, represented by 0.6. Drilling 20 sets of pressure relief boreholes was considered to represent a severe outburst risk, represented by 1.

Table 1. Collected sample data table.

Coal Type	Gas Desorption Δp	Strength Coefficient f	Gas Content $/m^3{\cdot}t^{-1}$	$K1$ Gas Release Rate $mL{\cdot}(g{\cdot}min^{0.5})^{-1}$		
2	16.86	0.31	7.13	0.4	0.27	0.29
2	17.83	0.30	10.62	0.42	0.21	0.24
2	17.78	0.35	7.52	0.22	0.16	0.18
3	17.09	0.32	11.44	0.11	0.25	0.21
...	...	...	...	...	...	...

Distance to Geological Structure/m	Burial Depth /m	Coal Thickness/m	Drilling Chip/kg·m^{-1}			Outburst Risk
32	366	0.8	3.7	3.4	3.9	1
0	358	2.2	3.1	3.6	3	0.6
42	352	2.2	3.5	3.4	3.8	0.1
62	370	2.5	3.2	3.2	3.2	0.6
...	...	...	...	...	...	...

3.1. Optimisation of the Transformer-SVM Model Using Different Algorithms

Firstly, the original data were split in a 7:3 ratio, with the first 70% used for the model optimisation. Various optimisation algorithms were employed to optimise the parameters of the Transformer-SVM model, including units_fc1, num_heads, key_dim, and learning_rate.

The optimisation process was conducted using a fitness function defined as $(1-\text{accuracy})$. A population of 30 individuals was selected, and the optimisation process was iterated 350 times, as illustrated in Figure 3.

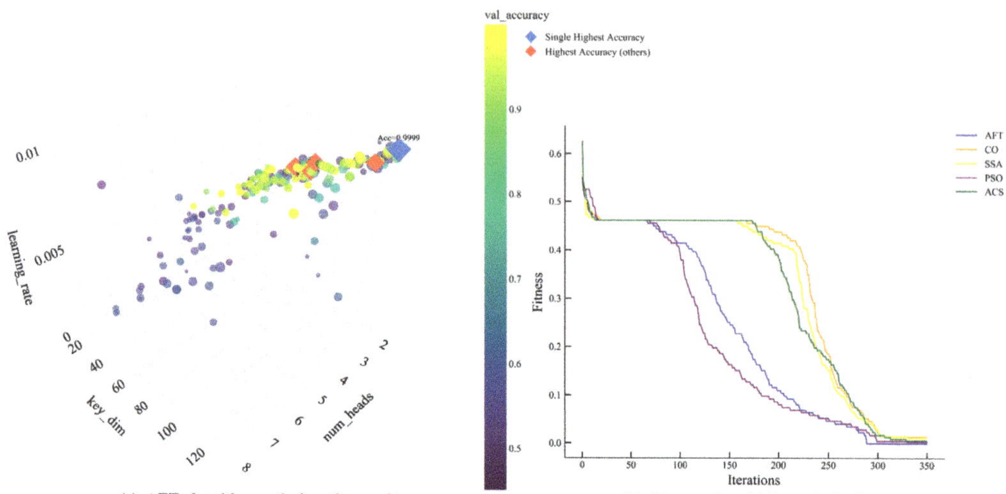

(a) AFT algorithm optimizes the results **(b) Fitness of multiple optimization algorithms**

Figure 3. Comparative results of various optimisation algorithms.

Figure 3a presents the optimisation results using the AFT algorithm, where the point with the highest accuracy reached 0.999. The size of the points in the figure represents units_fc1, the x-axis represents num_heads, the y-axis represents key_dim, the z-axis represents learning_rate, and the colour bar indicates the prediction accuracy. The final optimised parameters were determined as follows: [num_heads: 4, key_dim: 98, learning_rate: 0.009449, units_fc1: 64].

Figure 3b shows the relationship between the fitness function and the number of iterations for various optimisation algorithms. The blue curve represents the AFT algorithm, where the fitness value reached 0.00012 after 289 iterations, making it the algorithm with the smallest fitness value. This was followed by the PSO algorithm, with a fitness value of 0.0032, and then the CO, SSA, and ACS optimisation algorithms.

Through experimental comparisons, each algorithm was run ten times, and the optimal fitness values of the various optimisation algorithms were selected for a statistical analysis. The results are presented in Table 2, where the AFT algorithm demonstrated superior performance in terms of range, minimum, maximum, and standard deviation compared to the other algorithms. However, it exhibited a relatively higher variance. Following the AFT algorithm were the PSO and SSA algorithms, with the CO and ACS algorithms performing the least effectively.

Table 2. Comparative results of various optimisation algorithms.

Optimization Algorithms	Variance	Range	Minimum	Maximum	Standard Deviation
AFT	0.00004	0.0004	0.000110	0.00015	0.000011
CO	0.00001	0.003	0.013	0.016	0.000884
SSA	0.00003	0.003	0.006	0.009	0.000936
PSO	0.00002	0.0018	0.0032	0.005	0.000503
ACS	0.00013	0.012	0.016	0.028	0.003606

These results validate the efficiency and accuracy of the AFT algorithm in optimising the parameters of the Transformer-SVM model for coal and gas outburst predictions. Therefore, the AFT algorithm was chosen as the final intelligent optimisation algorithm.

3.2. AFT-Transformer-SVM Model for Coal and Gas Outburst Risk Prediction

To evaluate the performance of the model, 70% of the sample data were used as training data, while the remaining 30% were used as testing data to predict coal and gas outburst risks. Figure 4 illustrates the prediction results of the AFT-Transformer-SVM model. In the figure, the true values are represented by grey circles, while the prediction results are represented by red triangles. The "Predict Class" line indicates the predicted results, and the "True Class" line represents the actual results. In the sample data, the two lines overlap completely, indicating that the prediction results are 100% accurate. Figure 4a shows the results for the training set, and Figure 4b displays the results for the test set.

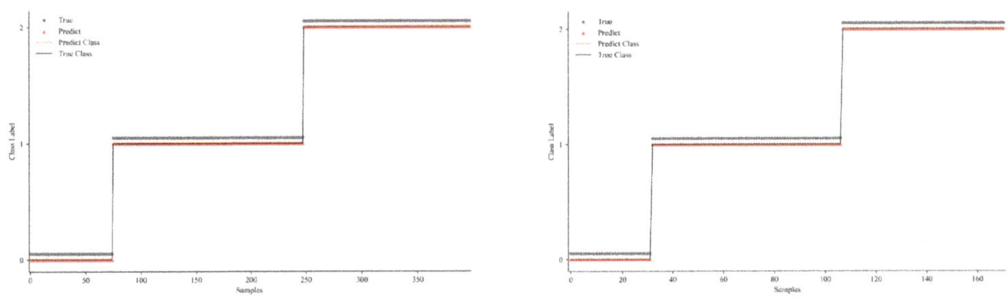

(**a**) train set predict results (**b**) test set predict results

Figure 4. Prediction results of the AFT-Transformer-SVM model.

Figure 5 presents the corresponding confusion matrices, where all predictions lie along the diagonal, signifying perfect prediction accuracy. The three colours in the figure represent three categories: 1, 2, and 3. The numerical values in the table indicate the number of instances in each category. Figure 5a corresponds to the results for the training set, while Figure 5b corresponds to the results for the test set.

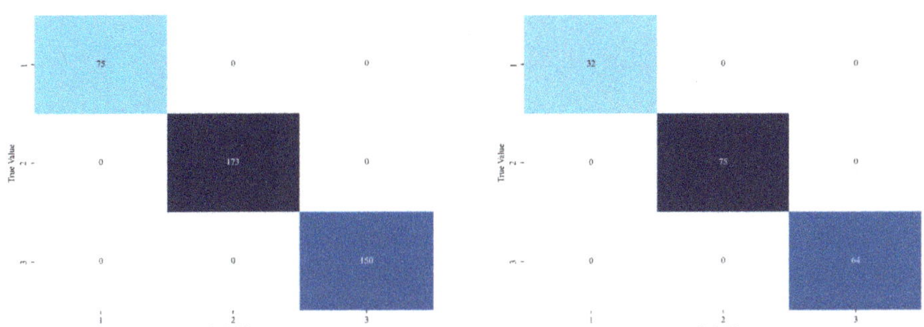

(**a**) train set predict results (**b**) test set predict results

Figure 5. Confusion matrix of the prediction results for the AFT-Transformer-SVM model.

Table 3 presents the sensitivity analysis of the prediction results, including eight parameters: recall, precision, F1-score, accuracy, sensitivity, specificity, AUC, and Kappa.

These metrics involve four key parameters: True Positive (TP), False Negative (FN), False Positive (FP), and True Negative (TN). TP represents instances where the actual label is positive, and the model predicts positive; FN represents instances where the actual label is positive, but the model predicts negative; FP represents instances where the actual label is negative, but the model predicts positive; and TN represents instances where the actual label is negative, and the model predicts negative [24–26].

Table 3. Sensitivity analysis of prediction results.

Recall/%	Precision/%	F1-Score/%	Accuracy/%	Sensitivity/%	Specificity/%	AUC/%	κ/%
100	100	100	100	100	100	100	100

$Recall = \frac{TP}{TP+FN}$, representing the proportion of positive samples correctly predicted as positive. $Percision = \frac{TP}{TP+FP}$, representing the proportion of correctly predicted positive samples among all predicted positive samples. $F1 = 2 \times \frac{Percision \times Recall}{Percision + Recall}$, combining precision and recall to indicate the proportion of perfect classification. $Accuracy = \frac{TP+TN}{All\ Samples}$, indicating the proportion of correctly classified samples, whether positive or negative. Sensitivity is similar to recall. $Specificity = \frac{TN}{TN+FP}$, representing the proportion of negative samples correctly classified as negative. AUC refers to the area under the ROC curve, representing the model's discriminative ability across different thresholds. $\kappa = \frac{p_0 - p_e}{1 - p_e}$, where p_0 denotes the observed accuracy, and p_e represents the expected accuracy under random guessing, indicating the model's accuracy after accounting for random agreement.

The prediction results shown in Table 3 indicate that the model has perfectly learnt the features of the sample data and performed accurate predictions.

Figure 6 illustrates the Polygon Area Metric (PAM) results of the AFT-Transformer-SVM model, with subplots a, b, and c representing the polygon area predictions for the three categories. In the figure, FM represents F_measure, or the F1-Score. K represents κ. SP denotes specificity. SE denotes sensitivity. CA represents accuracy. The prediction results for all three categories are entirely correct, confirming the feasibility of the model.

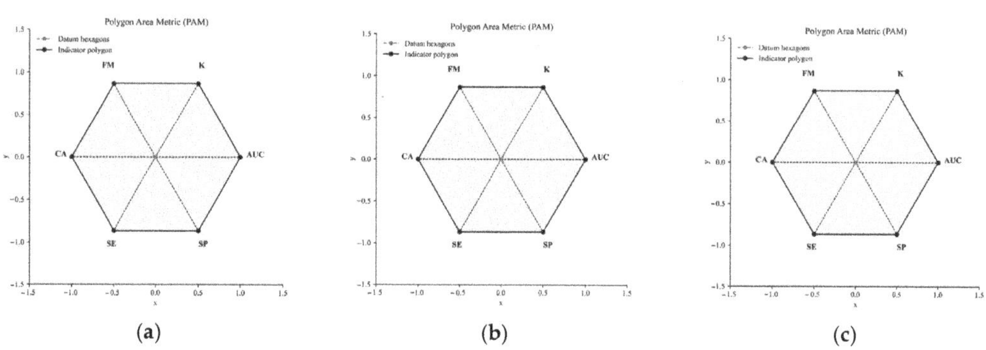

Figure 6. PAM results of AFT-Transformer-SVM predictions.

Figure 7 shows the ROC curve of the model's predictions. The actual curve starts from the point (0, 0), progresses to (0, 1), and ends at (1, 1), indicating that the model's predictions perfectly align with the actual results. The curves for Class 1, Class 2, and Class 3 overlap and converge at the top-right corner, further demonstrating the model's outstanding performance.

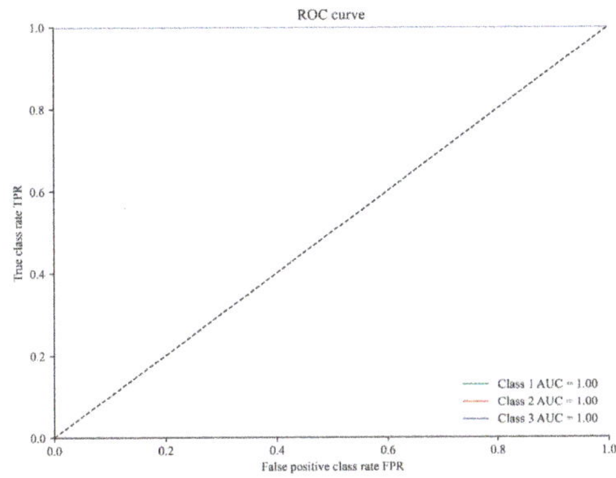

Figure 7. ROC curve of prediction results.

To interpret the model, a SHAP value analysis was conducted, and the results are shown in Figure 8. The figure illustrates three features extracted by the Transformer model. The SHAP analysis demonstrates clear feature distribution trends. For Feature 2, the relationship between the SHAP values and feature values is relatively monotonic: higher feature values (red) positively contribute to SHAP values, while lower feature values (blue) suppress SHAP values. This indicates that the model's treatment of Feature 2 is close to linear, where higher feature values have the greatest positive impact on the prediction results. For Feature 0, a reverse monotonic relationship is observed between the SHAP values and feature values: lower feature values (blue) positively contribute to the SHAP values, while higher feature values (red) suppress the SHAP values. The distribution trend is clear, indicating that the model processes Feature 0 in a nearly linear manner. For Feature 1, the SHAP value exhibits a certain level of complexity, where higher feature values (red) suppress the SHAP values, while middle SHAP values are not fully monotonic, showing both positive and negative SHAP values. This suggests that the model processes Feature 1 with some degree of nonlinearity.

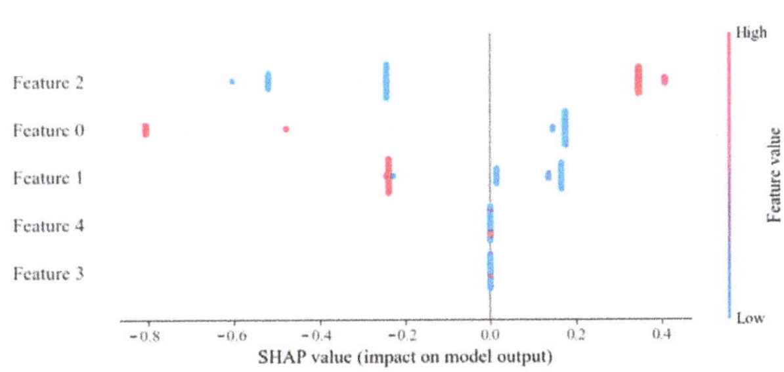

Figure 8. SHAP analysis of feature extraction in the AFT-Transformer-SVM model.

Figure 9 presents the SHAP analysis results for the original data. The figure shows the relationships between the feature values of various raw data variables and their SHAP values. Most variables exhibit highly nonlinear relationships, with the SHAP values distributed between −0.5 and 0.3, indicating a relatively small influence. Among these, gas content is the feature with the widest distribution and the highest degree of nonlinearity, followed by gas release rate, with other variables showing smaller contributions. A comparison between Figures 8 and 9 demonstrates that feature extraction reduces the nonlinearity of the original multidimensional data and increases the association between feature values and SHAP values.

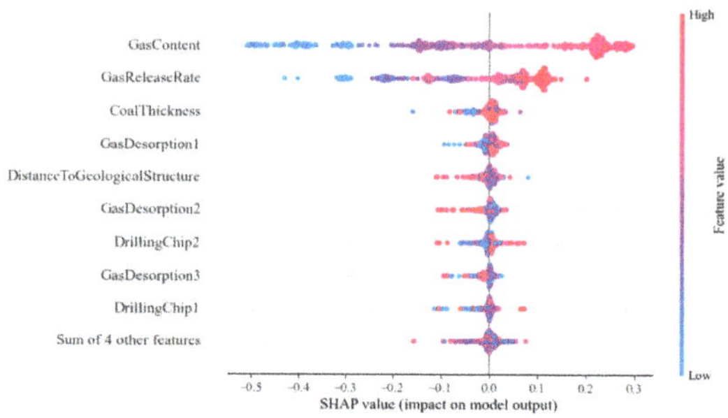

Figure 9. SHAP analysis of original sample features.

Figure 10 shows the waterfall plots for the training and testing samples. Figure 10a represents the training set features. In samples 0–100, Feature 3 and Feature 1 contribute strongly negatively to the model output (blue), while Feature 2 contributes positively (red). For samples 100–200, the positive contributions of Feature 3 and Feature 2 gradually increase, with the overall colour shifting towards blue, and Feature 1 shows a more substantial positive contribution. In samples 200–300, the contributions of Feature 1 and Feature 2 increase significantly (evidenced by the expansion of the red area), driving a marked rise in prediction values. In samples 300–400, the contributions of all three features increase (indicated by the red area), resulting in an upward trend in the model output. Figure 10b illustrates the feature contributions in the test set, which exhibit a similar overall trend to the training set. However, a noticeably varying region is observed within samples 0–50, where the model output shows an increase in this range.

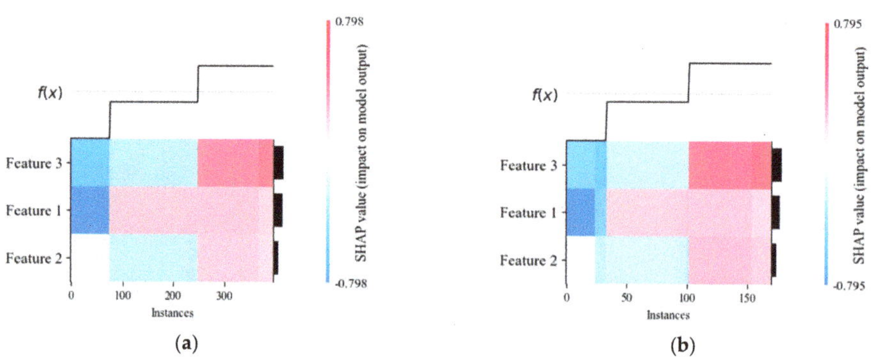

Figure 10. Waterfall plot of extracted sample features in training and testing sets.

3.3. Comparative Analysis of Different Algorithms

The XGBoost, KNN, RBFN, and Bayesian classifier algorithms were selected for comparison to analyse the prediction accuracy of different algorithms applied to coal and gas outburst sample data. The results are shown in Figure 11. The XGBoost algorithm achieved a training set accuracy of 93.47%, corresponding to Figure 11a, and a testing set accuracy of 82.46%, corresponding to Figure 11b. The KNN algorithm achieved a training set accuracy of 100%, corresponding to Figure 11c, and a testing set accuracy of 84.21%, corresponding to Figure 11d. The RBFN algorithm achieved a training set accuracy of 100%, corresponding to Figure 11e, and a testing set accuracy of 85.96%, corresponding to Figure 11f. The Bayesian classifier algorithm achieved a training set accuracy of 100%, corresponding to Figure 11g, and a testing set accuracy of 83.04%, corresponding to Figure 11h.

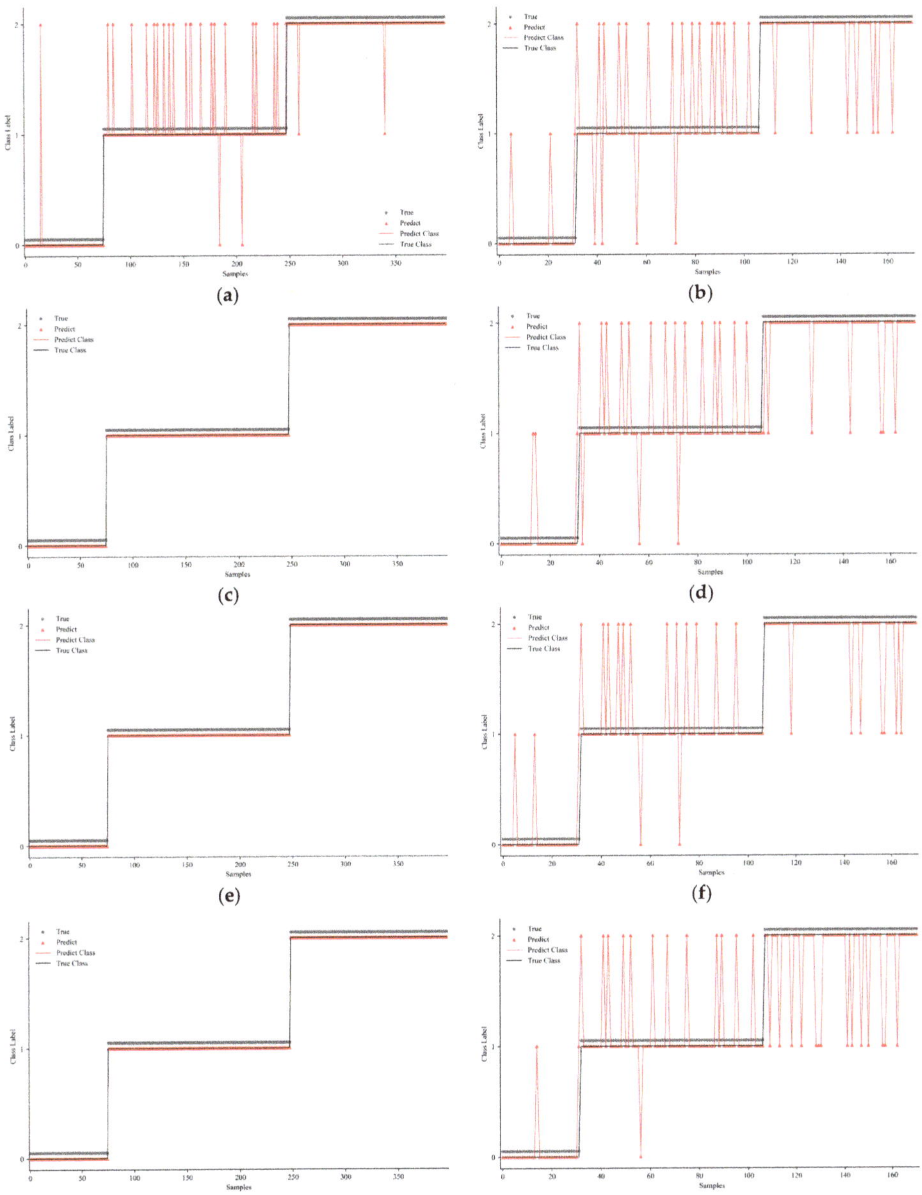

Figure 11. Prediction results of different comparison algorithms.

Figure 12 illustrates the ROC curves of different comparison algorithms. The overall AUC of the XGBoost algorithm was 0.94, with the AUCs of the three classes being 0.99,

0.89, and 0.94, represented in cyan, orange, and blue, respectively, as shown in Figure 12a. The KNN algorithm achieved an overall AUC of 0.93, with class-specific AUCs of 0.98, 0.88, and 0.95, as shown in Figure 12b. The RBFN algorithm achieved an overall AUC of 0.93, with class-specific AUCs of 0.97, 0.88, and 0.93, as shown in Figure 12c. The Bayesian classifier achieved an overall AUC of 0.93, with class-specific AUCs of 0.98, 0.89, and 0.91, as shown in Figure 12d.

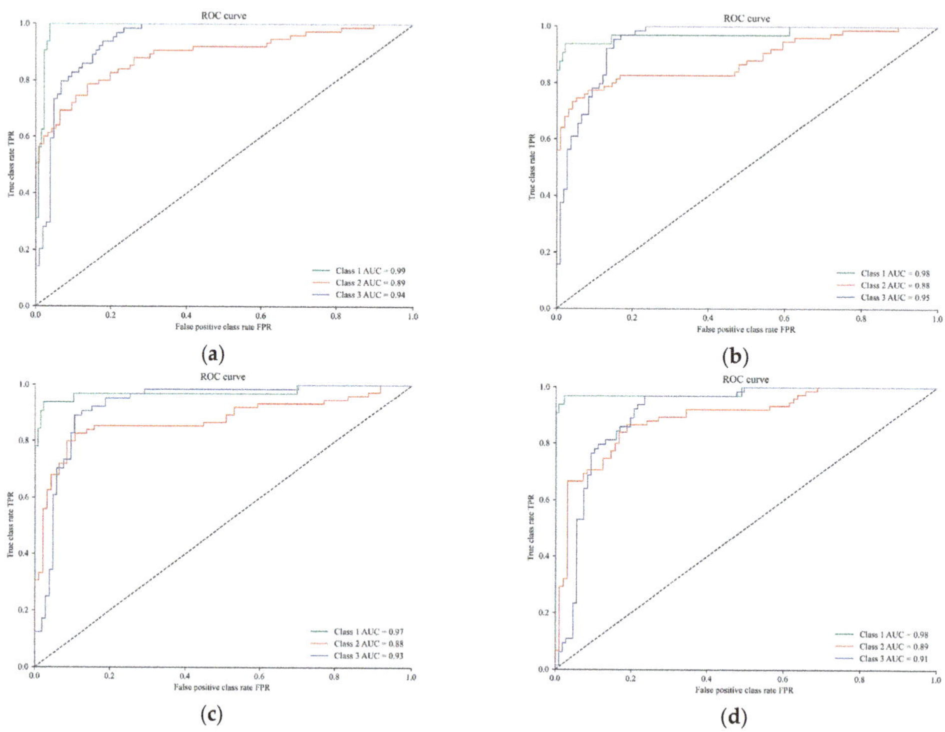

Figure 12. ROC curves of different comparison algorithms.

The comparative analysis of the four algorithms demonstrates that the RBFN algorithm achieved the highest prediction accuracy, followed by the KNN, Bayesian classifier, and, finally, the XGBoost algorithm. Although all four algorithms achieved 100% accuracy on the training set, none reached 100% accuracy on the testing set, further validating the superiority of the AFT-Transformer-SVM model.

Figure 13 presents the PAM (Polygon Area Metric) results of the four algorithms, corresponding to Figures 11 and 12, to analyse the sensitivity of the prediction accuracy of the different models. The corresponding data are summarised in Table 4. In Table 4, the comparative analysis of various data features reveals that all four algorithms performed poorly on the second and third classes of features while achieving better results on the first class of features. This finding indicates that the nonlinear characteristics of the data have not been fully extracted, and uncertainties remain in the data features.

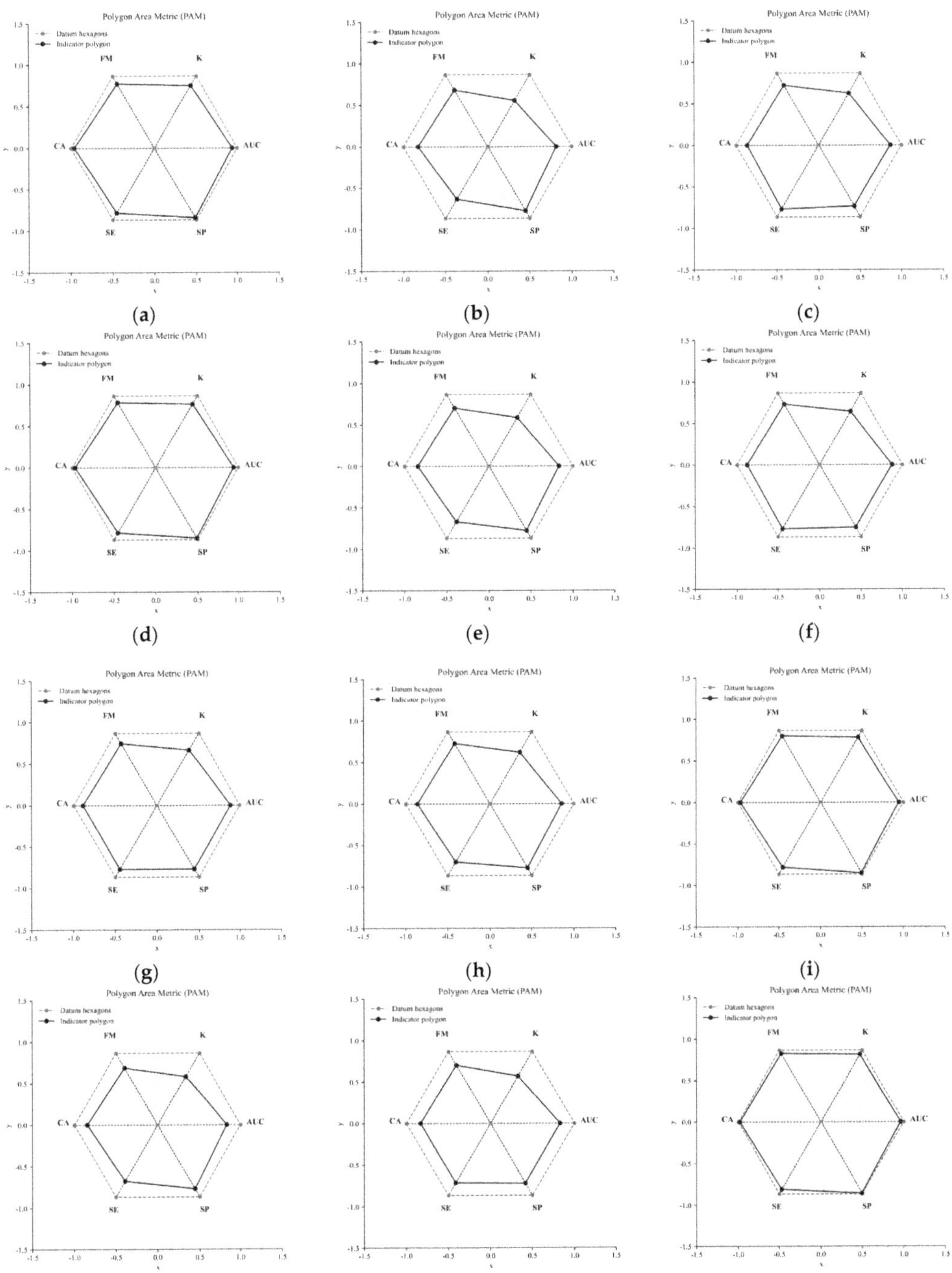

Figure 13. PAM results of different comparison algorithms.

Table 4. PAM values of different algorithms.

Algorithm	FM/%	κ/%	AUC/%	SP/%	SE/%	CA/%
XGBoost	89.23 78.57 83.21	86.70 63.85 72.07	93.87 81.46 87.05	97.12 89.58 85.05	90.62 73.33 89.06	95.91 82.46 86.55
KNN	90.62 81.12 84.44	88.47 67.61 74.34	94.23 83.46 87.99	97.84 89.58 86.92	90.62 77.33 89.06	96.49 84.21 87.72
RBFN	92.06 83.56 85.71	90.27 71.33 76.64	94.59 85.46 88.92	98.56 89.58 88.79	90.62 81.33 89.06	97.08 85.96 88.89
Bayesian classifiers	95.24 81.05 79.37	94.16 65.71 67.33	96.52 83.00 83.46	99.28 83.33 88.79	93.75 82.67 78.12	98.25 83.04 84.80

In Table 4, the PAM values for the first, second, and third classes of features are listed from top to bottom. Referring to Figure 12, the blue outer ring represents the ideal PAM value. None of the four algorithms achieved the optimal PAM value across all three classes. Specifically, the XGBoost algorithm achieved a PAM of 85.07% for the first class, 61.02% for the second class, and 70.27% for the third class. The KNN algorithm achieved a PAM of 86.05% for the first class, 64.79% for the second class, and 72.38% for the third class. The RBFN algorithm achieved a PAM of 88.07% for the first class, 68.62% for the second class, and 74.54% for the third class. The Bayesian classifier achieved a PAM of 92.52% for the first class, 63.63% for the second class, and 64.44% for the third class.

Through a comparative analysis with multiple algorithms, the AFT-Transformer-SVM algorithm demonstrated superior capabilities in extracting multidimensional nonlinear data features and achieving high prediction accuracy for coal and gas outbursts. The model achieved a testing set accuracy of 100%.

3.4. Data Validation for Other Mining Areas

To enhance the credibility of the model, based on the existing model, 66 sets of gas outburst test data from different time periods in a coal mine in Shaanxi, China, were selected as validation data. A validation experiment was designed. The prediction results of the model are shown in Figures 14–17. Among them, Figure 14 shows the prediction results of the AFT-Transformer-SVM model, where two test cases show slightly overestimated predictions, while the remaining values are predicted accurately, achieving an overall prediction accuracy of 96.97%, which is relatively high. Figure 15 shows the confusion matrix of the AFT-Transformer-SVM model's prediction results. From the figure, it can be seen that the model's prediction errors occur in two instances: one outburst risk point predicted as Class 2 and another outburst risk point predicted as Class 3, while the remaining values are all correctly predicted.

Figures 16 and 17 show the PAM results and ROC curve of the prediction results, respectively. In Figure 16a, the polygonal area is 96.42%; in Figure 16b, the polygonal area is 91.96%; and in Figure 16c, the polygonal area is 95.65%. The specific data can be referenced in Table 5. Classification 1 in Table 5 corresponds to Figure 16a, classification 2 corresponds to Figure 16b, and classification 3 corresponds to Figure 16c. Figure 17 presents the ROC curve of the prediction results, where all three classification results are close to the upper-left corner, indicating that the prediction results are relatively accurate.

Through the analysis, it has been found that the AFT-Transformer-SVM model achieved a prediction accuracy of 96.97% in a mining area in Shaanxi, China, which demonstrates a high level of accuracy and verifies the reliability of the model.

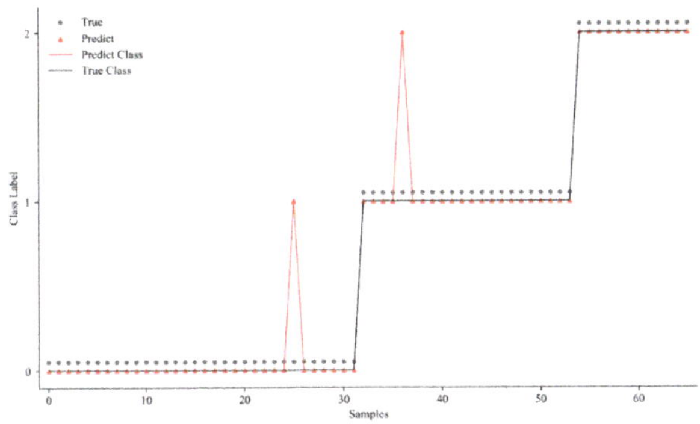

Figure 14. Prediction results of the AFT-Transformer-SVM model.

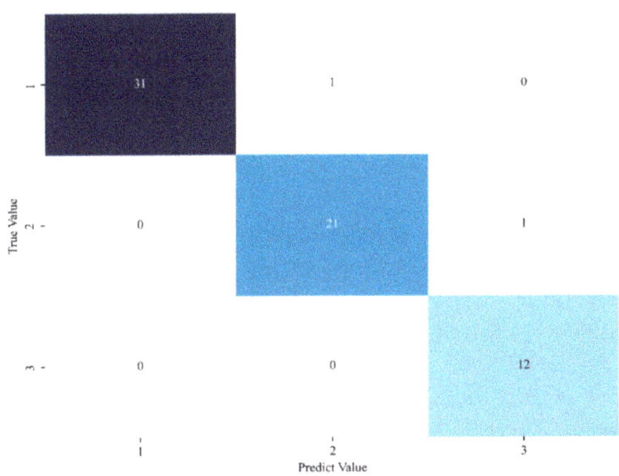

Figure 15. Confusion matrix of the prediction results for the AFT-Transformer-SVM model.

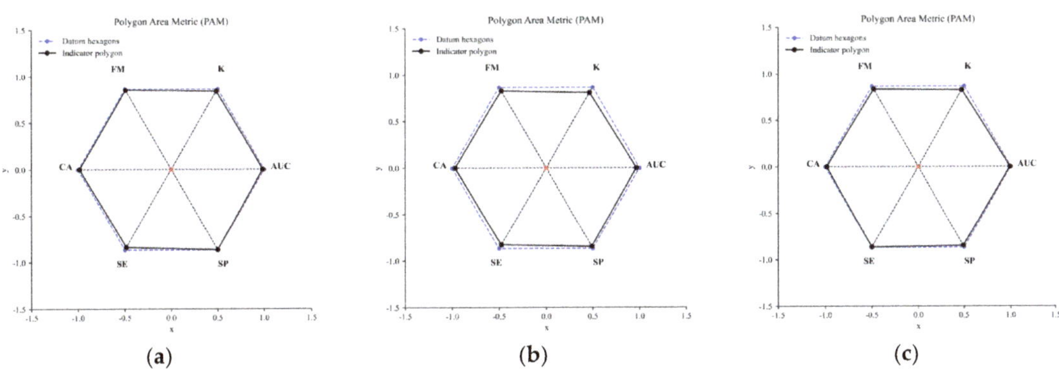

Figure 16. PAM results of AFT-Transformer-SVM predictions.

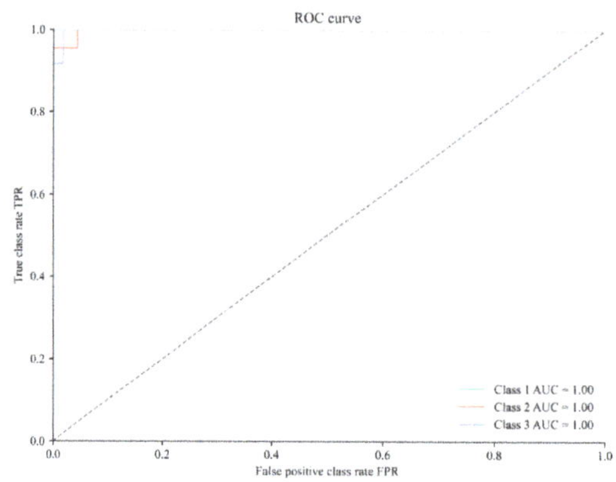

Figure 17. ROC curve of prediction results.

Table 5. PAM values of different algorithms.

Class	FM/%	κ/%	AUC/%	SP/%	SE/%	CA/%
1	98.41	96.96	98.44	100.00	96.88	98.48
2	95.45	93.18	96.59	97.73	95.45	96.97
3	96.00	95.07	99.07	98.15	100.00	98.48

3.5. Analysis of the Importance of Original Sample Data

PFI was employed to shuffle the original sample data and analyse the importance of the original features. Figure 18 illustrates the impact of different variables on the prediction results after shuffling. The variable with the greatest impact was gas content, followed by gas desorption, K1 gas release rate, drilling chip, distance to geological structure, burial depth, coal thickness, coal type, and strength coefficient. The box plots of the various parameters are marked in the figure. Table 6 provides the results of ten shuffles for each sample feature, listing the top five features with the highest importance. Among them, gas content had the highest feature importance, exhibiting the most significant impact on the prediction results, with the most apparent variations.

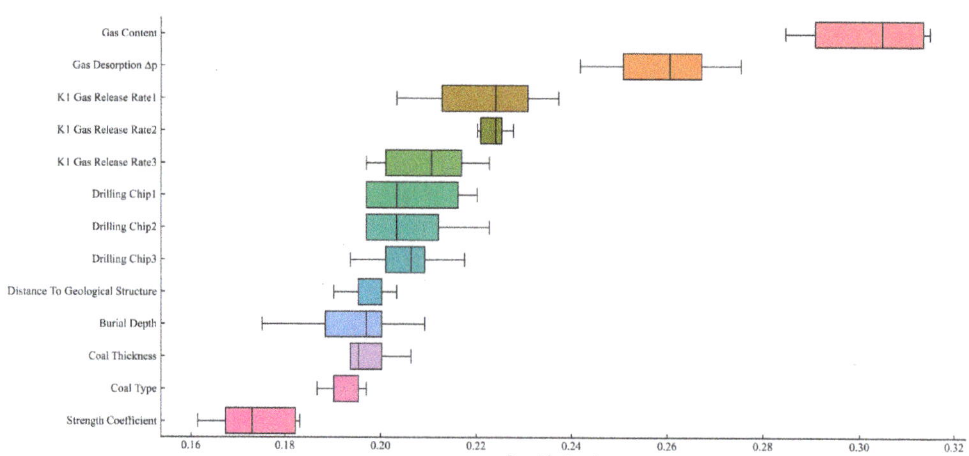

Figure 18. Box plots of feature importance based on PFI.

Table 6. Feature importance analysis after 10 shuffles.

Gas Content	0.20	0.39	0.40	0.28	0.30	0.29	0.30	0.30	0.31	0.29
Gas Desorption	0.26	0.26	0.24	0.24	0.26	0.25	0.26	0.24	0.27	0.25
Gas Release Rate1	0.23	0.21	0.22	0.22	0.23	0.23	0.23	0.20	0.21	0.22
Drilling Chip1	0.21	0.21	0.20	0.19	0.21	0.19	0.20	0.22	0.19	0.19
Distance to Geological Structure	0.20	0.19	0.19	0.20	0.20	0.20	0.20	0.20	0.20	0.20

Through this analysis, it was observed that in on-site sample testing data, the variable with the greatest impact on outburst risk was gas content, followed by gas desorption, K1 gas release rate, drilling chip, distance to geological structure, burial depth, coal thickness, coal type, and strength coefficient, which aligns with the characteristics of the real-world environment.

4. Conclusions

This study addresses the issue of gas outbursts in coal mine excavation faces by proposing a gas prediction and early warning method based on the interpretable AFT-Transformer-SVM model. A comprehensive framework for model construction, data analysis, and algorithm validation was developed. The findings of the study are as follows:

(1) This study innovatively integrates the AFT algorithm-optimised Transformer with the SVM model. By leveraging positional embedding, multi-head self-attention mechanisms, and fully connected layers, the model's ability to extract nonlinear and multidimensional data features of gas outbursts was significantly improved. The application of intelligent optimisation algorithms enabled the efficient parameter adjustment, ensuring the model's stability and accuracy in real-world operational environments.

(2) Based on SHAP and PFI analysis methods, the study conducted an in-depth examination of the impact of various features on model prediction, identifying the critical role of factors such as gas content, gas desorption rate, and drilling chip volume in assessing gas outburst risks. By incorporating multiple classification methods, including TP, TN, FP, and FN, the interpretability of the model was enhanced, providing a technical foundation for explaining gas prediction models in coal mines.

(3) A comparative analysis with traditional algorithms, such as XGBoost, KNN, RBFN, and Bayesian classifiers, demonstrated the significant advantages of the AFT-Transformer-SVM model in handling multidimensional nonlinear data and achieving high prediction accuracy. Experimental results indicated that the model achieved 100% prediction accuracy and excellent sensitivity on the test set, outperforming the comparison algorithms.

In conclusion, an effective method for predicting gas outburst risks in coal mines is provided by this study, which improves both the model accuracy and interpretability while offering valuable technical support for safe production in actual mining operations. On this basis, the model is embedded into the intelligent mine analysis platform, where it is periodically trained and used for the real-time prediction of coal and gas outburst hazards, integrated with on-site data. Future research will further combine theoretical analysis with field data to enrich real-time monitoring indicators, optimise monitoring methods, and advance the development of intelligent coal mine safety management.

Author Contributions: Conceptualisation, Y.W. (Yanping Wang) and J.D.; methodology, Z.Q.; software, Z.Q.; validation, Z.Q., Y.H. and Y.W. (Yiyang Wang); formal analysis, Y.C.; investigation, L.Z.; resources, Y.W. (Yanping Wang); data curation, Z.Q.; writing—original draft preparation, Y.W. (Yanping Wang); writing—review and editing, Y.W. (Yanping Wang); visualisation, Y.H.; supervision, J.D.; project administration, Z.Y.; funding acquisition, J.D. All authors have read and agreed to the published version of the manuscript.

Funding: This research was funded by the Key Research and Development Program of Shaanxi Province, grant numbers 2020GY-139 and 2022GY-150.

Informed Consent Statement: Informed consent was obtained from all subjects involved in the study.

Data Availability Statement: The data are not publicly available due to commercial confidentiality, as they contain information that could compromise the privacy of the research participants.

Conflicts of Interest: The authors declare no conflicts of interest.

References

1. Liang, Y.; Zheng, M.; Li, Q.; Mao, S.; Li, X.; Li, J.; Zhou, J. A review on prediction and early warning methods of coal and gas outburst. *J. China Coal Soc.* **2023**, *48*, 2976–2994. [CrossRef]
2. Zhang, C.; Wang, P.; Wang, E.; Xu, J.; Li, Z.; Liu, X.; Peng, S. Coal and gas outburst mechanism: Research progress and prospect inChina over the past 70 years. *Coal Geol. Explor.* **2023**, *51*, 59–94.
3. Black, D.J. Review of Coal and Gas Outburst in Australian Underground Coal Mines. *Int. J. Min. Sci. Technol.* **2019**, *29*, 815–824. [CrossRef]
4. Wang, E.; Zhang, G.; Zhang, C.; Li, Z. Research Progress and Prospect on Theory and Technology for Coal and Gas Outburst Control and Protection in China. *J. China Coal Soc.* **2022**, *47*, 297–322.
5. Zhu, Z.; Zhang, Y.; Wang, S.; Zhao, B. Effect of Plasma-Activated Water on the Settling Characteristics of Ultrafine Kaolinite. *Process Saf. Environ. Prot.* **2024**, *192*, 613–620. [CrossRef]
6. Nakajima, I.; Ujihira, M. *Considerations for Acoustic Emission and Gas Emission in Gas Outburst Processes*; VRMA: Morgantown, WV, USA, 1989.
7. Feyt, G.N. Selecting Methods for Preventing Sudden Outbursts Based on Comprehensive Criteria of the Loss of Stability and Avalanche Destruction of the Seam. *Nauchn. Soobshch.-Inst. Gorn. Dela A. A. Skochinskogo (USSR)* **1980**, *186*, 39–46.
8. Hanes, J.; Lama, R.D.; Shepherd, J. Research into the Phenomenon of Outbursts of Coal and Gas in Some Australian Collieries. In Proceedings of the 5th ISRM Congress, Melbourne, Australia, 10 April 1983.
9. Sato, K.; Fujii, Y. Source Mechanism of a Large Scale Gas Outburst at Sunagawa Coal Mine in Japan. *Pure Appl. Geophys.* **1989**, *129*, 325–343. [CrossRef]
10. Shepherd, J.; Rixon, L.K.; Griffiths, L. Outbursts and Geological Structures in Coal Mines: A Review. *Int. J. Rock Mech. Min. Sci. Geomech. Abstr.* **1981**, *18*, 267–283. [CrossRef]
11. Ji, P.; Shi, S.; Shi, X. Research on Early Warning of Coal and Gas Outburst Based on HPO-BiLSTM. *IEEE Trans. Instrum. Meas.* **2023**, *72*, 2529808. [CrossRef]
12. Zhu, J.; Zheng, H.; Yang, L.; Li, S.; Sun, L.; Geng, J. Evaluation of Deep Coal and Gas Outburst Based on RS-GA-BP. *Nat. Hazards* **2023**, *115*, 2531–2551. [CrossRef]
13. Xie, X.; Shu, X.; Fu, G.; Shen, S.; Jia, Q.; Hu, J.; Wu, Z. Accident Causes Data-Driven Coal and Gas Outburst Accidents Prevention: Application of Data Mining and Machine Learning in Accident Path Mining and Accident Case-Based Deduction. *Process Saf. Environ. Prot.* **2022**, *162*, 891–913. [CrossRef]
14. Wang, W.; Wang, H.; Zhang, B.; Wang, S.; Xing, W. Coal and Gas Outburst Prediction Model Based on Extension Theory and Its Application. *Process Saf. Environ. Prot.* **2021**, *154*, 329–337. [CrossRef]
15. Hanson, D.R.; Lawson, H.E. Using Machine Learning to Evaluate Coal Geochemical Data with Respect to Dynamic Failures. *Minerals* **2023**, *13*, 808. [CrossRef] [PubMed]
16. Zheng, X.; Lai, W.; Zhang, L.; Xue, S. Quantitative Evaluation of the Indexes Contribution to Coal and Gas Outburst Prediction Based on Machine Learning. *Fuel* **2023**, *338*, 127389. [CrossRef]
17. Li, Z.; Wang, E.; Ou, J.; Liu, Z. Hazard Evaluation of Coal and Gas Outbursts in a Coal-Mine Roadway Based on Logistic Regression Model. *Int. J. Rock Mech. Min. Sci.* **2015**, *80*, 185–195. [CrossRef]
18. Frid, V.; Vozoff, K. Electromagnetic Radiation Induced by Mining Rock Failure. *Int. J. Coal Geol.* **2005**, *64*, 57–65. [CrossRef]
19. Mathews, J.P.; Campbell, Q.P.; Xu, H.; Halleck, P. A Review of the Application of X-Ray Computed Tomography to the Study of Coal. *Fuel* **2017**, *209*, 10–24. [CrossRef]
20. Parsa, A.B.; Movahedi, A.; Taghipour, H.; Derrible, S.; Mohammadian, A. Toward Safer Highways, Application of XGBoost and SHAP for Real-Time Accident Detection and Feature Analysis. *Accid. Anal. Prev.* **2020**, *136*, 105405. [CrossRef] [PubMed]

21. Lerner, B.; Guterman, H.; Aladjem, M.; Dinstein, I.; Romem, Y. Feature Extraction by Neural Network Nonlinear Mapping for Pattern Classification. In Proceedings of the 13th International Conference on Pattern Recognition, Vienna, Austria, 25–29 August 1996; Volume 4, pp. 320–324.
22. Sun, L. Application Resrarch on the Coal and Gas Outburst Prediction Based on Gray Correlation Analysis and PSO-SVM. Master's Thesis, China University of Mining and Technology, Beijing, China, 2019.
23. Braik, M.; Ryalat, M.H.; Al-Zoubi, H. A Novel Meta-Heuristic Algorithm for Solving Numerical Optimization Problems: Ali Baba and the Forty Thieves. *Neural Comput. Appl.* **2022**, *34*, 409–455. [CrossRef]
24. Sokolova, M.; Japkowicz, N.; Szpakowicz, S. Beyond Accuracy, F-Score and ROC: A Family of Discriminant Measures for Performance Evaluation. In *AI 2006: Advances in Artificial Intelligence*; Sattar, A., Kang, B., Eds.; Lecture Notes in Computer Science; Springer: Berlin/Heidelberg, Germany, 2006; Volume 4304, pp. 1015–1021. ISBN 978-3-540-49787-5.
25. Zhu, W.; Zeng, N.F.; Wang, N. Sensitivity, Specificity, Accuracy, Associated Confidence Interval and ROC Analysis with Practical SAS. In Proceedings of the NESUG Proceedings: Health Care and Life Sciences, Baltimore, MD, USA, 14–17 November 2010.
26. Tharwat, A. Classification Assessment Methods. *Appl. Comput. Inform.* **2021**, *17*, 168–192. [CrossRef]

fire

Article

Anomaly Detection in Spatiotemporal Data from Fiber Optic Distributed Temperature Sensing for Outdoor Fire Monitoring

Haitao Bian [1,2], Xiaohan Luo [1], Zhichao Zhu [1], Xiaowei Zang [1,2,*] and Yu Tian [1]

1 College of Safety Science and Engineering, Nanjing Tech University, Nanjing 211816, China; bianhaitao@njtech.edu.cn (H.B.)
2 Jiangsu Key Laboratory of Hazardous Chemicals Safety and Control, Nanjing Tech University, Nanjing 211816, China
* Correspondence: nanozang@njtech.edu.cn

Abstract: Outdoor fire detection faces significant challenges due to complex and variable environmental conditions. Fiber Optic Distributed Temperature Sensing (FO-DTS), recognized for its high sensitivity and broad monitoring range, provides significant advantages in detecting outdoor fires. However, prediction models trained in laboratory settings often yield false and missed alarms when deployed in complex outdoor settings, due to environmental interferences. To address this issue, this study developed a fixed-power fire source simulation device to establish a reliable small-scale experimental platform incorporating various environmental influences for generating anomalous temperature data. We employed deep learning autoencoders (AEs) to integrate spatiotemporal data, aiming to minimize the impact of outdoor conditions on detection performance. This research focused on analyzing how environmental temperature changes and rapid fluctuations affected detection capabilities, evaluating metrics such as detection accuracy and delay. Results showed that, compared to AE and VAE models handling spatial or temporal data, the CNN-AE demonstrated superior anomaly detection performance and strong robustness when applied to spatiotemporal data. Furthermore, the findings emphasize that environmental factors such as extreme temperatures and rapid temperature fluctuations can affect detection outcomes, increasing the likelihood of false alarms. This research underscores the potential of utilizing FO-DTS spatiotemporal data with CNN-AE for outdoor fire detection in complex scenarios and provides suggestions for mitigating environmental interference in practical applications.

Keywords: outdoor fire detection; anomaly temperature detection; fiber optic distributed temperature sensing; spatiotemporal data; environmental interferences

1. Introduction

Outdoor fires are a common and destructive type of disaster worldwide, causing significant damage to ecosystems, human lives, and economic activities. Typical scenarios of outdoor fires include exterior buildings [1], electrical equipment [2], forests [3], grassland [4], and industrial facilities [5,6]. Large areas of forest and grassland are affected by wildfires annually, severely disrupting ecological balance and posing significant threats to human safety and property. Therefore, timely and effective outdoor fire warnings are crucial for minimizing property loss and protecting the environment [7–14].

Abnormal temperature detection, as an effective early fire-warning mechanism, achieves precise temperature monitoring and provides alerts through the use of various sensors and anomaly detection algorithms. Commonly used sensors include thermocouples, thermistors, infrared sensors, and infrared thermal cameras, all of which provide high-precision temperature measurements in diverse environments [15]. Researchers have developed a variety of anomaly detection methods to analyze sensor data and identify abnormal patterns that deviate from normal temperature fluctuations. A survey report conducted a comprehensive and systematic review of anomaly detection methods across almost two decades, covering approaches based on density, statistics, distance, clustering, ensemble algorithms, and machine learning [16]. These algorithms analyze temperature data collected by sensors to identify abnormal patterns deviating from normal states and trigger warning signals. Among these methods, machine learning techniques, known for their powerful feature extraction and pattern recognition capabilities, have proven particularly effective in handling complex and nonlinear temperature data, gradually becoming the dominant approach [17,18], significantly enhancing the accuracy and responsiveness of anomaly detection [19].

However, compared with indoor fires, outdoor fires present significant complexity and challenges. Outdoor fires tend to be more dispersed and spread more rapidly, heavily influenced by various environmental factors such as temperature, humidity, wind speed, and weather conditions [20,21]. The accuracy and sensitivity of abnormal temperature detection systems in practical applications are often affected by these environmental factors. For instance, natural conditions such as ambient temperature, humidity, wind speed, and weather can substantially impact the actual measurement of abnormal temperature rises [22–24]. These factors may cause temperature readings from sensors to inaccurately reflect abnormal heating events, as they are influenced by environmental interference. Such interference affects changes in data patterns and the extraction of abnormal features, causing detection models designed under laboratory conditions to deviate when used in practical applications. Because the temperature data fed into the model are impacted by environmental factors, this can lead to false negatives or false positives in fire warning systems [25].

Specifically, in fire detection for outdoor scenarios, such as forests, grassland, and outdoor equipment, high-temperature and high-humidity environments reduce thermal diffusion capacity, leading to more intense temperature rises and an increased fire risk. Conversely, under low-temperature and low-humidity conditions, rapid heat dissipation may obscure abnormal temperature increases during the early stages of a fire, resulting in delayed warnings. When wind speeds are high, temperature variations at the fire source become more dispersed and less concentrated, increasing the difficulty of detecting abnormal temperatures. Rainfall and snowfall can quickly lower the surface temperatures of the ground and equipment, concealing early signs of abnormal heating. Additionally, shifts in the monitored area between direct sunlight and shadow cause significant temperature fluctuations, masking the true temperature distribution and hindering the timely identification of early signs of abnormal heating indicative of fire events [26,27].

To mitigate the impact of environmental factors on abnormal temperature detection systems, researchers have proposed various improvement methods. Researchers have employed empirical correction methods, adjusting temperature monitoring thresholds based on prior experience and historical data analysis [28]. However, these methods are highly subjective with limited applicability and reliability in different environments. With advancements in machine learning, researchers have attempted to address these issues by improving data-driven models through improving their robustness to environmental factors and introducing composite data approaches [29,30]. In this study, the composite data were the spatiotemporal temperature datasets collected during outdoor fire simulations, capturing temperature variations across different times and spatial locations. This dataset reflected dynamic changes in monitored temperatures over time and space, offering a more comprehensive perspective for the model and enabling more accurate identification of abnormal temperature variations for fire warnings. By integrating both temporal and spatial data into the model's input, the spatiotemporal characteristics of the data are fully utilized, enhancing the model's performance and reducing false alarms and missed detection caused by environmental temperature fluctuations.

In this context, FO-DTS offers a promising solution to improve the reliability and accuracy of outdoor fire detection. Unlike traditional point sensors such as thermocouples and early single-point fiber optic sensors (e.g., fiber Bragg gratings), FO-DTSs provide continuous temperature distribution along the fiber, with measurement distances ranging from tens of meters to tens of kilometers. This makes FO-DTSs highly suitable for fire monitoring in large areas such as storage tanks, construction sites, forests, and grasslands [31]. The measurement process of an FO-DTS is based on optical scattering, making it immune to electromagnetic interference, which is ideal for use in harsh field environments, ensuring stability and reliability in complex surroundings [32]. Moreover, FO-DTS offers more flexible deployment, which enhances their suitability for various fire monitoring applications.

Integrating emerging sensor technologies and data-driven algorithms is crucial for developing reliable systems for abnormal temperature detection. The application of emerging sensor technologies provides more comprehensive temperature data at anomaly detection sites, leading to more accurate and reliable results. Simultaneously, data-driven methods, particularly deep learning techniques such as autoencoders (AEs), have achieved notable success in anomaly detection [33]. An AE is a neural network model used for unsupervised feature extraction. It learns compressed representations of the data to reconstruct the input data. An AE consists of two parts: an encoder that maps the input data to a compressed latent space, and a decoder that reconstructs the original input from the latent representation. During training, the AE aims to minimize the reconstruction errors between the input and the output, effectively capturing the underlying structure of the data [34,35]. When used for anomaly detection, the AE model is first trained on normal data to learn the normal patterns. Subsequently, for new data, if the reconstruction errors exceed a predefined threshold, the data point is considered anomalous. This approach is particularly effective for anomaly detection in complex, high-dimensional data, as traditional methods may perform poorly in such cases due to issues like data imbalance, high dimensionality, or the lack of labeled data [36–38]. This study proposes an AE-based method for detection of anomalous temperatures in spatiotemporal fiber optic distributed temperature sensor data, aiming to achieve early warning for incidents such as fires. This method improves the accuracy of the anomaly temperature detection system in the presence of environmental factors by integrating spatial and temporal dimensions, thereby reducing the risks of false alarms and missed detection.

This paper explores the effectiveness of an AE model that integrates two-dimensional spatiotemporal data for anomaly temperature detection, considering the specific characteristics of FO-DTS and the challenges posed by environmental factors. The objective is to achieve accurate detection of temperature anomalies in outdoor fire warning systems and mitigate the risks of false alarms and missed detections and investigate the environmental influences in the detection process. The structure of this paper is as follows. Section 2 introduces the methodology for detection of temperature anomalies and the experimental equipment setup. Section 3 presents the results and discussions from the perspectives of temporal, spatial, and spatiotemporal analysis. Section 4 concludes this study and outlines future research directions.

2. Methodology

The proposed system architecture is shown in Figure 1. The public meteorological information server provides weather information to the embedded system gateway via Ethernet. Temperature and humidity sensors send on-site temperature and humidity data to the embedded system gateway through the Modbus protocol. Then, the AT800 FO-DTS (Suzhou Agioe Technologies Co., Ltd., Suzhou, China) collects temperature data and forwards these to the embedded system gateway via the MQTT protocol. Subsequently, the embedded system gateway processes these data and stores them in the local MySQL database (Community Edition 8.0). Finally, the system employs AE models for anomaly detection, ensuring any abnormal conditions can be promptly identified. The AE models were implemented using PyTorch 1.12.1 with CUDA 11.6 for GPU acceleration on Ubuntu 22.04, and the python version is 3.9.

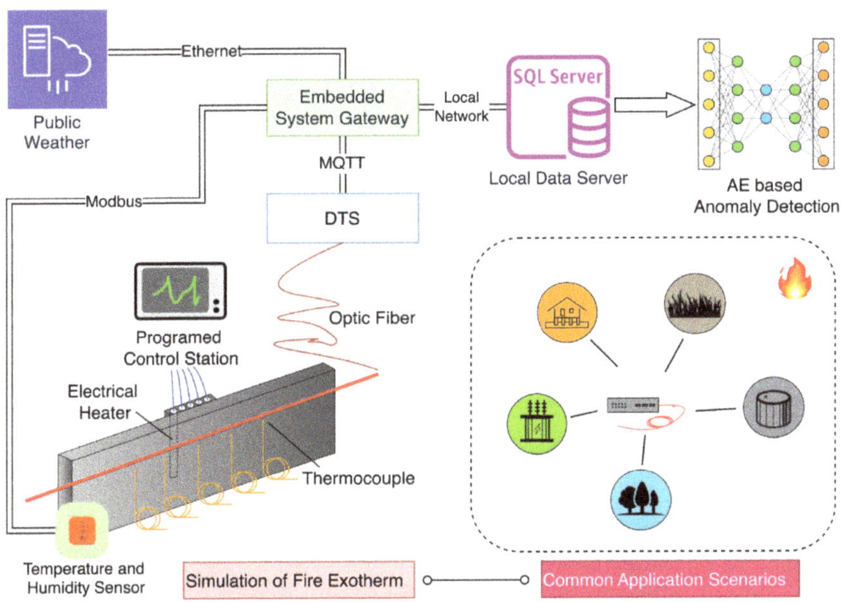

Figure 1. Architecture for anomalous temperature detection for outdoor fire warning systems.

2.1. Experiment

In the context of the time-consuming nature of outdoor fire experiments and safety hazards associated with using real fire sources, a constant power heating device can provide safe, controllable, and repeatable experimental conditions and accurately simulate the temperature change characteristics in the early stages of a fire [39]. Therefore, a constant-power electric heating module was used to heat a steel plate, simulating the rapid temperature rise process in the early stages of a fire. As shown in Figure 1, an experimental setup was designed to simulate the scenario of using an FO-DTS for abnormal temperature detection in an outdoor fire warning system, and to obtain temperature measurements under both normal and abnormal conditions. Outdoors, a constant-power fire-source simulation device plate was used to simulate the outdoor fire, with dimensions of 1200 mm × 345 mm × 15 mm. A customized heating module (100 mm × 100 mm) was installed on the back of the test plate. Five electrical heating rods (120 W) were embedded in the heating module, and a solid-state relay drive controlled the heating power with pulse width modulation generated from an embedded system. Five thermocouples were arranged on the front of the steel plate at 100 mm intervals, and the optical fiber was placed 5 mm away from the surface of the plate. Near this apparatus, temperature and humidity sensors were installed to record local environmental information. A thermocouple is a traditional point temperature sensor with very high accuracy and speed of response. It can accurately measure the temperature at a specific point without being limited by spatial resolution. Compared with thermocouples, the measurement results of an FO-DTS are far from the actual situation due to the spatial resolution, and it is difficult to distinguish the anomalies intuitively. Therefore, in this experiment, the main role of the thermocouple was to provide intuitive anomaly detection results.

The FO-DTS used in this experiment was based on Raman scattering and optical time-domain reflection. The spatial resolution of the FO-DTS was 500 mm and the precision of temperature measurement was 0.1 °C. The FO-DTS returned a temperature value every 100 mm along the fiber at intervals of 3 s. The thermocouples returned data every 3 s; since both had the same time sampling interval (both 3 s), it was easy to compare temperature data for the same moment. The other sensors were set to 30 s. All the above information was collected and saved in a local database.

The experiments were conducted outdoors in the summer in Nanjing, China. The experimental platform was fixed in an east-facing position so that it would be exposed to sunlight in the morning and hidden in the shade in the afternoon. Normal data were collected when the heating device was not working, and five different anomalous scenarios were created by heating the plate under different environmental conditions (all with a heating power of 120 W). The five scenarios covered morning, noon, afternoon, and evening, with temperatures ranging from 30.3 °C to 45.4 °C and humidity ranging from 32% to 81.6%, including both sunny and cloudy weather conditions. The patterns of temperature anomalies for the different scenarios are also shown in Figure 2. The environmental information for the different scenarios is presented in Table 1, where the time in the table indicates the time of the start of the experiment, i.e., the exact moment when the heating unit started working.

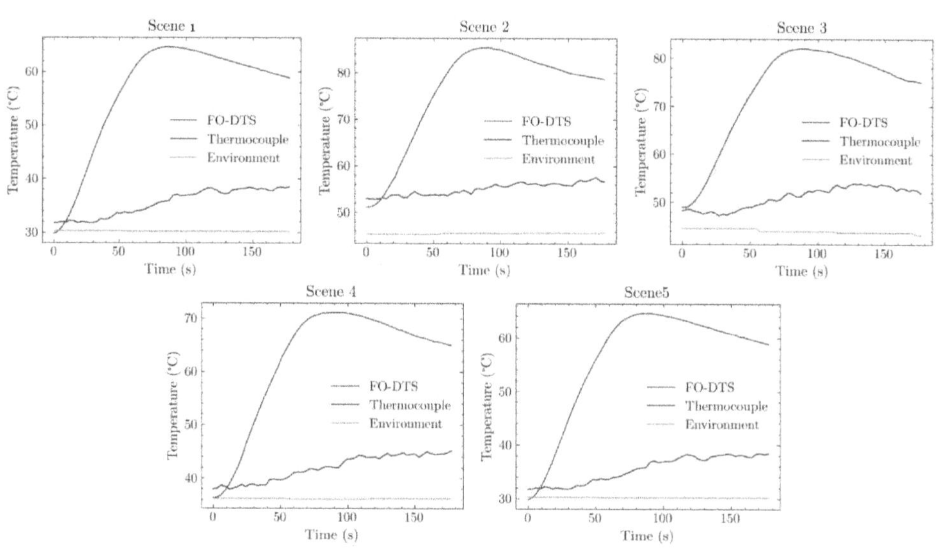

Figure 2. Abnormal temperature changes in different environmental scenarios.

Table 1. The details of the ambient information in anomaly scenes.

Scenes	Time	Temperature (°C)	Humidity (%)	Weather
1	12:16	37.9	54	Cloudy
2	9:27	45.4	32	Sunny
3	10:24	44.6	34.2	Sunny
4	15:04	36.2	57	Sunny
5	21:48	30.3	81.6	Cloudy

2.2. Detection of Temperature Anomalies via FO-DTS

Different from other temperature sensors, an FO-DTS offers both spatial and temporal information. Consequently, when an anomaly occurs, the collected field data can be analyzed from two perspectives, as illustrated in Figure 3, where the orange dots represent continuous measurement points along the FO-DTS sensing optical fiber.

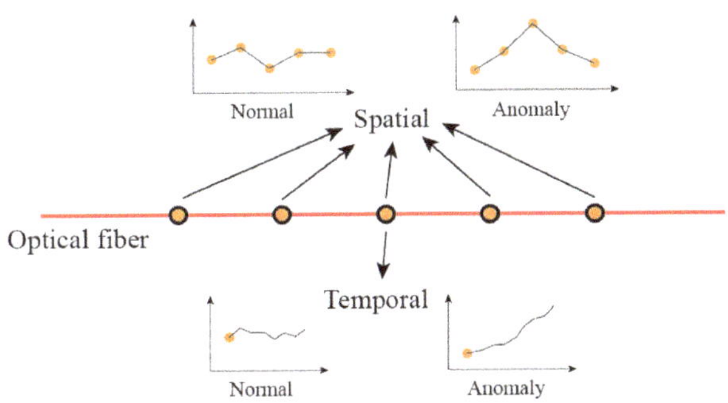

Figure 3. Two perspectives for detecting anomalous temperatures via FO-DTS.

From the spatial perspective, due to the existence of spatial resolution, the system may not precisely match the actual temperatures at certain scales. When detecting a relatively small hotspot, several continuous points along the FO-DTS record the temperature, with points closer to the hotspot typically registering higher temperatures. However, if a hotspot extends beyond the spatial resolution of the system, the FO-DTS can accurately report the temperature according to the thermocouple. Thus, for anomaly detection from the spatial perspective, the continuous temperature at a particular time can be used to either estimate the hotspot temperature or evaluate the system state. The environmental conditions at different positions within a small range of fiber are uniform; therefore, the temperature from continuous measurement points changes synchronously under environmental influence. Anomalies in temperature at a specific location cause spatial variations in the measured values, which is fundamental to detecting anomalous temperatures from the spatial perspective.

From the temporal perspective, temperature variations at a specific location are influenced by environmental effects. Distinguishing the characteristics of temperature changes under normal conditions from those under anomalous conditions enables effective detection of anomalies. However, environmental fluctuations often complicate this distinction, leading to potential false alarms. For instance, minimal measurement fluctuations pose significant challenges in detecting temperature anomalies, necessitating the use of algorithms to enhance detection accuracy.

2.3. Detection of Anomalous Temperatures with AE

The imbalance between normal data (i.e., data collected when no fire occurs) and abnormal data (i.e., data collected during a fire) presents challenges for traditional supervised learning methods. Traditional models may overfit the normal data, thereby weakening their ability to detect abnormal data [40,41]. In contrast, as an unsupervised learning method, AEs are particularly well suited for scenarios with data imbalance because they do not require a large amount of abnormal data for training. An AE can be trained with only normal data, learning to reconstruct normal patterns. For new data, if the reconstruction errors exceed a set threshold, the data point is considered anomalous [36]. Moreover, the temperature data obtained using the FO-DTS contains both spatial and temporal dimensions, with many measurement points and timestamps for each dimension. Traditional methods may perform poorly in this context due to the curse of dimensionality or the lack of labeled data. In contrast, an AE can effectively capture the underlying structure of high-dimensional data by learning compressed representations, thereby improving the accuracy of anomaly detection. Therefore, these characteristics make AEs an ideal choice for handling imbalanced, high-dimensional temperature data.

Specifically, an AE is a data reconstruction model consisting of an encoder and a decoder. The original data X are first passed through the encoder, which extracts their features $(f : X \rightarrow F)$. The extracted information is then delivered into the decoder to attempt to restore the original data $(g : F \rightarrow \hat{X})$. The loss between output $\hat{X}$ and input X is used to optimize the algorithm, enabling it to capture the input features and generate an output that closely resembles the original data (Equation (1)).

$$f, g = arg\,\min_{f,g} \|X - g[f(X)]\|^2 \tag{1}$$

The convolutional autoencoder (CNN-AE) replaces the fully connected layers of a traditional autoencoder with convolutional layers, enabling it to more effectively handle input data with spatial structures. For the input a, linear layers process the data via linear calculation, as $z = W^T a + b$, where w and b are weight vector and bias. For convolution layers, the input is processed by cross-correlation, as $z = K \otimes a + b$, where K is the kernel. The cross-correlation can be described via Equation (2):

$$y_{i,j} = \sum_{u=1}^{U} \sum_{v=1}^{V} K_{u,v} X_{i+u-1,j+v-1}, \ K \in R^{U*V} \tag{2}$$

The VAE operates as an encoding–decoding framework that utilizes variational inference to model the probability distribution of the input data by approximating it with a simpler distribution family. The VAE's encoder outputs the approximate posterior distribution of the latent variable z, which can be described as $q(z|x; \varnothing)$, and the decoder is the likelihood of input x, which is $p(x|z; \theta)$. For simplicity, $q(z|x; \varnothing)$ is commonly postulated as Gaussian distribution, so the latent variable z or the outputs of encoder are mean μ and variance σ^2. To ensure z had independent randomness, a reparameterization trick expressed via Equation (3) was applied:

$$z = \mu + \sigma \odot \epsilon, \epsilon \sim \mathcal{N}(0, I) \tag{3}$$

In the detection of temperature anomalies, normal data are sent for training the AE, and the algorithm can learn the features of the normal conditions. When an anomaly occurs, the AE, unfamiliar with the anomaly's features, generates a larger loss between the output and the input. If this loss exceeds a predefined threshold, an anomaly is detected. Normal data are split into training and test sets. The threshold is commonly set as a certain percentile of this loss array or according to kernel density estimation corresponding to a certain confidence level. Consequently, the threshold setting is somewhat subjective and adjustable, affecting the sensitivity of anomaly detection. To compare different conditions equally, a test set is used to balance the performance. After a default threshold is determined, it is used for detection within the test set, which includes the datasets of temperature anomalies simulated by electrical heating. The results can be used to evaluate the performance of the algorithms. Due to the ambiguous boundary between normal and anomalous states, the loss from anomalies may fall below the set threshold, while loss from the normal state may exceed it. Algorithms that more accurately extract normal features are often better at identifying anomalies and can more quickly return anomaly results. In this study, the delay in temperature anomaly detection—defined as the time taken for the algorithm to return anomaly results after heating began—served as a key metric for evaluating the detection performance.

In this study, we conducted anomaly detection from temporal, spatial, and spatiotemporal perspectives using various AE models. The workflow for this process is shown in Figure 4. To ensure that the data accurately reflected temperature variations during the occurrence and development of the fire, we employed a rigorous data selection strategy.

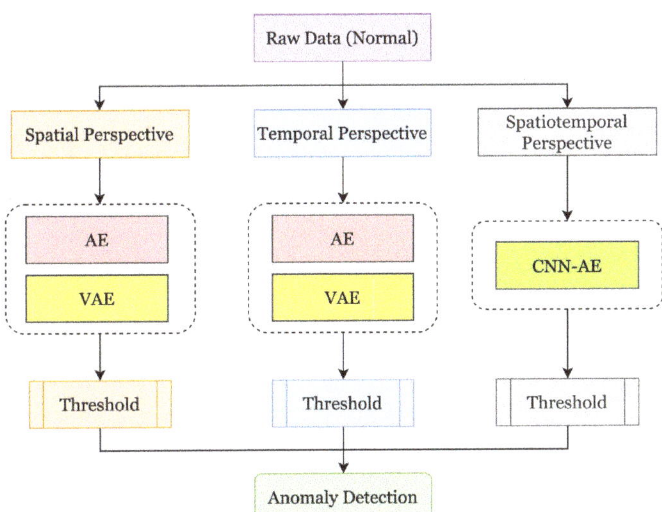

Figure 4. The workflow of temperature anomaly detection in this study.

2.3.1. Temporal Perspective Anomaly Detection

Time series data are one of the most commonly used types of data used in current research, widely applied in fields such as anomaly detection, forecasting, and pattern recognition [42]. In this study, within the time series dimension, we selected the temperature data of all timestamps from the measurement point at the middle position of the experimental board's FO-DTS. This point was located at the heating source and represented the most significant temperature change. By focusing on a fixed spatial point, the influence of spatial variables on time series features can be minimized. Then, a sliding window technique was used to capture the dynamic features of the time series data, where each sliding window contained temperature values from multiple consecutive time points. The data from these sliding windows were proportionally divided into training and test sets.

For anomaly detection in time series data, this study used an AE and a VAE. First, the temperature data from all timestamps at the middle measurement point were used as input to form sliding window sequences. Then, the AE and VAE models were trained using time series data from normal conditions. The AE model learned the normal patterns by minimizing the reconstruction errors between the input and output, while the VAE model introduced a probabilistic generative model and was trained by maximizing the log likelihood of the data. During training, the AE and VAE models both learned to extract features from the normal data and reconstruct the input data. For new data, the reconstruction errors (AE) or reconstruction loss (VAE) was calculated. If the reconstruction errors or loss exceeded a predefined threshold, the data point was considered anomalous. The 95th percentile of the reconstruction errors from the training set was selected as the threshold to ensure a low false-positive rate.

2.3.2. Spatial Perspective Anomaly Detection

In the spatial dimension, we selected all measurement points along the entire monitoring path of the FO-DTS. Each ID represented the temperature measurement points at different spatial locations at the same time, with each ID corresponding to 10 temperature feature points. In the early stages of a fire, temperature increases are often confined to a specific area. The collection of multi-point spatial data can significantly enhance the sensitivity and reliability of fire detection systems. Temperature differences across different locations can be substantial, with higher temperatures closer to the heating source and lower temperatures further away. This difference can affect the model's learning process, causing temperature data from certain locations to be either overemphasized or overlooked. Therefore, we normalized the temperature data collected along the FO-DTS path to eliminate dimensional differences and ensure that all data were compared and processed on the same scale. For each ID corresponding to 10 temperature points, we extracted spatial features by calculating statistics such as the difference, mean, and standard deviation between adjacent temperature points to enhance the AE model's ability to capture local anomalies.

For the spatial dimension data, this study used an AE and a VAE for anomaly detection. First, the temperature data from all measurement points along the FO-DTS monitoring path were used as input, forming multiple spatial vectors, each containing 10 features. The data from all IDs together formed a complete spatial feature matrix. The AE and VAE models were trained using the spatial data under normal conditions. The AE model learned the normal spatial temperature distribution by minimizing the reconstruction errors between the input and output, while the VAE introduced a probabilistic generative model and was trained by maximizing the log likelihood of the data. During training, the AE and VAE models both learned to extract features from the normal data and reconstruct the input data. For new data, the reconstruction errors (AE) or reconstruction loss (VAE) for the spatial vector corresponding to each ID was calculated. If the reconstruction errors or loss for a particular ID exceeded the predefined threshold, it was considered that an anomaly existed at that location. The 95th percentile of the reconstruction errors from the training set was selected as the threshold to ensure a low false-positive rate.

2.3.3. Spatiotemporal Perspective Anomaly Detection

To comprehensively consider both the temporal and spatial dimensions and fully capture the temperature changes during the occurrence and development of a fire, this study combined the spatial data from all measurement points along the entire monitoring path with the time series data captured using the sliding window technique, forming a spatiotemporal dataset containing both temporal and spatial information. For each timestamp, the temperature data from all measurement points were combined into a spatial vector containing 10 features. Each spatial vector represented the temperature values at 10 different locations for that timestamp. For each sliding window, spatial vectors from multiple timestamps were stacked together, forming a 2D sample with the shape [window length, spatial feature number]. These samples were proportionally divided into training and test sets.

For the spatiotemporal data, we used a convolutional autoencoder (CNN-AE) for anomaly detection. The collected samples used as model input contains both temporal and spatial information. The CNN-AE model was trained using spatiotemporal data under normal conditions. The encoder part extracted local features through multiple convolutional layers, while the decoder part reconstructed the input data using deconvolution layers, with the goal of minimizing the reconstruction errors between the input and output. Fully connected layers were added between the encoder and decoder to further compress and expand the features, ensuring that the model was able to learn higher-level abstract representations. During training, the mean squared error (MSE) was used as the loss function, and model parameters were optimized through the backpropagation algorithm. The 95th percentile of the reconstruction errors from the training set was selected as the threshold. For new data, the reconstruction errors were calculated. If the reconstruction error exceeded the predefined threshold, the data point was considered anomalous.

3. Results and Discussion

3.1. Detection of Anomalous Temperatures from the Temporal Perspective

3.1.1. Temporal Model Design and Setting

To capture the spatiotemporal properties in the time series data, we employed a sliding window technique, where each window contained temperature values from multiple time points. Temperature values returned from measurement point 5 in the middle of the FO-DTS were chronologically programmed into a sliding window to ensure that the model captured the characteristics of temperature changes over short periods of time. The input lengths of temperature sequences were 10, representing a 30-s interval, totaling 303,911 samples after division of normal data. All normal data were also separated into the training set and part of the test set, in a proportion of 98:2, comprising 297,832 and 6079 samples, respectively. The detailed size of each dataset is described in the attachment.

Before inputting data into algorithms, it is common to use normalization to standardize data that are in different orders of magnitude. In this experiment, the data were preprocessed with standard normalization (standardscaler). Normalization can improve the convergence speed and model accuracy of neural networks. Because the normalization process is based on the training set, if the distribution state of the new data in the test set is different, the normalization will fail. In this process, temperatures from different positions in the FO-DTS were in a similar distribution state, i.e., the values did not vary widely. An AE and VAE were used; the number of epochs was 20 and the batch size was 64. Adam was used as the optimizer and the learning rate was set to 0.001. ReLU was used as the activation function. The details of the AE networks are shown in Table 2. For the loss function, the AE used MSE, and the VAE combined the reconstruction loss and the Kullback–Leibler divergence. The 95th percentile of the reconstruction errors from the training set was selected as the threshold for detections of anomalous temperatures. The test set consisted of 2% of the normal data and the entire data from the five anomalous scenarios, a total of 9634 samples.

3.1.2. Anomaly Detection Performance Using Temporal Data

In the process of detection, if the algorithms fail to identify an anomaly, the missed detection is marked as false negative (FN), while classifying a normal event as an anomaly is marked as a false positive (FP). Specifically, to compare the anomaly detection performance across models, the delays in detecting temperature anomalies delays were collected.

Table 2. The brief structure of the (**a**) AE and (**b**) VAE model.

(a)		
Type	**Layer**	**Description**
Encoder	(0) Linear	in-features = 10, out-features = 128
	(1) ReLU	activation function
	(2) BatchNorm1d	in-features = 128, out-features = 128
	(3) Dropout	activation function
	(4) Linear	in-features = 128, out-features = 256
	(5) ReLU	activation function
	(6) BatchNorm1d	in-features = 256, out-features = 256
	(7) Dropout	$p = 0.2$
	(8) Linear	in-features = 256, out-features = 64
Decoder	(0) Linear	in-features = 64, out-features = 256
	(1) ReLU	activation function
	(2) BatchNorm1d	in-features = 256, out-features = 256
	(3) Dropout	$p = 0.2$
	(4) Linear	in-features = 256, out-features = 128
	(5) ReLU	activation function
	(6) BatchNorm1d	in-features = 128, out-features = 128
	(7) Dropout	$p = 0.2$
	(8) Linear	in-features = 128, out-features = 10

(b)		
Type	**Layer**	**Description**
Encoder	(dense1) Linear	in-features = 10, out-features = 128
	(dense2) Linear	in-features = 128, out-features = 256
	(dense3_mu) Linear	in-features = 256, out-features = 64
	(dense3_logvar) Linear	in-features = 256, out-features = 64
Decoder	(dense4) Linear	in-features = 64, out-features = 256
	(batch_norm) BatchNorm1d	in-features = 256, out-features = 256
	(dropout) Dropout	$p = 0.2$
	(dense5) Linear	in-features = 256, out-features = 128
	(recon) Linear	in-features = 128, out-features = 10

The overall performance of the algorithm ignoring the scenarios and the delay time of the algorithm's detection rate under different scenarios are shown in Table 3. The VAE significantly outperformed the AE in terms of recognition rate, with a recognition rate of 91.87%. In terms of detection latency, VAE also showed better performance in most scenarios, especially in Scenario 2 and Scenario 3, where the detection latency was 39 and 30 s, respectively, while the AE returned FNs in several scenarios. Unlike a standard AE, a VAE is inherently robust to noise and less prone to overfitting due to its use of variational inference and the reparameterization trick. This makes it better suited to handling variability in data compared with a basic autoencoder. Although these advantages mean that the VAE model can play an important role in data generation and the detection of temperature anomalies, if the normal and the anomalous data are similar enough or some other factors make the anomaly appear like a normal condition, the VAE cannot provide the expected results, e.g., relatively long detection delays were observed in Scenarios 1, 4, and 5.

Table 3. Results of temperature anomaly detection by the AE and VAE from a temporal perspective.

Model	Accuracy	Detection Delay (s)				
		Scene 1	Scene 2	Scene 3	Scene 4	Scene 5
AE	73.04%	FN	69	FN	FN	FN
VAE	91.87%	102	39	30	93	150

In order to compare different experimental scenarios, the ambient temperatures one hour before and after the experiment are shown in Figure 5a. At the beginning of heating, Scenario 2 and Scenario 3 both represented the hottest environment, but the ambient temperature of Scene 2 increased slowly, which did not have much effect on the anomalous temperature detection. Scenarios 1 and 3 included a decreasing trend in temperature, but the decreasing trend in Scenario 1 was relatively intense, so the delay in Scenario 1 was significantly higher than that in Scenario 3. At the beginning of heating, Scenario 3 and Scenario 2 both represented the hottest environment, but the ambient temperature in Scenario 3 was decreasing all the time. As shown in Figure 5b, although the temperature returned by the FO-DTS increased after heating, before the experiment, the temperature dropped dramatically with the environment.

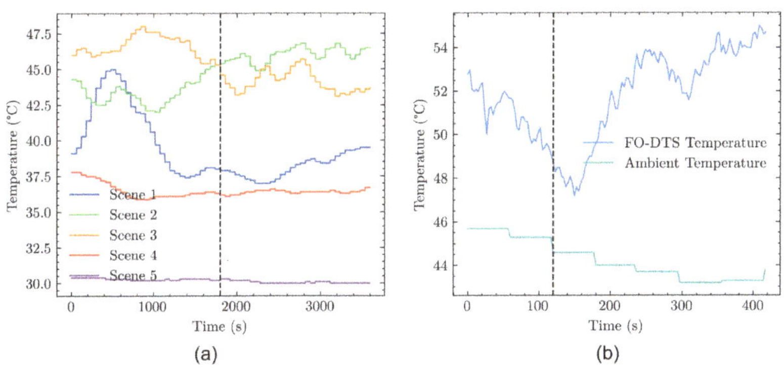

Figure 5. (**a**) Ambient temperature of different scenarios, (**b**) Ambient temperature for Scenario 3 and temperature from FO-DTS. Dashed line marks the start time of heating.

3.2. Detection of Temperature Anomalies Detection from a Spatial Perspective

3.2.1. Spatial Model Design and Setting

In the spatial aspect, measurements from points along the whole monitoring path of the FO-DTS were taken to obtain the temperature data. Each ID represented different spatial locations of temperature measurement points at the same moment, and each ID corresponds to 10 temperature point featured in this study.

Before training the algorithms, spatial data were processed via standard normalization as described in Section 3.1.1. The algorithms were AE and VAE; ReLU was used as the activation function. The number of epochs was 20 and the batch size was 64. Adam was used as the optimizer and the learning rate was set to 0.001. Network details were the same as stated in Section 3.1.1. The loss function for AE was MSE, and VAE used a combination of reconstruction loss and the Kullback–Leibler divergence. A total of 303,983 data samples were obtained after division of the normal data. All the normal data were also separated into the training set and part of the test set, in a proportion of 98:2, comprising 297,903 and

6080 samples, respectively. The test set consisted of 2% of the normal data and the entire data from the five anomalous scenarios, a total of 9680 samples.

3.2.2. Anomaly Detection Performance Using Spatial Data

In the experiment, the anomaly detection threshold was determined using the 95th percentile of the reconstruction errors from the training set to assess the accuracy of the test set. This ensured that all algorithms were evaluated at the same level for anomalous temperature detection performance. As shown in Table 4, The overall recognition rate of VAE was 91.14%, which was slightly lower than the 92.58% achieved by the AE, but the difference was not significant. Although the overall recognition rate of the VAE was slightly lower than that of the AE, its detection speed was significantly faster than that of the AE model. The AE model had FNs in several scenarios, while the VAE model did not have FNs in any of the five scenarios, showing its higher stability.

Table 4. Results of temperature anomaly detection by the AE and VAE from a spatial perspective.

Model	Accuracy	Detection Delay (s)				
		Scene 1	Scene 2	Scene 3	Scene 4	Scene 5
AE	92.58%	FN	234	FN	FN	FN
VAE	91.14%	12	186	3	9	81

Generally, all of the rest had false negatives except Scenario 2, while Scenario 2 had the comparatively longest delay in detecting temperature anomalies. Various scenarios appeared to affect the likelihood of false negatives in detection of anomalous temperatures. As discussed in Section 2.2, from the spatial perspective, the AE was trained to recognize the characteristics of normal data that the temperature would keep consistent over a small range, and if the temperature had fluctuations in different positions, an anomaly event might happen. Since the VAE performed better in the spatial dimension, in order to understand the characteristics of the false positive samples in this test, we conducted an analysis to study the error distribution of the ten measurement points in the spatial dimension. Figure 6 shows that the variance of the measurement points in the central positions was relatively large.

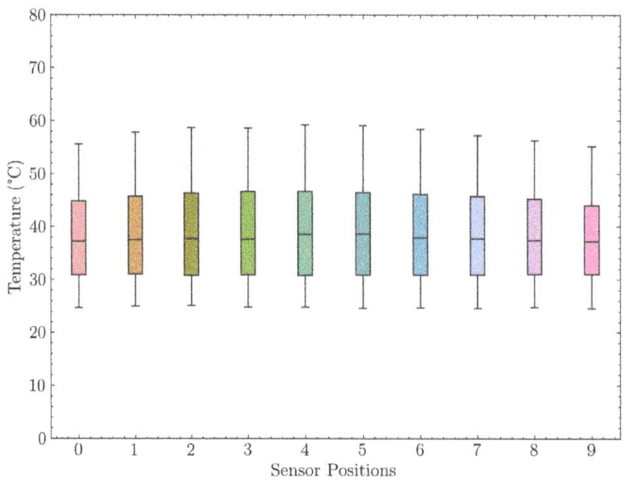

Figure 6. Box plots of VAE-based false positive samples from a spatial perspective.

3.3. Temperature Anomaly Detection from a Spatiotemporal Perspective

3.3.1. Spatiotemporal Model Design and Setting

It is difficult to find an algorithm that performs well across all five scenes from a single perspective. Considering that the FO-DTS is a spatiotemporal measurement sensor, combining both spatial and temporal information may be a better approach. In this spatiotemporal dimension experiment, a 10 by 10 2D window was constructed, which represented a sliding window of 10 consecutive timestamps. At each timestamp, there were temperature values from 10 different measurement points, and each column represented the temperature variation at a specific measurement point across multiple timestamps.

Unlike the algorithms mentioned above, which process only one-dimensional data, handling two-dimensional inputs requires additional techniques. CNNs are inherently designed to work with multi-dimensional data, particularly images. Therefore, CNN-AE, which incorporates convolutional layers, can directly process spatiotemporal inputs. In contrast, traditional autoencoders, which use linear layers for computation, can only process data in one-dimension at a time. This study employed CNN-AE to handle two-dimensional inputs. CNN-AE is widely used; the kernel sizes in its convolutional layers are represented as two-dimensional arrays.

Datasets were reconstructed and for each input, the spatial dimensions were 10 and the temporal dimensions were 10. A total of 303,911 samples were obtained after division of the normal data. All the normal data were also separated into the training set and part of the test set, in a proportion of 98:2, including 297,832 and 6079 samples, respectively. As described in Sections 3.1.1 and 3.2.1, data were processed by standard normalization. The number of epochs was 20 and the batch size was 32. Adam was used as the optimizer and the learning rate was set to 0.001. MSE was set as the loss function and ReLU was used as the activation function. The details of the networks are shown in Table 5. The threshold for detection of temperature anomalies was set as the 97th percentile of the training set's loss value. The test set consisted of 2% of the normal data and the entire data from the five anomalous scenarios, a total of 9634 samples.

Table 5. Brief structure of the proposed CNN-AE model.

Type	Layer	Description
Encoder	(0) Conv2d	1, 64, kernel size = (3, 3), stride = (1, 1), padding = (1, 1)
	(1) ReLU	activation function
	(2) Conv2d	64, 128, kernel size = (3, 3), stride = (1, 1), padding = (1, 1)
	(3) ReLU	activation function
	(4) MaxPool2d	kernel size = 2, stride = 2, padding = 0, dilation = 1
Decoder	(0) ConvTranspose2d	128, 64, kernel size = (2, 2), stride = (2, 2)
	(1) ReLU	activation function
	(2) ConvTranspose2d	64, 1, kernel size = (3, 3), stride = (1, 1), padding = (1, 1)

3.3.2. Anomaly Detection Performance Using Spatiotemporal Data

The convolutional layers of the CNN-AE effectively extract spatial features, and its recurrent structure can capture dynamic changes in time series. This combination of spatial and temporal features makes the CNN-AE more robust to environmental disturbances; the model achieved high detection rates across all five scenarios. As shown in Table 6, the average accuracy of the CNN-AE was 85.08%, with relatively stable performance across different scenes. Particularly in Scenarios 1 to 4, the CNN-AE demonstrated high detection rates, showing good adaptability to different environmental conditions. Moreover, the CNN-AE exhibited consistent detection latency across these four scenarios, reflecting good temporal stability. Although the detection latency in Scenario 5 was slightly longer, the CNN-AE still maintained relatively low latency compared with the other models, especially under complex environmental conditions.

Table 6. Results of temperature anomaly detection by the CNN-AE from a spatiotemporal perspective.

Model	Accuracy	Detection Delay (s)				
		Scene 1	Scene 2	Scene 3	Scene 4	Scene 5
CNN-AE	85.08%	60	60	48	75	108

The determination of anomalies depends on the comparison between the reconstructed loss and the threshold, and the results of the reconstruction error over time for the five different scenes are shown in Figure 7 for further analysis. Overall, CNN-AE demonstrated robust performance across all scenes, effectively identifying anomalies and quickly returning to normal levels after an anomaly occurred.

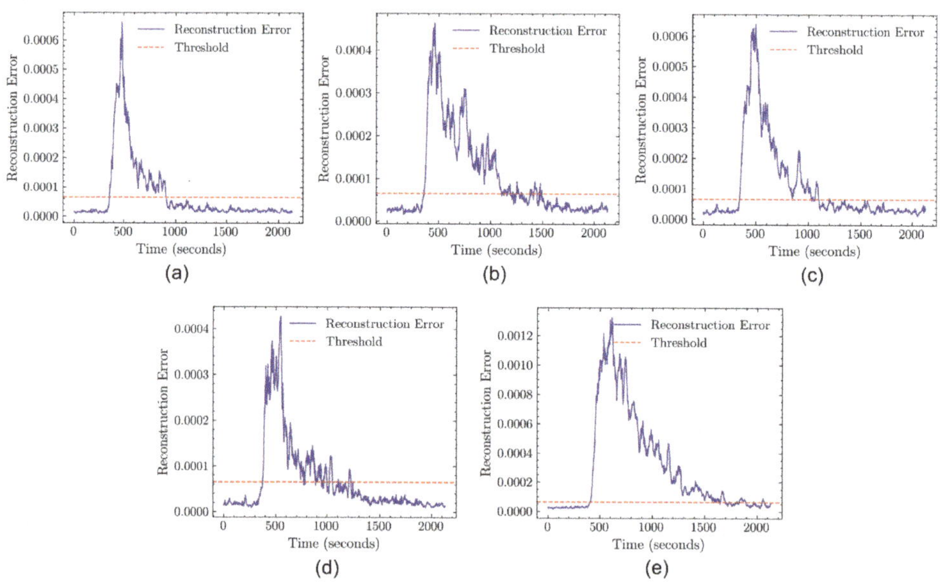

Figure 7. CNN-AE for spatiotemporal data anomaly detection: (**a**) Scenario 1 reconstruction error; (**b**) Scenario 2 reconstruction error; (**c**) Scenario 3 reconstruction error; (**d**) Scenario 4 reconstruction error; (**e**) Scenario 5 reconstruction error.

3.4. Discussion

Currently, research on AE-based fire detection revolves around how to exploit the properties of AE to improve the performance of fire warning systems. These studies involve feature extraction and dimensionality reduction, anomaly detection, multimodal data fusion, and real-time and computational resource optimization. Researchers have fused data from different types of sensors and employed AE for comprehensive analysis to further enhance the accuracy and reliability of fire detection [36,37,43]. However, these approaches also face some challenges; for complex data distributions, more complex network structures may be required to learn features effectively; training data may lead to false predictions if they do not adequately cover all normal operating conditions; handling multimodal data increases the complexity of the system and requires solving the problems of data synchronization and calibration and, while pursuing a fast response, a certain degree of accuracy to improve the adaptability, accuracy, and response speed of the system in different environments.

Compared with the existing studies, this paper puts more emphasis on the influence of environmental factors on fire detection and explores the detection performance of different AE models. Meanwhile, this paper not only considers the changes in time series and spatial dimensions, but also integrates the spatiotemporal data of FO-DTS to construct a more comprehensive dataset, which helps to more accurately capture the characteristics of the fire occurrence and improve the adaptability and detection accuracy of the system.

Despite demonstrating the potential for detection of temperature anomalies using an FO-DTS and an AE, this study emphasizes that these methods should extend beyond theoretical exploration to support practical engineering applications. For deep learning methods, datasets are always the foundation of the algorithms. As presented in this study, after the installation of the FO-DTS, the temperature data returned from the sensor can serve as the training set for the algorithms. Theoretically, to mitigate algorithmic bias, it is advisable to collect data under varied conditions to ensure a balanced representation across different scenarios. Specifically, when constructing the dataset for the monitored equipment, temperature data should be collected in different weather conditions and at various times of the day, as abnormal increases in temperature can occur under any conditions, and the algorithms should learn patterns from all possible scenarios. Temperature trends can differ across seasons, and our study only considered summer conditions. We recommend preparing different datasets for each season so that the algorithms can focus more on the temperature characteristics relevant to the current time. Additionally, the datasets can be incrementally updated, allowing the algorithms to be retrained regularly to capture new patterns.

Determining the size of the input remains a challenge as the optimal parameters vary across different devices and scenarios, necessitating further research. In the context of the scenes discussed in this study, the spatial size of the input should at least be larger than the spatial resolution, and a temporal size greater than 60 would be preferable. If the spatial size of the input is smaller than the spatial resolution of the FO-DTS, it cannot provide accurate information about the spatial distribution of temperature, as spatial resolution represents the minimum length necessary for the FO-DTS to differentiate temperature measurements. Additionally, we suggest that the spatial size of the input should not be excessively large, as this can complicate spatial positioning. Regarding the temporal size, short time series are susceptible to random error interference and are unable to reflect accurately changes over time. Therefore, we recommend a larger temporal size to effectively reveal trends in temperature.

When the algorithm starts running, it serves as part of an outdoor fire alarm system to ensure the safety of life and property. Once the system reports an anomaly, it can coordinate and initiate appropriate emergency measures. Through these efforts, personnel can fully utilize the information provided by the FO-DTS to connect this spatiotemporal sensor with the fire alarm system. With the help of deep learning methods, reducing the rate of false alarms will enhance personnel's trust in the system, helping to avoid fatigue caused by excessive alerts.

4. Conclusions

This study presents an enhanced method based on FO-DTS and autoencoders to address the challenges of poor performance of traditional data-driven models in outdoor fire warning, which are susceptible to environmental factors leading to false and missing alarms in fire warning systems. An experimental setup with an electrical heating platform was constructed to simulate abnormal temperature rises, and the performance of various types of autoencoders, including AE, VAE, and CNN-AE, was tested. The results indicate that CNN-AE, which uses spatiotemporal data, offers superior detection rates and robustness, maintaining stable performance in complex environments. Although AE and VAE demonstrate slightly higher overall accuracy in some cases when using temporal or spatial dimensions, they exhibit lower performance in detection delay and anomaly detection rate, both of which are critical indices in anomaly detection for outdoor early fire warning. Environmental conditions such as temperature, humidity, and weather significantly impact detection results, with false alarms more likely to occur under high ambient temperatures and rapid temperature fluctuations.

Future research will focus on expanding datasets that incorporate diverse environmental factors to enhance the model's generalization capability and effectively evaluate the algorithm's performance across various scenarios. Additionally, the study will aim to further optimize the CNN-AE architecture to improve its accuracy and robustness in practical applications. Furthermore, integrating measurement data with environmental information and developing a model for the comprehensive detection of anomalies in complex scenarios could be considered.

Supplementary Materials: The following supporting information can be downloaded at: https://www.mdpi.com/article/10.3390/fire8010023/s1.

Author Contributions: Conceptualization, H.B. and X.Z.; Formal analysis, X.L. and Z.Z.; Investigation, X.L., Z.Z. and H.B.; Methodology, H.B. and X.Z.; Project administration, H.B.; Software, X.L., Z.Z. and Y.T.; Supervision, X.Z.; Validation, X.L. and X.Z.; Visualization, X.L., Z.Z. and Y.T; Writing—original draft, H.B., X.L. and Z.Z.; Writing—review and editing, H.B. and X.Z. All authors have read and agreed to the published version of the manuscript.

Funding: This research was funded by the National Natural Science Foundation of China (No. 52004134) and the Postgraduate Research and Practice Innovation Program of Jiangsu Province 2024 (No. SJCX24_0528).

Data Availability Statement: The original contributions presented in this study are included in the article/Supplementary Materials. Further inquiries can be directed to the corresponding author.

Conflicts of Interest: The authors declare no conflicts of interest.

References

1. Meacham, B.J. Fire safety of existing residential buildings: Building regulatory system gaps and needs. *Fire Saf. J.* **2023**, *140*, 103902. [CrossRef]
2. Xie, D.; Chen, Q.; Zhu, Y. Optimal layout scheme of fire detectors and operation condition monitoring technology in urban integrated substation. *Measurement* **2024**, *236*, 115093. [CrossRef]
3. Hu, T.; Xu, Z.; Yu, C.; Dou, X.; Zhang, Y.; Sun, L. Impacts of different forest fire management policies and fuel treatment models on forest fire risk in boreal forest of China. *Ecol. Indic.* **2024**, *169*, 112806. [CrossRef]
4. Desai, A.; Heilman, W.E.; Skowronski, N.S.; Clark, K.L.; Gallagher, M.R.; Clements, C.B.; Banerjee, T. Features of turbulence during wildland fires in forested and grassland environments. *Agric. For. Meteorol.* **2023**, *338*, 109501. [CrossRef]
5. Wu, Y.; Sun, J.; Yang, G.; Cui, L.; Wang, Z.; Wang, M. Research on digital twin based temperature field monitoring system for LNG storage tanks. *Measurement* **2023**, *215*, 112864. [CrossRef]

6. Zhao, S.; Duan, Y.; Roy, N.; Zhang, B. A deep learning methodology based on adaptive multiscale CNN and enhanced highway LSTM for industrial process fault diagnosis. *Reliab. Eng. Syst. Saf.* **2024**, *249*, 110208. [CrossRef]
7. Linn, R.R.; Reisner, J.M.; Colman, J.; Winterkamp, J. Studying wildfire behavior using FIRETEC. *Int. J. Wildland Fire* **2002**, *11*, 233–246. [CrossRef]
8. Himoto, K.; Tanaka, T. Development and validation of a physics-based urban fire spread model. *Fire Saf. J.* **2008**, *43*, 477–494. [CrossRef]
9. Zhang, Q.; Zhu, J.; Dong, Y.; Zhao, E.; Song, M.; Yuan, Q. 10-minute forest early wildfire detection: Fusing multi-type and multi-source information via recursive transformer. *Neurocomputing* **2025**, *616*, 128963. [CrossRef]
10. Moritz, M.A.; Batllori, E.; Bradstock, R.A.; Gill, A.M.; Handmer, J.; Hessburg, P.F.; Leonard, J.; McCaffrey, S.; Odion, D.C.; Schoennagel, T.; et al. Learning to coexist with wildfire. *Nature* **2014**, *515*, 58–66. [CrossRef]
11. Luo, Y.-x.; Li, Q.; Jiang, L.-r.; Zhou, Y.-h. Analysis of Chinese fire statistics during the period 1997–2017. *Fire Saf. J.* **2021**, *125*, 103400. [CrossRef]
12. Yoshioka, H.; Himoto, K.; Kagiya, K. Large Urban Fires in Japan: History and Management. *Fire Technol.* **2020**, *56*, 1885–1901. [CrossRef]
13. Chowdhury, E.H.; Hassan, Q.K. Operational perspective of remote sensing-based forest fire danger forecasting systems. *ISPRS J. Photogramm. Remote Sens.* **2015**, *104*, 224–236. [CrossRef]
14. Amiro, B.D.; Cantin, A.; Flannigan, M.D.; de Groot, W.J. Future emissions from Canadian boreal forest fires. *Can. J. For. Res.* **2009**, *39*, 383–395. [CrossRef]
15. Prasad, D.; Nath, V. An Overview of Temperature Sensors. In *Proceeding of the Second International Conference on Microelectronics, Computing & Communication Systems (MCCS 2017)*; Nath, V., Mandal, J.K., Eds.; Springer: Singapore, 2019; pp. 777–784.
16. Wang, H.; Bah, M.J.; Hammad, M. Progress in Outlier Detection Techniques: A Survey. *IEEE Access* **2019**, *7*, 107964–108000. [CrossRef]
17. Chatterjee, P.; Kadir, S.U.; Srivastava, A.; Laszka, A. Chapter Seventeen—Grid resilience against wildfire with machine learning: Machine learning based detection, localization and mitigation of the impact of forest fires on power grids. In *Big Data Application in Power Systems*, 2nd ed.; Arghandeh, R., Zhou, Y., Eds.; Elsevier Science: Amsterdam, The Netherlands, 2024; pp. 393–417.
18. Ji, W.; Li, G.-Q.; Zhu, S.; Li, J.; Qi, H.; Wang, Y. Machine learning-driven real-time identification of large-space building fires and forecast of temperature development. *Expert Syst. Appl.* **2024**, *255*, 124758. [CrossRef]
19. Pang, G.; Shen, C.; Cao, L.; Hengel, A.V.D. Deep Learning for Anomaly Detection: A Review. *ACM Comput. Surv.* **2021**, *54*, 38. [CrossRef]
20. Chen, Y.; Morton, D.C.; Randerson, J.T. Remote sensing for wildfire monitoring: Insights into burned area, emissions, and fire dynamics. *One Earth* **2024**, *7*, 1022–1028. [CrossRef]
21. Toledo-Jaime, C.; Díaz-Avalos, C.; Chaudhuri, S.; Serra, L.; Juan, P. Understanding wildfire occurrence and size in Jalisco, Mexico: A spatio-temporal analysis. *For. Ecol. Manag.* **2024**, *573*, 122349. [CrossRef]
22. Toullier, T.; Dumoulin, J. Bias and bottlenecks study in outdoor long term thermal monitoring by infrared thermography: Leveraging opportunistic data for temperature estimation. *Infrared Phys. Technol.* **2024**, *141*, 105471. [CrossRef]
23. Dastour, H.; Ahmed, M.R.; Hassan, Q.K. Analysis of forest fire patterns and their relationship with climate variables in Alberta's natural subregions. *Ecol. Inform.* **2024**, *80*, 102531. [CrossRef]
24. Carrasco-Escaff, T.; Garreaud, R.; Bozkurt, D.; Jacques-Coper, M.; Pauchard, A. The key role of extreme weather and climate change in the occurrence of exceptional fire seasons in south-central Chile. *Weather Clim. Extrem.* **2024**, *45*, 100716. [CrossRef]
25. Junior, O.S.; Coninck, J.C.P.; Magrin, F.G.S.; Ganacim, F.I.S.; Pombeiro, A.; Fernandes, L.G.; Romaneli, E.F.R. Impacts of Atmospheric and Load Conditions on the Power Substation Equipment Temperature Model. *Energies* **2023**, *16*, 4295. [CrossRef]
26. Liu, S.-S.; Li, P.-W.; Lan, W.-h.; Lin, W.-j. High-temperature high-humidity and electrical static discharge stress effects on GaN p–i–n UV sensor. *Mater. Sci. Eng. B* **2005**, *121*, 29–33. [CrossRef]
27. Xing, H.; Fang, K.; Yao, Q.; Zhou, F.; Ou, T.; Liu, J.; Zhou, S.; Jiang, S.; Chen, Y.; Bai, M.; et al. Impacts of changes in climate extremes on wildfire occurrences in China. *Ecol. Indic.* **2023**, *157*, 111288. [CrossRef]
28. Diao, W.; Geng, Y. An Automatic Threshold Adjustment Algorithm for Thermal Fault Extraction. In Proceedings of the 2023 6th International Conference on Software Engineering and Computer Science (CSECS), Chengdu, China, 22–24 December 2023; pp. 1–5.
29. Yang, S.; Huang, Q.; Yu, M. Advancements in remote sensing for active fire detection: A review of datasets and methods. *Sci. Total Environ.* **2024**, *943*, 173273. [CrossRef]
30. Saleh, A.; Zulkifley, M.A.; Harun, H.H.; Gaudreault, F.; Davison, I.; Spraggon, M. Forest fire surveillance systems: A review of deep learning methods. *Heliyon* **2024**, *10*, e23127. [CrossRef]

31. Gasser, J.; Warpelin, D.; Bussières, F.; Extermann, J.; Pomarico, E. Distributed temperature sensor combining centimeter resolution with hundreds of meters sensing range. *Opt. Express* **2022**, *30*, 6768–6777. [CrossRef]
32. Bao, Y.; Huang, Y.; Hoehler, M.S.; Chen, G. Review of Fiber Optic Sensors for Structural Fire Engineering. *Sensors* **2019**, *19*, 877. [CrossRef]
33. Neloy, A.A.; Turgeon, M. A comprehensive study of auto-encoders for anomaly detection: Efficiency and trade-offs. *Mach. Learn. Appl.* **2024**, *17*, 100572. [CrossRef]
34. Li, P.; Pei, Y.; Li, J. A comprehensive survey on design and application of autoencoder in deep learning. *Appl. Soft Comput.* **2023**, *138*, 110176. [CrossRef]
35. Bank, D.; Koenigstein, N.; Giryes, R. Autoencoders. *arXiv* **2020**, arXiv:2003.05991.
36. Xu, Z.; Guo, Y.; Saleh, J.H. Advances Toward the Next Generation Fire Detection: Deep LSTM Variational Autoencoder for Improved Sensitivity and Reliability. *IEEE Access* **2021**, *9*, 30636–30653. [CrossRef]
37. Üstek, İ.; Arana-Catania, M.; Farr, A.; Petrunin, I. Deep Autoencoders for Unsupervised Anomaly Detection in Wildfire Prediction. *Earth Space Sci.* **2024**, *11*, e2024EA003997. [CrossRef]
38. Qian, J.; Song, Z.; Yao, Y.; Zhu, Z.; Zhang, X. A review on autoencoder based representation learning for fault detection and diagnosis in industrial processes. *Chemom. Intell. Lab. Syst.* **2022**, *231*, 104711. [CrossRef]
39. Chu, C.; Zhu, Z.; Bian, H.; Jiang, J. Design of self-heating test platform for sulfide corrosion and oxidation based on Fuzzy PID temperature control system. *Meas. Control* **2021**, *54*, 1082–1096. [CrossRef]
40. Choi, K.; Yi, J.; Park, C.; Yoon, S. Deep Learning for Anomaly Detection in Time-Series Data: Review, Analysis, and Guidelines. *IEEE Access* **2021**, *9*, 120043–120065. [CrossRef]
41. He, H.; Garcia, E.A. Learning from Imbalanced Data. *IEEE Trans. Knowl. Data Eng.* **2009**, *21*, 1263–1284. [CrossRef]
42. Malhotra, P.; Ramakrishnan, A.; Anand, G.; Vig, L.; Agarwal, P.; Shroff, G.M. LSTM-based Encoder-Decoder for Multi-sensor Anomaly Detection. *arXiv* **2016**, arXiv:1607.00148.
43. Khan, Z.A.; Hussain, T.; Ullah, F.U.M.; Gupta, S.K.; Lee, M.Y.; Baik, S.W. Randomly Initialized CNN with Densely Connected Stacked Autoencoder for Efficient Fire Detection. *Eng. Appl. Artif. Intell.* **2022**, *116*, 105403. [CrossRef]

fire

MDPI

Article

Thermal Reaction Process and Thermokinetic Characteristics of Coking Coal Oxidation

Ruoyu Bao [1,2], Changkui Lei [3,*], Chengbo Wang [3] and Fubao Zhou [1]

1 China Academy of Safety Science and Technology, Beijing 100012, China
2 Information Research Institute of the Ministry of Emergency Management, Beijing 100029, China
3 School of Safety and Emergency Management Engineering, Taiyuan University of Technology, Taiyuan 030024, China
* Correspondence: leichangkui@tyut.edu.cn

Abstract: The coal–oxygen composite reaction is a complex physicochemical reaction process, and different heating rates have a great influence on this reaction. In order to reveal the influence of different heating rates on the coal–oxygen composite reaction of coking coal, the TG-DSC experimental method was adopted to analyze the hysteresis effect of the characteristic temperature, inflection point temperature, and peak temperature under different heating rates. Furthermore, the KAS method was employed to calculate the apparent activation energy, and the Málek method was utilized to infer the most probable mechanism functions and determine the compensation effects at different stages of the coal oxidation process. The results show that with an increase in heating rate, the temperature values corresponding to each characteristic temperature point increase, the characteristic temperature exhibits a hysteresis phenomenon, and the heat flow rate and heat flux rate also show an increasing trend. The apparent activation energy gradually increases in Stages II and III, with a maximum value of 198.7 kJ/mol near the ignition point T_3, which first increases and then gradually decreases in Stage IV, where the maximum value is around the temperature point T_4 of the maximum mass loss rate, which is 170.02 kJ/mol. The variation trend in the pre-exponential factor is consistent with the apparent activation energy, and the dynamic compensation effect is greater in Stage IV. The three different oxidation stages have different mechanism functions: a three-dimensional diffusion mode is present in Stages II and III, which is ultimately transformed into an accelerated form α-t curve with E_1 and $n = 1$ in Stage IV.

Keywords: coal oxidation; thermodynamic parameters; apparent activation energy; mechanism function; compensation effect

1. Introduction

Coal spontaneous combustion is one of the most common disasters in the process of coal mining, transportation, storage, and use [1–3]. Coal spontaneous combustion not only causes a lot of waste of coal resources and economic losses, but also causes death and injury in the production process, and the discharged gas further aggravates environmental pollution [4–6]. Therefore, it is very important to carry out relevant research and technological developments regarding the process of coal oxidation and spontaneous combustion to prevent coal spontaneous combustion disasters. Currently, with the progress of science and technology, the scope and amount of coal mining are increasing year by year, which makes the prevention and control of the spontaneous combustion of coal a top priority in ensuring safe coal production [7].

To find more efficient methods of inhibiting spontaneous coal combustion, domestic and foreign scholars have conducted a large number of experimental studies on the factors influencing spontaneous coal combustion. The influence degree of different factors on coal spontaneous combustion was analyzed through looking at complex coal–oxygen reactions. Wang et al. [8] proposed that there are a large number of pores in coal; the pores provide good conditions for interactions between active groups and oxygen in coal, which are the cause of the spontaneous combustion of coal. They therefore put forward the theory of group interaction. Yang et al. [9] determined the characteristic temperature of coal spontaneous combustion using a series of parameters, such as heating rate, oxygen consumption rate, gas production rate, the single gas index, and the complex gas index, and analyzed the corresponding relationship between the characteristic temperature and each parameter. Yang et al. [10] researched the heat release characteristics of the spontaneous combustion of coal and established a prediction model of the shortest spontaneous ignition period of coal based on the heat release characteristics. Qi et al. [11] divided the fast and slow oxidation stages of coal–oxygen recombination at low temperatures and found that the kinetic parameters of the two stages showed little difference at low oxygen concentrations.

Thermogravimetry (TG or TGA) is the main method used to study the coal–oxygen recombination reaction [12,13]. Wang et al. [14] employed a thermal analysis of the experimental data obtained using the random nucleation growth model (RNGM) and volume model (VM), obtained the kinetic parameters and mechanism of combustion reactions, and found the kinetic characteristics that lead to sample combustion. Ren et al. [15] found the combustion and kinetic mechanisms of waste plastic blends with 0%, 10%, 20%, 40%, and 100% plastic and coal proportions through thermogravimetric analysis. Qi et al. [16] and Xu et al. [17] researched the oxidation and heat release characteristics of coal using the cross-point temperature method, determined each characteristic temperature in the spontaneous combustion process of coal, and calculated the activation energy.

The spontaneous combustion of coal involves very complex physical and chemical reactions, in which the functional groups in coal play a decisive role. Zhao et al. [18] used X-ray diffraction and Fourier transform infrared spectroscopy to determine the microscopic characteristics of coal in the process of high-temperature oxidation. The mineral structure and functional groups of coal were determined to ensure that the original structural characteristics of coal were obtained. Sonibare et al. [19] and Okolo et al. [20] compared and analyzed the differences in aromatic structure, aliphatic hydrocarbon, and microcrystalline structure in different coal molecular structures using various techniques. MoraKanska et al. [21] researched the evolution characteristics of hydrogen bonds and oxygen structures during coal pyrolysis. Deng et al. [22] and Cao et al. [23] investigated the changes in functional groups before and after coal combustion and analyzed the key functional groups of coal spontaneous combustion. Chen et al. [24] studied the spontaneous combustion of bituminous coal using thermogravimetric experiments and in situ Fourier transform infrared spectroscopy. Through the semi-quantitative calculation of alkyl side bond length, the aromatic condensation degree and oxygen enrichment degree were obtained. Zhou et al. [25] researched the evolution law of microscopic functional groups during coal–oxygen recombination at low temperatures and analyzed the correlation between weight changes and functional group evolution.

Many scholars have analyzed the kinetics of the coal–oxygen recombination reaction using the apparent activation energy and pre-exponential factors. Ozbas et al. [26] assumed that the low-temperature oxidation of coal was a first-order reaction model to analyze the thermogravimetric data, and calculated the activation energy and pre-exponential factors in the reaction process of coal with different particle sizes using the Arrhenius kinetic model. Iliyas et al. [27] predicted the reaction rate of sulfur-containing minerals at different temperatures through thermal analysis and dynamic simulation and verified the autocatalysis that occurs in coal during the oxidation stage. Kaljuvee et al. [28] showed that the apparent activation energy is related to the reactive functional groups in coal, and the higher the content of the reactive functional groups, the higher the apparent activation energy. Rotaru et al. [29] studied the differences in the apparent activation energies of different coal samples.

The oxidation and spontaneous combustion process of coal is a complex oxidation kinetic process. Understanding and mastering the kinetics of coal oxidation is the key to understanding the mechanism of coal spontaneous combustion. Therefore, we take the coking coal of the working face 4502 of Shaqu Coal Mine No. 1 as the research object, investigate the thermal reaction process and thermodynamic characteristics of oxidation combustion of coking coal under different heating rates, and explore the influence of heating rate on the characteristic temperature and heat release effect of coking coal. Further, the KAS method was adopted to calculate the apparent activation energy of coal samples, and the Málek method was employed to determine the most probable mechanism functions and kinetic compensation effects of different reaction stages.

2. Experiment and Method

2.1. Coal Sample

The experimental coal sample is from the working face 4502 of Shaqu Coal Mine No. 1 in Shanxi Province, China, and consists of coking coal. The coal sample was removed from the mine and transported to the laboratory under anaerobic conditions, and then it was ground into coal powder of 100–120 mesh by a coal crusher after removing the oxidized and deteriorated part of the surface.

Proximate analyzer 5E-MAG6700 and element analyzer Elementar Vario EL III were employed to conduct proximate and element analysis tests on coal, and the results are shown in Table 1.

Table 1. Proximate and elemental analysis of coal.

Coal Sample	Proximate Analysis				Elemental Analysis				
	M_{ad}	A_{ad}	V_{ad}	FC_{ad}	C_{ad}	H_{ad}	O_{ad}	N_{ad}	S_{td}
Working face 4502	3.00	18.54	10.31	68.15	78.63	9.21	9.21	1.42	0.41

Note: M_{ad}, A_{ad}, V_{ad}, and FC_{ad} represent the content of moisture, ash, volatile, and fixed carbon in coal, respectively. C_{ad}, H_{ad}, O_{ad}, N_{ad}, and S_{td} represent the carbon, hydrogen, oxygen, nitrogen, and sulfur content of coal, respectively.

2.2. Experiment

The instrument used in the experiment is the synchronous thermal analyzer SDTQ600. The weight of the experimental coal sample is 3 ± 1 mg. The coal is heated from room temperature to 800 °C with heating rates of 10 °C/min, 15 °C/min, and 20 °C/min under air atmosphere, and the inlet flow rate is set to 100 mL/min.

3. Results and Discussion

3.1. Mass and Heat Variation

3.1.1. Characteristic Temperature

To analyze the reaction law of the coal oxidation process under different heating rates and divide the characteristic temperature points, the TG-DSC curve was adopted to analyze the entire coal–oxygen composite reaction. The heating rates selected in the experiment were 10 °C/min, 15 °C/min, and 20 °C/min, respectively. The division of characteristic temperature stages under different heating rates is shown in Figure 1.

From Figure 1, we can observe that the TG-DTG curves of experimental coal samples at different heating rates show almost the same change trend, but due to different heating rates, there are some differences in the characteristic temperature points and mass change rates in the whole reaction process. According to the classification method mentioned in the literature review, the characteristic temperature is selected based on the TG-DTG curve, and T_1 is the temperature corresponding to the minimum value of the TG curve in the early stage of the coal–oxygen composite reaction, which corresponds to the temperature at which the mass change rate is 0. At this temperature, the chemisorption of oxygen by coal is enhanced, offsetting the mass reduction caused by water evaporation and gas desorption, and the coal mass reaches a minimum value. After exceeding the temperature, the chemisorption of oxygen by coal continues to increase, manifested macroscopically as an increase in coal mass. T_2 is the temperature with the highest mass. At this temperature, due to the high temperature, the coal–oxygen recombination effect is enhanced, the oxidation gases and the alkane gases generated by molecular fracture increase, and the mass increase caused by the chemical adsorption of oxygen by coal reaches a balance with the release of gas products. After exceeding the temperature, the reaction consumption of coal continues to increase and enters the rapid reaction weight loss stage. Therefore, the macroscopic performance at this temperature is that the mass reaches the maximum value; T_3 is the ignition point temperature of the coal sample and is the tangent of the intersection point of the maximum peak vertical line on the DTG curve and the TG curve. The tangent line corresponds to the intersection point of the pyrolysis temperature T_3 on the TG curve. T_4 is the temperature corresponding to the maximum point of the mass change rate of the coal sample in the combustion stage. T_5 is the burnout temperature at the end of the reaction. At this point, the DTG curve value is close to 0, the whole reaction process is over, and the quality of the coal sample also does not change significantly. The characteristic temperature points under different heating rates are presented in Figure 1 and Table 2.

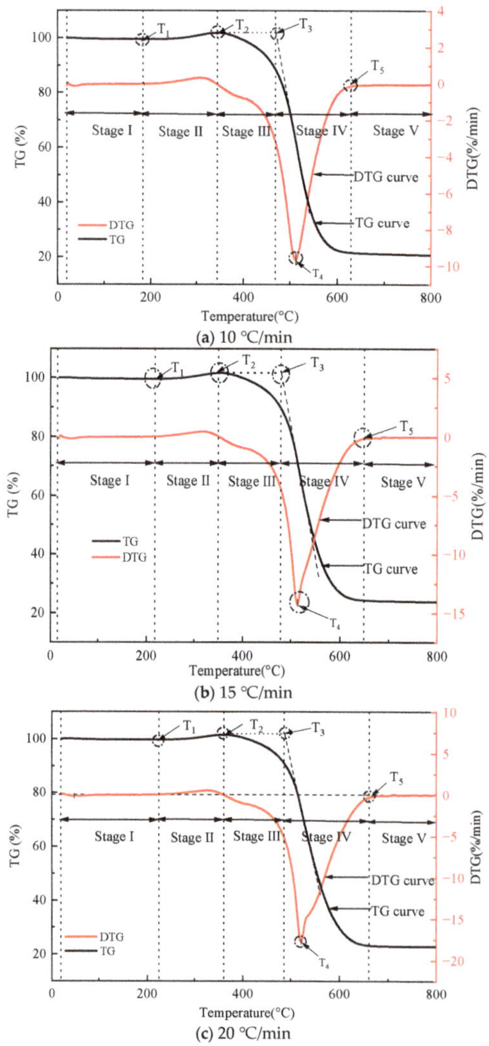

Figure 1. Characteristic temperature points and stage division.

Table 2. Characteristic temperature points at different heating rates.

Heating Rate (°C/min)	T_1 (°C)	T_2 (°C)	T_3 (°C)	T_4 (°C)	T_5 (°C)
10	189.48	343.19	471.91	512.19	605.45
15	190.83	351.49	479.85	514.20	625.38
20	207.14	358.65	487.47	522.22	644.81

3.1.2. Stage Characteristics

The reaction process of the coal sample from low-temperature oxidation to the end of the burnout reaction can be divided into five stages (I–V) according to the characteristic temperature point.

(1) Water evaporation and minimal mass loss stage (Stage I: T_0–T_1)

The temperature range of this stage is from the initial temperature to T_1, which is mainly due to the mass loss of coal samples during low-temperature oxidation. In this stage, the TG curve presents a trend of slightly increasing at first and then decreasing, which is because the physical adsorption of gas by coal samples occurred at the beginning of the stage, resulting in a slight increase in coal weight within a very short time range. Then, as the water contained in the coal sample began to evaporate with increasing temperature, and as the temperature rose, the thermal movement of molecules accelerated, which weakened the binding effect of intermolecular interaction on gas molecules, and the adsorbed gas began to desorb. As a result, the TG curve showed a downward trend, and the quality of coal samples decreased. According to the DTG curve, the mass reduction rate at this stage gradually decreased, which is due to the limited adsorption capacity and water content of the gas. With the continuous desorption of the gas and the continuous evaporation of water, the mass loss rate gradually decreased. When the mass change rate was 0, the coal sample weight was the minimum value before the composite reaction.

(2) Oxygen absorption and mass gain stage (Stage II: T_1–T_2)

At this stage, the temperature range is from T_1 to T_2, and the chemical adsorption of oxygen by coal mainly occurred. Different from the physical adsorption process, the adsorption of gas by coal at this stage is selective, mainly due to the adsorption of oxygen in the environment by the active groups in coal. In the coal–oxygen composite process, unstable carbon and oxygen compounds are formed and gas products are released. However, the adsorption amount of oxygen is greater than the release amount of gas products, resulting in increasing weight. According to the DTG curve, the weight increase rate first increases and then decreases, which is because the temperature in the early stage is low, the strength of coal–oxygen recombination is weak, and the amount of gas products released is small. As the temperature rises, the functional groups in coal are continuously activated, the strength of coal–oxygen recombination is increased, the amount of gas products released is increased, and the unstable carbon and oxygen compounds formed in the early stage may undergo thermal decomposition, weakening the increasing trend of coal mass [30].

(3) Slow chemical reaction stage (Stage III: T_2–T_3)

The temperature range of this stage is from T_2 to T_3, mainly involving the thermal decomposition of the active groups in the coal sample and slow mass loss. The main active groups in coal are the aromatic ring, naphthene aliphatic hydrocarbon, and oxygen-containing functional groups. Because the structure of aromatic rings and naphthene is stable and difficult to break during the heating process of coal, it is not considered. Aliphatic hydrocarbon and oxygen-containing functional groups begin to participate in the coal–oxygen reaction at low temperatures, and they are the main active functional groups in the coal–oxygen complex reaction [31]. As the temperature gradually rises, the aliphatic hydrocarbon begins to undergo pyrolysis reaction with the active groups of oxygen-containing functional groups, and the carboxyl and hydroxyl groups contained in aliphatic hydrocarbon continue to react, and the C=O, -COOH transition groups are in a complex process of constant consumption and generation in the coal–oxygen composite reaction. In addition, the overall content of C=O and -COOH decreased with rising temperature, and some macromolecules in coal began to undergo a pyrolysis reaction, resulting in a slow decline in coal mass at this stage [32,33].

(4) Combustion stage (Stage IV: T_3–T_5)

The temperature range of this stage is from T_3 to T_5, which mainly causes the thermal decomposition of macromolecular structure in coal and the violent coal–oxygen compound effect. Due to the rapid increase in the release of volatile substances in the thermal decomposition of macromolecular structure and the gas products in the coal-oxygen composite process, the quality of the gas decreases rapidly [34]. According to the DTG curve, the weight loss velocity increases rapidly at first and then decreases sharply. This is because in the initial stage, thermal decomposition plays a dominant role, the release of volatile substances and the combustion of fixed carbon rapidly reduce the weight, and the coal–oxygen composite reaction aggravates the weight reduction. Then, when the temperature exceeds a certain level, the residual combustible substances in the coal are relatively difficult to react, and the weight loss rate decreases rapidly, and the weight loss rate tends to zero at the end of the reaction.

(5) Burnout stage (Stage V: >T_5)

The temperature range of this stage is T_5 to the end of the experiment, at which time the whole coal–oxygen composite reaction is close to the end, and the main active groups and macromolecular structure in the coal after full combustion of the experimental coal sample are gone [24]. In this stage, the quality of coal hardly changes, and the residual quality after the reaction is almost similar to the ash quality in the proximate analysis of coal samples. At the same time, the DTG curve values almost tend to 0, and the coal mass shown by the TG curve hardly changes.

At each stage of the coal–oxygen recombination reaction, mass changes will be presented on a macro level along with the internal reaction. The mass loss and gain ratio of coal at each stage in the thermogravimetric experiment is shown in Table 3.

Table 3. Mass variation of coal samples at different stages.

Heating Rate (°C/min)	Mass Variation at Each Stage (%)				
	I	II	III	IV	Residual Mass
10	−0.79	1.96	−11.31	−78.66	21.34
15	−0.76	1.73	−11.12	−75.86	24.14
20	−0.70	1.68	−9.30	−74.85	25.15

According to the comparison of the mass loss ratio of each stage at different heating rates, the mass ratio remaining at the end of the coal–oxygen composite reaction is almost the same on the whole. The mass loss ratio of coal sample decreases with the increase in heating rate in Stage I. The mass gain ratio in Stage II shows a trend of first increasing and then decreasing, and the maximum temperature rise rate of mass gain ratio is 10 °C/min. The mass loss ratio of coal sample decreases with the increasing heating rate in Stage III, indicating that the slower the heating rate, the more fully the reactive groups contained in coal can react, and the greater the mass loss ratio of the coal sample. The coal sample shows a trend of gradually increasing mass loss ratio after Stage IV; the heating rate with the largest mass loss ratio is 10 °C/min, while the heating rate with the smallest residual mass ratio is also 10 °C/min, indicating that the coal sample is most fully burned at this heating rate.

3.1.3. Hysteresis Effect of Thermogravimetry

The whole coal–oxygen compound reaction is a complicated process, and the mass changes of the reaction at different heating rates are not the same. Therefore, the TG-DTG curve can more clearly analyze the influence degree of each heating rate on the coal–oxygen complex reaction, as shown in Figure 2.

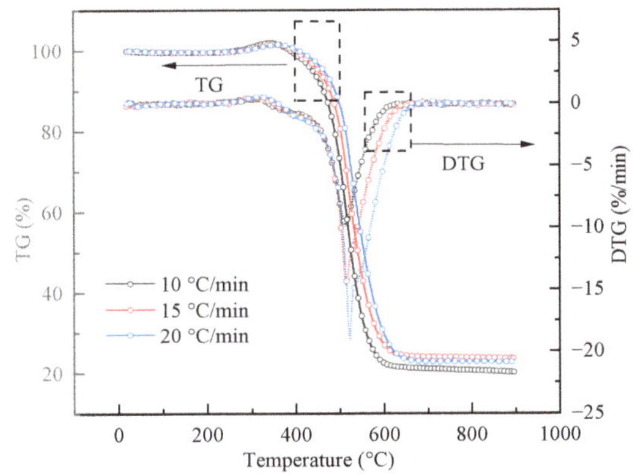

Figure 2. TG-DTG curves at different heating rates.

It can be found from Figure 2 that the change trend of the TG-DTG curve of the coal–oxygen complex reaction under different heating rates is almost the same, but with the increase in heating rate, the reaction gradually lags. The mass change is not obvious in Stages I and II, which is due to the coal–oxygen recombination reaction being relatively gentle in a low-temperature environment, and the mass change rate of coal in these two stages is almost the same from the DTG curve. With the increase in temperature, the mass change is more obvious when reaching Stages III and IV. On account of the increase in heating rate, the temperature change increases at the same time, the active groups and volatile molecules contained in coal react more vigorously at high temperatures, and the degree of coal–oxygen recombination reaction is more intense. The weight loss rate of the DTG curve is also gradually increasing. The maximum mass loss rate presented by T_4 also increases with the increase in heating rate. When the heating rate is 20 °C/min, the maximum mass loss rate (T_4) presents a maximum value of −17.8%/min. It can also be seen from the TG curve that when the heating rate is 10 °C/min, the mass ratio of the coal sample remaining after the whole reaction is the lowest in Stage V. These show that the higher the heating rate, the less sufficient the active groups and volatile reactions contained in the interior of the coal, the less mass loss ratio presented, and the greater the remaining mass ratio [35]. Furthermore, in terms of its rate of mass change, at the same temperature, the higher the heating rate, the greater the rate of mass change. This is because an increase in heating rate will lead to a rapid increase in temperature difference between the ambient temperature and the coal sample temperature, as well as an increase in ambient air pressure, which is more conducive to the diffusion of oxygen to the coal surface, the progress of the coal oxygen reaction, and the formation of carbon oxides, resulting in an increase in the rate of mass change with an increase in heating rate.

Meanwhile, it can be observed from Figure 2 that the corresponding temperature values of each characteristic temperature point also increase with the increasing heating rate, and the characteristic temperature has a backward phenomenon; in particular, the temperature point has a more obvious backward trend in Stages III and IV. This is because as the heating rate increases, the reaction time at each stage gradually shortens, resulting in insufficient coal–oxygen recombination reaction in the previous stage, leading to some activated groups inside the coal sample not participating in the reaction [36]. At this time, heterogeneous reactions occur inside the coal sample with a higher heating rate, resulting in a lag phenomenon in the entire oxidation recombination reaction process. Finally, as the heating rate increases, each characteristic temperature point also increases, and this change is more obvious around T_4 (characteristic temperature of maximum mass loss rate).

With the increasing heating rate, the temperature point at which the whole coal–oxygen recombination reaction ends also shifts backwards.

3.1.4. Heat Effect

To analyze the influence of different heating rates on the temperature corresponding to the extreme point of heat flow rate, DSC curves at different heating rates were made, as shown in Figure 3.

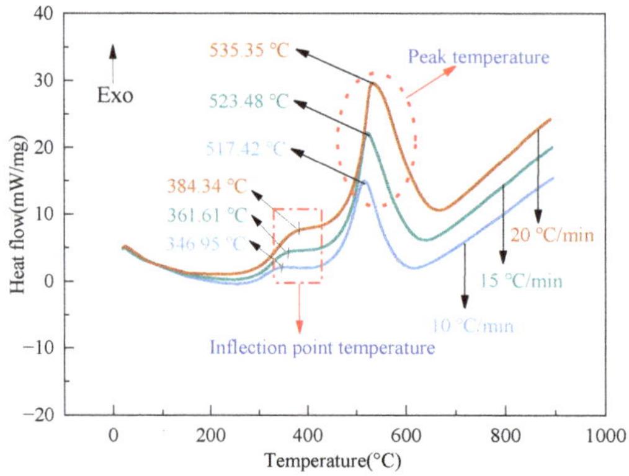

Figure 3. DSC curves at different heating rates.

The characteristic temperature points of DSC curves at different heating rates are shown in Figure 3, and the DSC curves of coal samples under different heating rates are almost the same. In the initial stage (0–200 °C), the whole reaction occurs relatively slowly, and the energy released is small, and the moisture and gas volatilization in the coal sample also need to consume a certain amount of heat, so the overall heat release is small. In the second stage (200–400 °C), when the temperature continues to rise, the water is gradually consumed in the process of continuous evaporation, and the energy released by the whole system also increases with the increasing temperature, because the active groups (aliphatic hydrocarbons and oxygen-containing functional groups) that are difficult to react at low temperatures are more active at this stage. As a result, the types and quantities of active groups involved in the reaction in the coal molecules increase, resulting in an increase in the intensity of the coal–oxygen composite reaction, and the heat released begins to increase [37]. When the temperature reaches a certain value, the first inflection point temperature appears at this time, and then the growth rate of coal heat release rate shows a relatively lower phenomenon because the unstable carbon and oxygen compounds and macromolecular structures in coal crack at high temperatures, which requires more energy and absorbs part of the reaction heat release, resulting in a lower growth rate of heat release. In the third stage (400–700 °C), when the main thermal decomposition is completed, oxygen quickly occupies the surface and reacts with the volatile molecules and fixed carbon contained in the coal, releasing a lot of heat, and the heat release rate increases rapidly; at this time, the heat flow extreme value of the entire curve will appear. However, with the volatilization of coal combustible substances at high temperatures, the remaining combustible substances in coal become more stable as they reach the later stage of the reaction. Some of the combustible substances will absorb heat release in the reaction process, and the heat release rate of coal decreases rapidly. When the combustible substances in coal are completely burned, the heat release rate of coal approaches 0. These are consistent with the thermogravimetric results. In Stage I, the coal oxidation reaction rate is slower and the heat release is less. As the reaction progresses, the reaction rate further accelerates, the heat release increases, and the mass rapidly decreases. In Stage V, the coal sample burns completely, the heat release decreases, and the mass change approaches 0.

In addition, from the analysis of inflection point temperature and peak temperature under different heating rates in Figure 3, it can be observed that different heating rates have a certain influence on the heat released by the same coal sample in the coal–oxygen composite reaction. The variation of DSC characteristic temperature points under different heating rates is shown in Figure 4.

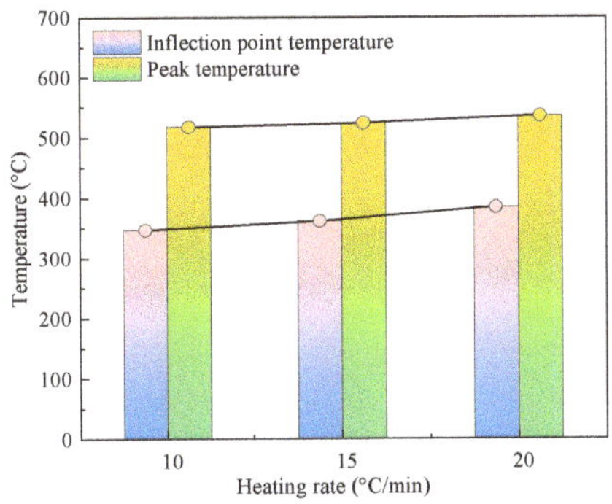

Figure 4. Characteristic temperature variation of DSC curve at different heating rates.

According to Figure 4, it can be seen that the inflection point temperatures at the heating rates of 10 °C/min, 15 °C/min, and 20 °C/min are 346.95 °C, 361.61 °C, and 384.34 °C, respectively, indicating that the temperature at the corresponding inflection point also lags as the heating rate increases. In addition, through the analysis of Figures 3 and 4, the heat flow rates corresponding to the inflection point temperatures at the heating rates of 10 °C/min, 15 °C/min, and 20 °C/min are 2.18 mW/mg, 4.63 mW/mg, and 7.94 mW/mg, respectively, and the heat flow rate increases with the increasing heating rate. By comparing the peak width and maximum heat release rate temperature of volatile and fixed carbon combustion at different heating rates, it can be seen from Figures 3 and 4 that the heat release peak at this stage becomes wider with the heating rate increasing, the peak shape changes from sharp to blunt, and the heat release of the coal–oxygen composite reaction also gradually increases. When the heating rate is 10 °C/min, 15 °C/min, and 20 °C/min, the corresponding peak temperatures are 517.42 °C, 523.48 °C, and 535.35 °C, respectively, and the corresponding heat flow rates are 15.06 mW/mg, 22.34 mW/mg, and 29.63 mW/mg, respectively. Meanwhile, it can be found that as the heating rate increases, the inflection point temperature and peak temperature relatively lag behind, but the heat flow rate and heat flow increase accordingly, indicating that the higher the heating rate, the stronger the coal–oxygen recombination reaction, and the greater the heat release.

3.2. Kinetic Characteristics

3.2.1. Kinetic Methods

Due to the complexity of the coal molecular structure, the competitive reaction and parallel reaction of various functional groups in coal exist at the same time, and the apparent activation energy required for the same conversion at different heating rates is the same. Therefore, the influence of different conversion rates on the coal–oxygen recombination reaction is analyzed from the calculation of the apparent activation energy and its changing law. The apparent activation energy generally refers to the energy required for a molecule to change from its normal state to an active state that is prone to chemical reactions.

When solving the kinetic parameters, the kinetic parameters calculated by the same curve using different solving methods are not the same, so a more appropriate method should be selected to solve the kinetic parameters. At present, the commonly used kinetic methods are divided into the isothermal method and the non-isothermal method. The multi-scanning rate method in the non-isothermal method is based on the analysis and calculation of the experimental data measured under different scanning rates to obtain the kinetic parameters and can calculate the apparent activation energy according to the coal sample data with the same conversion rate under different heating rates, so it is also called the equal conversion method. It is mainly represented by the Flynn–Wall–Ozawa (FWO) method, Kissen–Akahira–Sunose (KAS) method, and Friedman method.

The advantage of the non-isothermal method is that it does not need to determine the mechanism function first compared with the isothermal method, and can directly calculate the apparent activation energy of the same conversion under different heating rates. Compared with the FWO method and Friedman method, the KAS method has more accurate calculation results and a simple calculation process. Therefore, the KAS method is selected to calculate and analyze the kinetic parameters of coal samples under three different heating rates after comprehensive consideration [38,39].

$$\ln\left(\frac{\beta}{T^2}\right) = \ln\left(\frac{A \cdot R}{E_a \cdot G(\alpha)}\right) - \frac{E_a}{R \cdot T} \tag{1}$$

where A refers to the pre-exponential factor, min^{-1}; E_a is the apparent activation energy, J/mol; R is the gas universal constant, 8.314 J/(mol·K); $G(\alpha)$ is the integral of the mechanism function; α is the conversion rate, which is defined as follows:

$$\alpha = \frac{w_0 - w}{w_0 - w_1} \tag{2}$$

where w_0, w_1, and w are the start, end, and mass corresponding to t at a certain time of this stage, respectively, mg.

3.2.2. Apparent Activation Energy

Because of the small mass loss generated by the coal sample in Stages I and V, the kinetic analysis of coal samples mainly focuses on Stage II, Stage III and Stage IV. According to the KAS method, the apparent activation energy required for the reaction with the same conversion rate α under different heating rates is the same, so the linear correlation curve is drawn between $\ln(\frac{\beta}{T^2})$ and 1/T, with $\ln(\frac{\beta}{T^2})$ as the Y axis and 1/T as the X axis for linear fitting. The apparent activation energy E_a is calculated by the fitted slope, and the pre-exponential factor ln A is calculated by the intercept.

The conversion rates of the three stages under the heating rates of 10 °C/min, 15 °C/min, and 20 °C/min were calculated by Equation (2), and the apparent activation energy was calculated according to Equation (1), while the curves of conversion rate and temperature and the linear correlation between $\ln(\frac{\beta}{T^2})$ and 1/T are shown in Figures 5–7. The calculation results of apparent activation energy are shown in Tables 4–6.

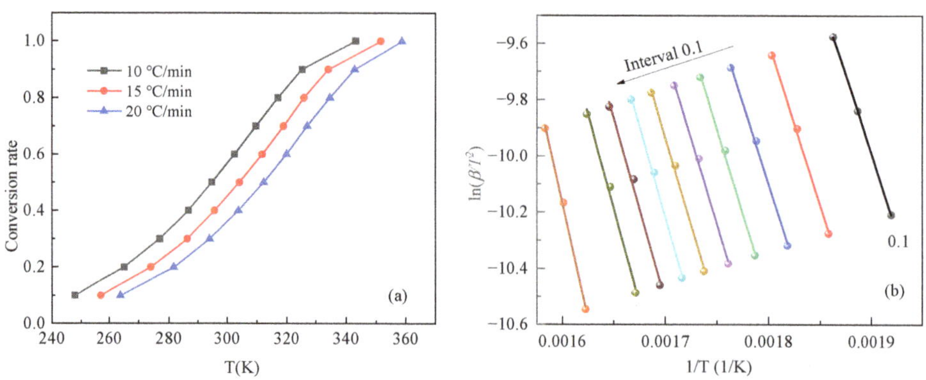

Figure 5. Relationship curves in Stage II: (**a**) conversion rate and temperature; (**b**) linear correlation between $\ln(\frac{\beta}{T^2})$ and 1/T.

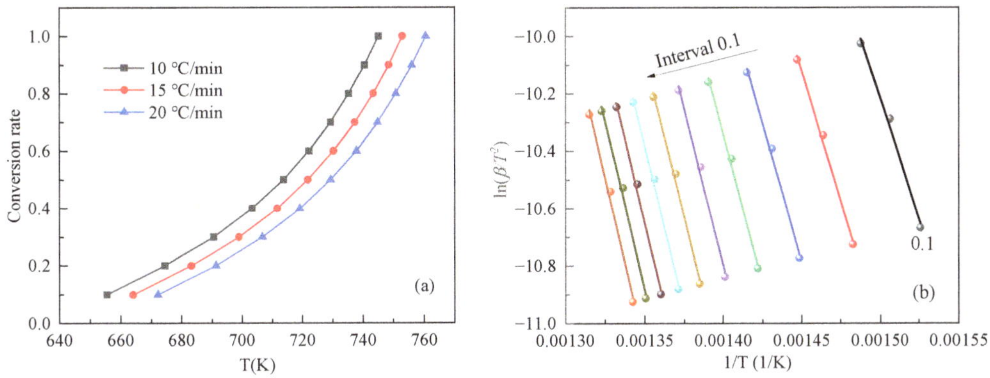

Figure 6. Relationship curves in Stage III: (**a**) conversion rate and temperature; (**b**) linear correlation curve $\ln(\frac{\beta}{T^2})$ with $1/T$.

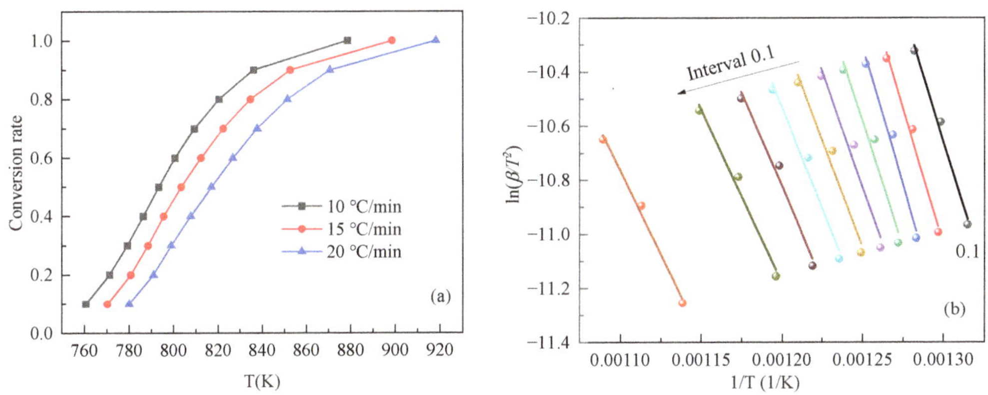

Figure 7. Relationship curves in Stage IV: (**a**) conversion rate and temperature; (**b**) linear correlation curve $\ln(\frac{\beta}{T^2})$ with $1/T$.

Table 4. Calculation results of apparent activation energy for Stage II.

Conversion Rate	Linear Fitting Results	E_a (kJ/mol)	R^2
0.1	y = −10,979x + 10.87167	91.29	0.99999
0.2	y = −11,510x + 11.11828	95.70	0.99783
0.3	y = −11,581x + 10.74511	96.29	0.99769
0.4	y = −11,916x + 10.94634	99.07	0.99630
0.5	y = −12,115x + 10.95963	100.73	0.99699
0.6	y = −12,470x + 11.26598	103.68	0.99624
0.7	y = −12,831x + 11.5945	106.68	0.99637
0.8	y = −12,907x + 11.4306	107.31	0.99249
0.9	y = −13,434x + 11.98012	111.70	0.99143
1.0	y = −16,245x + 15.82241	135.07	0.99723

Table 5. Calculation results of apparent activation energy for Stage III.

Conversion Rate	Linear Fitting Equation	E_a (kJ/mol)	R^2
0.1	y = −16,922x + 15.16613	140.69	0.99325
0.2	y = −17,964x + 15.9234	149.36	0.99457
0.3	y = −19,535x + 17.53494	162.42	0.995

Table 5. *Cont.*

Conversion Rate	Linear Fitting Equation	E_a (kJ/mol)	R^2
0.4	y = −20,781x + 18.75295	172.78	0.99458
0.5	y = −21,695x + 19.57657	180.38	0.99415
0.6	y = −22,147x + 19.82337	184.14	0.99372
0.7	y = −22,872x + 20.49755	190.17	0.99399
0.8	y = −23,279x + 20.77722	193.55	0.99465
0.9	y = −23,542x + 20.89851	195.74	0.99316
1.0	y = −23,899x + 21.17233	198.70	0.99274

Table 6. Calculation results of apparent activation energy for Stage IV.

Conversion Rate	Linear Fitting Equation	E_a (kJ/mol)	R^2
0.1	y = −19,755x + 15.0253	164.25	0.98869
0.2	y = −19,850x + 14.77119	165.04	0.98530
0.3	y = −20,402x + 15.19492	169.63	0.98067
0.4	y = −18,789x + 12.90239	156.21	0.96409
0.5	y = −17,230x + 10.71005	143.26	0.96623
0.6	y = −15,850x + 8.76239	131.78	0.97053
0.7	y = −14,818x + 7.24984	123.20	0.97570
0.8	y = −13,915x + 5.87322	115.69	0.97910
0.9	y = −12,916x + 4.31641	107.39	0.98391
1.0	y = −12,430x + 2.90876	103.34	0.99199

From the calculation results of the apparent activation energy values in the above tables, we can find the correlation between apparent activation energy and conversion rate in the three stages. Furthermore, the relationship between apparent activation energy and temperature can be obtained based on the corresponding relationship between conversion rate and temperature in each stage, as shown in Figures 8 and 9.

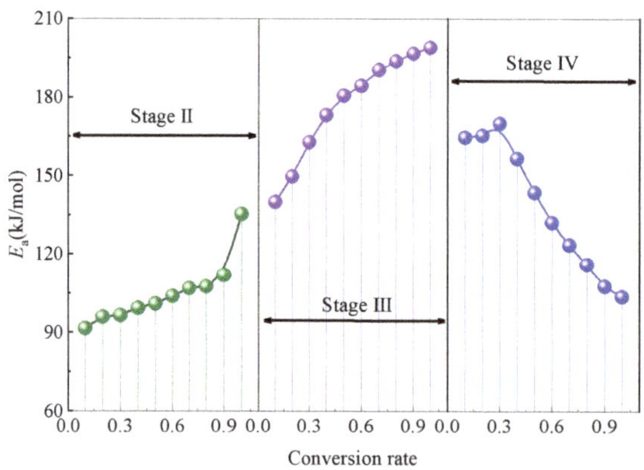

Figure 8. Relationship between apparent activation energy and conversion rate at Stages II, III and IV.

As can be seen from Figure 8, the apparent activation energy of Stage II gradually increases with the increase in conversion rate. In the stage of oxygen absorption and weight gain, coal mainly undergoes chemical adsorption of oxygen, and it is mainly that the active groups in coal adsorb the oxygen in the environment and form unstable carbon and oxygen compounds in the coal–oxygen composite process. However, due to the low temperature in early Stage II, the reaction degree is weak, and the mass changes are small; as the temperature increases, the DTG curve shows a trend of first increasing and then decreasing, the functional groups in coal undergo continuous oxidation reaction, the strength of coal–oxygen recombination increases and the release of gas products increases. Moreover, the unstable carbon and oxygen compounds formed in the early stage may cause thermal decomposition, which weakens the increasing trend of coal mass. Due to various chemical reactions such as the thermal decomposition of carbon and oxygen compounds and oxygen adsorption, the apparent activation energy required at this stage shows a significant increasing trend after the conversion rate of 0.8. In addition, due to the low temperature range at this stage, the number of active functional groups involved in the reaction is relatively small, requiring less energy and resulting in a lower apparent activation energy [40]. As heat accumulates and the temperature rises, the energy of the reaction system increases. Functional groups that were previously difficult to react with oxygen at low temperatures are gradually activated and participate in the reaction. As these functional groups are relatively stable, the energy required for the reaction is high. At the same time, an increase in the types and quantities of functional groups participating in the reaction also leads to an increase in the required energy, ultimately resulting in an increase in apparent activation energy with increasing conversion rate.

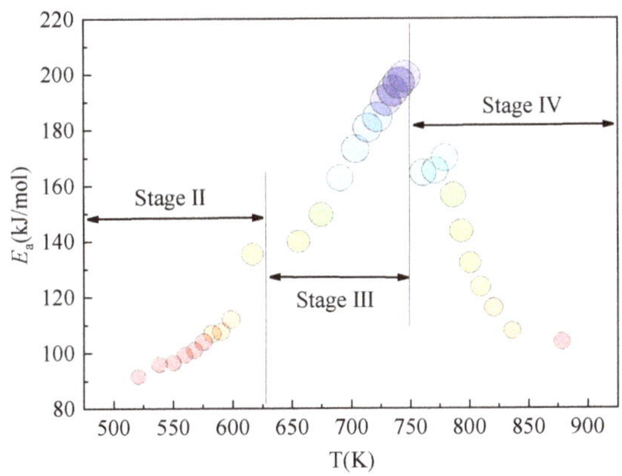

Figure 9. Relationship between apparent activation energy and temperature at Stages II, III and IV.

The calculated values of the apparent activation energy of Stage III shown in Figure 8 present an overall increasing trend with the increase in conversion rate, and compared to the apparent activation energy curve of Stage II, it can be clearly observed that the apparent activation energy required at this stage is larger. This because the oxygen-containing functional groups contained in the coal sample at this stage begin to react, and the oxygen-containing functional groups undergo a chemical reaction. Moreover, with the increase in temperature, the large molecules contained in the coal sample begin to break down into smaller molecular structures in the later stage [41,42].

As the conversion rate increases, the required apparent activation energy of Stage IV in Figure 8 shows a trend of first increasing and then decreasing, reaching the maximum value of 170.02 kJ/mol at 506.13 °C. This is because the initial pyrolysis reaction is still in progress at this stage, the volatile macromolecules contained in the coal are pyrolyzed into small molecules, a large amount of gas is released, and the number of active functional groups participating in the reaction gradually increases. However, as the temperature rises, the reaction of active functional groups contained in the coal sample gradually weakens, and in the later stage, the oxygen concentration will gradually decrease with the progress of the reaction, and the number of active functional groups involved in the reaction will decrease [43]. Therefore, the apparent activation energy curve at this stage shows a trend of first increasing and then decreasing.

As can be seen from Figure 9, the trend of apparent activation energy with the increase in temperature is almost the same as that of apparent activation energy with the increase in conversion rate, as shown in Figure 8. This is because the change of conversion rate from 0 to 1 at each stage shows an increasing trend with the increase in temperature. Therefore, the changing trend of apparent activation energy presented by them is also basically the same.

3.2.3. Kinetic Compensation Effect

To more clearly analyze the linear relationship between the pre-exponential factor ln A and the apparent activation energy E_a, the kinetic compensation effect is employed for analysis. The mathematical expression of the kinetic compensation effect is as follows:

$$\ln A = a \cdot E_a + b \tag{3}$$

where a and b are compensation parameters.

The kinetic compensation effect is commonly presented in various stages of the whole reaction system, and the values obtained by different mechanism functions in the same stage of the same reaction system are not the same. Therefore, the Málek method was adopted to validate and optimize the kinetic modes of each stage, in order to determine the most probable mechanism function of the oxidation reaction in each stage. The Málek method is an approach of deducing the kinetic mode $G(\alpha)$ or $f(\alpha)$ by defining the function $y(\alpha)$, which is combined with the reaction rate Equation (4) and the Coats–Redfern Equation (5) to obtain Equation (6).

$$\frac{d\alpha}{dt} = A\exp\left(-\frac{E_a}{RT}\right)f(\alpha) \tag{4}$$

$$\int_0^{\alpha} \frac{d\alpha}{f(\alpha)} = G(\alpha) = \frac{ART^2}{E_a\beta}\exp(-\frac{E_a}{RT}) \tag{5}$$

$$G(\alpha) = \frac{RT^2}{E_a\beta}\frac{d\alpha}{dt} \cdot \frac{1}{f(\alpha)} \tag{6}$$

When $\alpha = 0.5$:

$$G(0.5) = \frac{RT_{0.5}{}^2}{E_a\beta}(\frac{d\alpha}{dt})_{0.5} \cdot \frac{1}{f(0.5)} \tag{7}$$

where $T_{0.5}$ and $(\frac{d\alpha}{dt})_{0.5}$ represent the temperature and reaction rate when α is equal to 0.5, respectively.

Combine Equations (6) and (7) to obtain Equation (8):

$$y(\alpha) = \left(\frac{T}{T_{0.5}}\right)^2 \cdot \frac{\left(\frac{d\alpha}{dt}\right)}{\left(\frac{d\alpha}{dt}\right)_{0.5}} = \frac{f(\alpha) \cdot G(\alpha)}{f(0.5) \cdot G(0.5)} \tag{8}$$

Substitute experimental data into the left-hand equation and compare it with the standard curve on the right-hand side of the equation. If the experimental data curve overlaps with the standard curve, or if all the experimental data points fall on a standard curve, then $G(\alpha)$ or $y(\alpha)$ corresponding to the standard curve is the most probable kinetic mechanism function. Common kinetic mechanism functions are shown in Table 7.

When inferring the most probable mechanism function of different stages in the coal oxidation process according to the Málek method, it is necessary to calculate the theoretical value of the mechanism function first, then calculate the experimental value of different stages, and finally conduct a comparative analysis of the two to judge the apparent kinetic reaction model of different stages in the coal oxidation process. Málek theoretical values of different mechanism functions are shown in Figure 10.

Table 7. Common kinetic mechanism functions.

Function Number	Function Name	Mechanism	Integral Form $G(\alpha)$	Differential Form $f(\alpha)$
1	Parabola principle	One-dimensional diffusion	α^2	$\frac{1}{2}\alpha^{-1}$
2	Valensi equation	Two-dimensional diffusion	$\alpha + (1-\alpha)\ln(1-\alpha)$	$[-\ln(1-\alpha)]^{-1}$
3	Jander equation	Two-dimensional diffusion, $n = 1/2$	$\left[1-(1-\alpha)^{\frac{1}{2}}\right]^{\frac{1}{2}}$	$4(1-\alpha)^{\frac{1}{2}}\left[1-(1-\alpha)^{\frac{1}{2}}\right]^{\frac{1}{2}}$
4		Two-dimensional diffusion, $n = 2$	$\left[1-(1-\alpha)^{\frac{1}{2}}\right]^{2}$	$(1-\alpha)^{\frac{1}{2}}\left[1-(1-\alpha)^{\frac{1}{2}}\right]^{-1}$
5		Three-dimensional diffusion, $n = 1/2$	$\left[1-(1-\alpha)^{\frac{1}{3}}\right]^{\frac{1}{2}}$	$6(1-\alpha)^{\frac{2}{3}}\left[1-(1-\alpha)^{\frac{1}{3}}\right]^{\frac{1}{2}}$
6	Ginstling–Brounshtein equation	Three-dimensional diffusion, cylindrical symmetry	$1-\frac{2}{3}\alpha-(1-\alpha)^{\frac{2}{3}}$	$\frac{3}{2}\left[(1-\alpha)^{-\frac{1}{3}}-1\right]^{-1}$
7	Anti-Jander equation	Three-dimensional diffusion, $n = 2$	$\left[(1+\alpha)^{\frac{1}{3}}-1\right]^{2}$	$\frac{3}{2}(1+\alpha)^{\frac{2}{3}}\left[(1+\alpha)^{\frac{1}{3}}-1\right]^{-1}$
8	Avrami–Erofeev equation	$n = 1/4$, $m = 4$	$[-\ln(1-\alpha)]^{\frac{1}{4}}$	$4(1-\alpha)[-\ln(1-\alpha)]^{\frac{3}{4}}$
9		$n = 1/3$, $m = 3$	$[-\ln(1-\alpha)]^{\frac{1}{3}}$	$3(1-\alpha)[-\ln(1-\alpha)]^{\frac{2}{3}}$
10		$n = 1/2$, $m = 2$	$[-\ln(1-\alpha)]^{\frac{1}{2}}$	$2(1-\alpha)[-\ln(1-\alpha)]^{\frac{1}{2}}$
11	Mample single line principle, level 1	Random nucleation and subsequent growth, A1, F1, S-shaped α-t curves, $n = 1$, $m = 1$	$-\ln(1-\alpha)$	$1-\alpha$
12	Power function principle	$n = 1/2$	$\alpha^{\frac{1}{2}}$	$2\alpha^{\frac{1}{2}}$
13		$n = 1$	α	1
14		$n = 2$	α^2	$\frac{1}{2}\alpha^{-1}$
15	Exponential principle	E_1, $n = 1$, accelerated form α-t curve	$\ln\alpha$	α
16	Shrinkage cylinder	$n = 1/2$	$1-(1-\alpha)^{\frac{1}{2}}$	$2(1-\alpha)^{\frac{1}{2}}$
17		$n = 2$	$2\left[1-(1-\alpha)^{\frac{1}{2}}\right]$	$(1-\alpha)^{\frac{1}{2}}$
18	Reaction order	$n = 2$	$1-(1-\alpha)^2$	$\frac{1}{2}(1-\alpha)^{-1}$
19		$n = 3$	$1-(1-\alpha)^3$	$\frac{1}{3}(1-\alpha)^{-2}$
20		$n = 4$	$1-(1-\alpha)^4$	$\frac{1}{4}(1-\alpha)^{-3}$
21	Level 2	Decelerated α-t curve	$(1-\alpha)^{-1}$	$(1-\alpha)^2$
22	Reaction order	Chemical reaction	$(1-\alpha)^{-1}-1$	$(1-\alpha)^2$
23	Level 2/3	Chemical reaction	$(1-\alpha)^{-\frac{1}{2}}$	$2(1-\alpha)^{\frac{3}{2}}$
24	Level 3	Three-level	$(1-\alpha)^{-2}$	$\frac{1}{2}(1-\alpha)^3$

The experimental data of the three stages are substituted into the left side of Equation (8) to calculate the experimental numerical curve, and the comparison between the experimental values and the theoretical values of the function is shown in Figure 11. Therefore, it can be concluded that the data points of Stage II are most likely to overlap with the theoretical value of Function 5 (Jander equation). Similarly, the most probable mechanism function of Stage III is most likely to be Function 7 (Anti-Jander equation), and the most probable mechanism function of Stage IV is most likely to be Function 15 (Exponential principle).

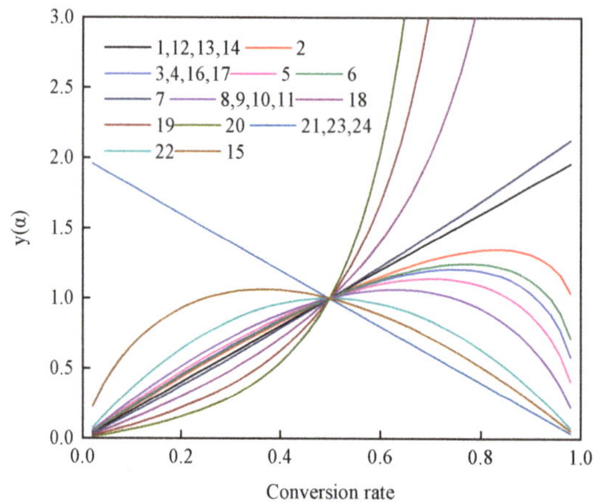

Figure 10. Málek theoretical values of different mechanism functions.

As can be seen from Figure 11, the experimental values presented in the three stages coincide with the inferred theoretical values of the functions. Moreover, the most probable mechanism functions at each stage are summarized in Table 8. The kinetic reaction mechanism of different stages of the coal oxidation process is different. The reaction mechanism in Stage II is a three-dimensional diffusion mode with $n = 1/2$, which transformed into a three-dimensional diffusion mode with $n = 2$ in Stage III, and further transformed into an accelerated form α-t curve with E_1 and $n = 1$ in Stage IV. It is mainly due to the different reaction modes and reaction rates between coal and oxygen at each stage. In the oxygen absorption and mass gain stage (Stage II), the chemical adsorption of coal on oxygen mainly occurs, which is greater than the gas release of the coal oxidation reaction from the perspective of its quality change, and the chemical adsorption of coal on oxygen is dominant, and the chemical reaction intensity is weak. The diffusion pattern at this stage indicates that the reaction rate is proportional to the square root of the concentration gradient, and the reaction rate is limited by the diffusion rate of oxygen into the coal interior. As the reaction progresses and enters Stage III, the three-dimensional diffusion mode changes, with the exponent n increasing to 2. This means that the reaction rate is proportional to the square of the concentration gradient, and this change reflects the changes in physical or chemical properties during coal oxidation, such as the transformation of coal structure or the accumulation of oxidation products, thereby affecting the diffusion mode of oxygen in coal. Furthermore, in Stage IV, the reaction mechanism shifts to an accelerated form of the α-t curve, characterized by activation energy E_1 and exponent $n = 1$. The α-t curve of this stage indicates that the reaction rate is proportional to time, indicating that the coal oxygen reaction has entered an acceleration period. This acceleration is due to the macromolecular structure in coal being thermally decomposed to produce a large number of active groups, and the coal oxidation reaction is strengthened and more intense, and the chemical reaction occupies a dominant position. The oxidation reaction generates more heat, leading to an increase in temperature and thus increasing the reaction rate. The existence of activation energy E_1 indicates that the reaction requires overcoming a certain energy barrier in order to proceed, and in Stage IV, this energy barrier is more easily overcome, resulting in an accelerated reaction rate.

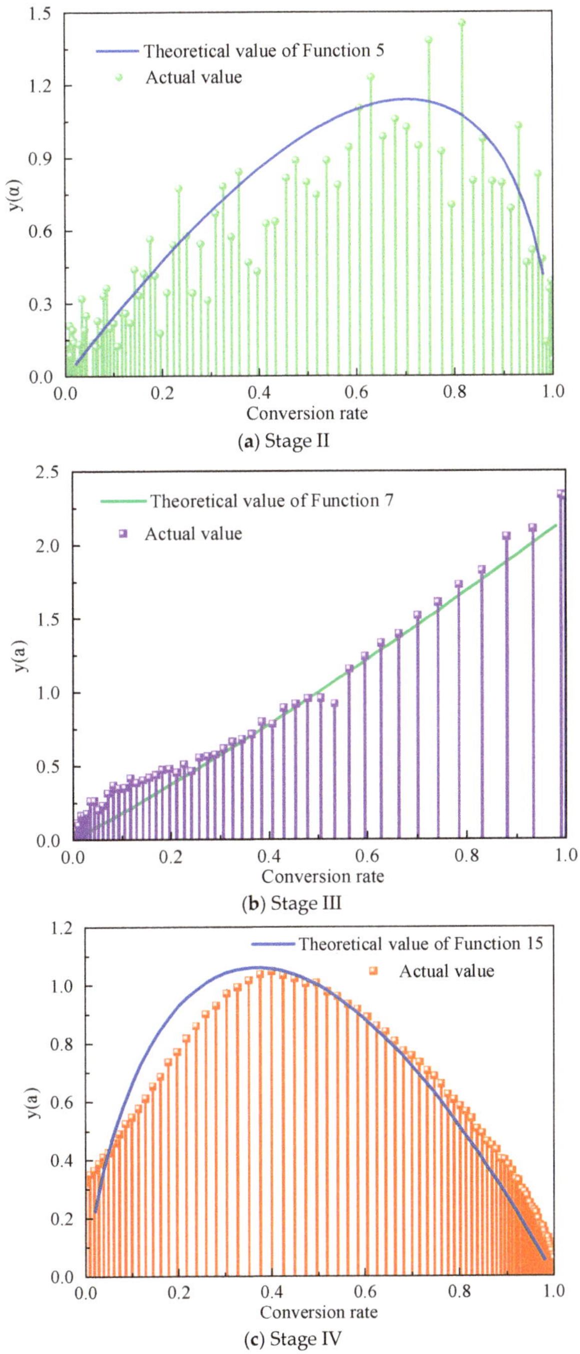

Figure 11. Málek method for determining mechanism functions at different stages.

Table 8. Most probable mechanism functions of Stages II, III, and IV.

Stage	Function Number	Function Name	Mechanism	$G(\alpha)$	$f(\alpha)$
Stage II	5	Jander equation	Three-dimensional diffusion, $n = 1/2$	$\left[1-(1-\alpha)^{\frac{1}{3}}\right]^{\frac{1}{2}}$	$6(1-\alpha)^{\frac{2}{3}}\left[1-(1-\alpha)^{\frac{1}{3}}\right]^{\frac{1}{2}}$
Stage III	7	Anti-Jander equation	Three-dimensional diffusion, $n = 2$	$\left[(1+\alpha)^{\frac{1}{3}}-1\right]^{2}$	$\frac{3}{2}(1+\alpha)^{\frac{2}{3}}\left[(1+\alpha)^{\frac{1}{3}}-1\right]^{-1}$
Stage IV	15	Exponential principle	E_1, $n = 1$, accelerated form α-t curve	$\ln\alpha$	α

The most probable mechanism function in Table 8 can be substituted into Equation (3) to calculate the pre-exponential factor ln A. To facilitate the calculation, 10 points with a conversion rate from 0.1 to 1 and an interval of 0.1 are selected for calculation, and the calculation results are shown in Table 9.

Table 9. Calculation of pre-exponential factor ln A.

Conversion Rate	Stage II	Stage III	Stage IV
0.1	11.585	11.128	18.843
0.2	12.244	13.272	18.235
0.3	12.100	15.722	18.396
0.4	12.497	17.525	15.748
0.5	12.665	18.787	13.190
0.6	13.122	19.373	10.854
0.7	13.593	20.342	8.915
0.8	13.548	20.864	7.006
0.9	14.266	21.191	4.625
1	18.610	21.651	0.829

The linear correlation of the compensation effect of ln $A - E_a$ is shown in Figure 12, and the arrows represent the calculation sequence of ln A.

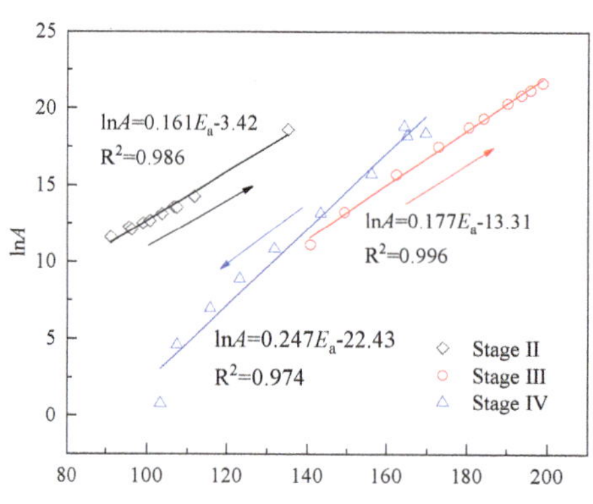

Figure 12. Linear relationship of ln $A - E_a$ compensation effect.

After linear fitting in Figure 12, the expressions of the ln $A - E_a$ compensation effect in the three stages were ln $A = 0.161E_a - 3.42$ (Stage II), ln $A = 0.177E_a - 13.31$ (Stage III), and ln $A = 0.247E_a - 22.43$ (Stage IV), respectively, and the correlation degree of fitting in the three stages was greater than 0.97. In addition, the pre-exponential factor ln A gradually increases with the increase in the apparent activation energy E_a. This is because the pre-exponential factor represents the effective collision frequency of molecules. The higher the pre-exponential factor, the higher the effective collision frequency of molecules, and the greater the possibility of a reaction, which leads to the increase in the energy required for the reaction and the increase in the apparent activation energy.

4. Conclusions

(1) Each characteristic temperature point increases with the increasing heating rate, and the mass loss rate also shows an increasing trend with the increase in heating rate; it reaches the maximum value when the heating rate is 20 °C/min, at −17.8%/min. In addition, in the whole coal–oxygen composite reaction, the heat flow rate also shows an increasing trend with the increase in heating rate; it reaches the maximum value of 29.63 mW/mg when the heating rate is 20 °C/min. Moreover, the inflection point temperature and peak temperature lag behind with the increase in heating rate.

(2) The KAS method was employed to calculate the apparent activation energy of three reaction stages, and there are certain differences in their changes at each stage. As the coal–oxygen reaction gradually accelerates, chemical reaction gradually becomes dominant in the coal oxidation process, and the apparent activation energy in Stages II and III gradually increases with the increase in conversion rate, while in stage IV, it shows a trend of first increasing and then decreasing, which is closely related to the fact that the decomposition of active substances in the coal–oxygen reaction begins to increase and eventually deplete, requiring less energy.

(3) The kinetic compensation effect of the three stages were determined by deriving the corresponding most probable mechanism function of each stage and identifying the correlation between the pre-exponential factors and the apparent activation energy. The three different oxidation stages have different mechanism functions, reflecting different oxidation characteristics, which further reflects the complexity and dynamism of the coal oxidation process. The pre-exponential factor ln A is positively correlated with the apparent activation energy, and the linear relationship is very obvious, which is consistent with the law of mass reduction and heat release increase in the process of the coal oxidation reaction.

Author Contributions: Conceptualization, R.B. and C.L.; methodology, R.B., C.L, C.W. and F.Z.; software, R.B.; validation, R.B., C.L., C.W. and F.Z.; formal analysis, R.B.; investigation, R.B. and C.L.; resources, R.B.; data curation, R.B.; writing—original draft preparation, R.B.; writing—review and editing, R.B. and C.W.; visualization, R.B. and C.L.; supervision, F.Z.; project administration, F.Z.; funding acquisition, F.Z. All authors have read and agreed to the published version of the manuscript.

Funding: This research was funded by the National Natural Science Foundation of China (grant numbers 52104225 and 52204229), and the Scientific and Technological Innovation Programs of Higher Education Institutions in Shanxi (grant number 2021L054).

Institutional Review Board Statement: Not applicable.

Informed Consent Statement: Not applicable.

Data Availability Statement: Data will be made available on request.

Conflicts of Interest: The authors declare no conflicts of interest.

References

1. Wang, H.; Fan, C.; Li, J.; Wu, Y.; Xing, S.; Wang, W. A Field Study of Coal Fire Areas Re-Burning Behavior Assessment and Related Carbon Emissions. *Fire* **2022**, *5*, 186. [CrossRef]
2. Stracher, G.B.; Taylor, T.P. Coal fires burning out of control around the world: Thermodynamic recipe for environmental catastrophe. *Int. J. Coal Geol.* **2004**, *59*, 7–17. [CrossRef]
3. Yang, Y.; Niu, H.; Wang, H.; Bu, Y.; Liu, Y. An experimental investigation of how coal's secondary oxidation process is impacted by primary oxidation temperature and ventilation. *Fuel* **2023**, *354*, 129198. [CrossRef]
4. Wang, F.; Ji, Z.; Wang, H.; Chen, Y.; Wang, T.; Tao, R.; Su, C.; Niu, G. Analysis of the Current Status and Hot Technologies of Coal Spontaneous Combustion Warning. *Processes* **2023**, *11*, 2480. [CrossRef]
5. Nyashina, G.S.; Kurgankina, M.A.; Strizhak, P.A. Environmental, economic and energetic benefits of using coal and oil processing waste instead of coal to produce the same amount of energy. *Energy Convers. Manag.* **2018**, *174*, 175–187. [CrossRef]

6. Du, X.; Sun, D.; Li, F.; Tong, J. A Study on the Propagation Trend of Underground Coal Fires Based on Night-Time Thermal Infrared Remote Sensing Technology. *Sustainability* **2022**, *14*, 14741. [CrossRef]
7. Hu, X.; Su, M.; Dong, H.; Chen, X.; Wang, F.; Luo, C.; Yu, S. Study on Coal Spontaneous Combustion Dangerous Zone Under Different Air Leakage in the Shallow Buried Coal Seam Fissure. *Combust. Sci. Technol.* **2023**, *196*, 4811–4835. [CrossRef]
8. Wang, H.; Dlugogorski, B.Z.; Kennedy, E.M. Theoretical analysis of reaction regimes in low-temperature oxidation of coal. *Fuel* **1999**, *78*, 1073–1081. [CrossRef]
9. Yang, Y.; Fei, J.; Luo, Z.; Wen, H.; Wang, H. Experimental study on characteristic temperature of coal spontaneous combustion. *J. Therm. Anal. Calorim.* **2023**, *148*, 10011–10019. [CrossRef]
10. Yang, Y.; Li, Z.; Tang, Y.; Liu, Z.; Ji, H. Fine coal covering for preventing spontaneous combustion of coal pile. *Nat. Hazards* **2014**, *74*, 603–622. [CrossRef]
11. Qi, G.; Wang, D.; Zheng, K.; Xu, J.; Qi, X.; Zhong, X. Kinetics characteristics of coal low-temperature oxidation in oxygen-depleted air. *J. Loss Prevent. Proc.* **2015**, *35*, 224–231. [CrossRef]
12. Varol, M.; Atimtay, A.T.; Bay, B.; Olgun, H. Investigation of co-combustion characteristics of low quality lignite coals and biomass with thermogravimetric analysis. *Thermochim. Acta* **2010**, *510*, 195–201. [CrossRef]
13. Deng, J.; Ren, L.; Ma, L.; Lei, C.; Wei, G.; Wang, W. Effect of oxygen concentration on low-temperature exothermic oxidation of pulverized coal. *Thermochim. Acta* **2018**, *667*, 102–110. [CrossRef]
14. Wang, Q.; Wang, G.; Zhang, J.; Lee, J.; Wang, H.; Wang, C. Combustion behaviors and kinetics analysis of coal, biomass and plastic. *Thermochim. Acta* **2018**, *669*, 140–148. [CrossRef]
15. Ren, S.; Zhang, J. Thermogravimetric analysis of anthracite and waste plastics by iso-conversional method. *Thermochim. Acta* **2013**, *561*, 36–40. [CrossRef]
16. Qi, X.; Li, Q.; Zhang, H.; Xin, H. Thermodynamic characteristics of coal reaction under low oxygen concentration conditions. *J. Energy Inst.* **2017**, *90*, 544–555. [CrossRef]
17. Xu, Y.; Wang, D.; Wang, L.; Zhong, X.; Chu, T. Experimental research on inhibition performances of the sand-suspended colloid for coal spontaneous combustion. *Saf. Sci.* **2012**, *50*, 822–827. [CrossRef]
18. Zhao, J.; Deng, J.; Chen, L.; Wang, T.; Song, J.; Zhang, Y.; Shu, C.; Zeng, Q. Correlation analysis of the functional groups and exothermic characteristics of bituminous coal molecules during high-temperature oxidation. *Energy* **2019**, *181*, 136–147. [CrossRef]
19. Sonibare, O.O.; Haeger, T.; Foley, S.F. Structural characterization of Nigerian coals by X-ray diffraction, Raman and FTIR spectroscopy. *Energy* **2010**, *35*, 5347–5353. [CrossRef]
20. Okolo, G.N.; Neomagus, H.W.J.P.; Everson, R.C.; Roberts, M.J.; Bunt, J.R.; Sakurovs, R.; Mathews, J.P. Chemical–structural properties of South African bituminous coals: Insights from wide angle XRD–carbon fraction analysis, ATR–FTIR, solid state 13 C NMR, and HRTEM techniques. *Fuel* **2015**, *158*, 779–792. [CrossRef]
21. Kotyczka-Morańska, M.; Tomaszewicz, M. Comparison of the first stage of the thermal decomposition of Polish coals by diffuse reflectance infrared spectroscopy. *J. Energy Inst.* **2018**, *91*, 240–250. [CrossRef]
22. Deng, J.; Wang, K.; Zhang, Y.; Yang, H. Study on the kinetics and reactivity at the ignition temperature of Jurassic coal in North Shaanxi. *J. Therm. Anal. Calorim.* **2014**, *118*, 417–423. [CrossRef]
23. Cao, W.; Cao, W.; Peng, Y.; Qiu, S.; Miao, N.; Pan, F. Experimental study on the combustion sensitivity parameters and pre-combusted changes in functional groups of lignite coal dust. *Powder Technol.* **2015**, *283*, 512–518. [CrossRef]
24. Chen, X.; Ma, T.; Zhai, X.; Lei, C. Thermogravimetric and infrared spectroscopic study of bituminous coal spontaneous combustion to analyze combustion reaction kinetics. *Thermochim. Acta* **2019**, *676*, 84–93. [CrossRef]
25. Zhou, C.; Zhang, Y.; Wang, J.; Xue, S.; Wu, J.; Chang, L. Study on the relationship between microscopic functional group and coal mass changes during low-temperature oxidation of coal. *Int. J. Coal Geol.* **2017**, *171*, 212–222. [CrossRef]
26. Ozbas, K.E. Effect of particle size on pyrolysis characteristics of Elbistan lignite. *J. Therm. Anal. Calorim.* **2008**, *93*, 641–649. [CrossRef]
27. Iliyas, A.; Hawboldt, K.; Khan, F. Thermal stability investigation of sulfide minerals in DSC. *J. Hazard. Mater.* **2010**, *178*, 814–822. [CrossRef]
28. Kaljuvee, T.; Keelman, M.; Trikkel, A.; Petkova, V. TG-FTIR/MS analysis of thermal and kinetic characteristics of some coal samples. *J. Therm. Anal. Calorim.* **2013**, *113*, 1063–1071. [CrossRef]
29. Rotaru, A. Thermal analysis and kinetic study of Petroşani bituminous coal from Romania in comparison with a sample of Ural bituminous coal. *J. Therm. Anal. Calorim.* **2012**, *110*, 1283–1291. [CrossRef]
30. Niu, H.; Sun, Q.; Li, S.; Sun, S.; Bu, Y.; Yang, Y.; Mao, Z.; Tao, M. Study on the thermal release characteristics and the correlation transformation mechanism of microscopic active groups of oxidized coal combustion in a deep mined-out area. *Sci. Total Environ.* **2023**, *890*, 164354. [CrossRef]
31. Zhang, Y.; Zhai, F.; Shu, P.; Wang, K.; Deng, J.; Duan, Z. Study on evolution characteristics of thermal contribution functional groups in low temperature oxidation process of bituminous coal. *Fuel* **2023**, *341*, 127683. [CrossRef]
32. Li, J.; Lu, W.; Li, J.; Yang, Y.; Li, Z. Mutual conversion of active sites and oxygen-containing functional groups during low-temperature oxidation of coal. *Energy* **2023**, *272*, 127151. [CrossRef]
33. Lu, W.; Li, J.; Li, J.; He, Q.; Hao, W.; Li, Z. Oxidative kinetic characteristics of dried soaked coal and its related spontaneous combustion mechanism. *Fuel* **2021**, *305*, 121626. [CrossRef]

34. Wang, W.; Zhu, H.; Huang, H.; Zhao, H.; Pan, R. Study on the thermal kinetics and microscopic characteristics of oxidized coal. *Environ. Sci. Pollut. Res.* **2023**, *30*, 85953–85967. [CrossRef] [PubMed]
35. Zhai, X.; Ge, H.; Shu, C.; Obracaj, D.; Wang, K.; Laiwang, B. Effect of the heating rate on the spontaneous combustion characteristics and exothermic phenomena of weakly caking coal at the low-temperature oxidation stage. *Fuel* **2020**, *268*, 117327. [CrossRef]
36. Li, J.; Lu, W.; Li, J.; Yang, Y.; Li, Z. Towards understanding of internal mechanism of coal reactivity enhancement after thermal decomposition at low temperature. *Fuel* **2023**, *337*, 127118. [CrossRef]
37. Pan, R.; Li, C.; Chao, J.; Hu, D.; Jia, H. Thermal properties and microstructural evolution of coal spontaneous combustion. *Energy* **2023**, *262*, 125400. [CrossRef]
38. Li, Q.; Xiao, Y.; Wang, C.; Deng, J.; Shu, C. Thermokinetic characteristics of coal spontaneous combustion based on thermogravimetric analysis. *Fuel* **2019**, *250*, 235–244. [CrossRef]
39. Starink, M.J. Activation energy determination for linear heating experiments: Deviations due to neglecting the low temperature end of the temperature integral. *J. Mater. Sci.* **2007**, *42*, 483–489. [CrossRef]
40. Deng, J.; Li, Q.; Xiao, Y.; Wen, H. The effect of oxygen concentration on the non-isothermal combustion of coal. *Thermochim. Acta* **2017**, *653*, 106–115. [CrossRef]
41. Lu, W.; Gao, A.; Liang, Y.; He, Z.; Li, J.; Sun, Y.; Song, S.; Meng, S. Stable and highly efficient HMDS terminated m-Cresol inhibitor for inhibiting coal spontaneous combustion. *Energy* **2023**, *282*, 128915. [CrossRef]
42. Li, J.; Lu, W.; Li, J.; Zhang, Q.; Zhuo, H. Thermodynamics of oxygen-containing intermediates and their role in coal spontaneous combustion. *Energy* **2022**, *260*, 124989. [CrossRef]
43. Miao, G.; Li, Z.; Yang, J.; Yang, Y.; Liu, H. Microstructure evolution and higher-molecular-weight gas emission during the low temperature oxidation of coal. *Energy* **2023**, *282*, 128424. [CrossRef]

fire

MDPI

Article

Study on the Influence Mechanism of the Key Active Structure of Coal Molecules on Spontaneous Combustion Characteristics Based on Extraction Technology

Jun Guo [1,2], Yunfei Wu [1,*], Yin Liu [1,3], Guobin Cai [1], Dailin Li [1] and Yan Jin [1,4]

1 College of Safety Science and Engineering, Xi'an University of Science and Technology, Xi'an 710054, China; guojun@xust.edu.cn (J.G.); liuyin@xust.edu.cn (Y.L.); 19220214086@stu.xust.edu.cn (G.C.); 22220226172@stu.xust.edu.cn (D.L.); 19220214099@stu.xust.edu.cn (Y.J.)
2 Key Laboratory of Western Mine and Hazard Prevention, Ministry of Education of China, Xi'an 710054, China
3 College of Energy Engineering, Xi'an University of Science and Technology, No. 58, Yanta Mid. Rd., Xi'an 710054, China
4 Taizhou Petroleum Branch, Sinopec Sales Co., Ltd., Taizhou 318000, China
* Correspondence: 23220226169@stu.xust.edu.cn

Abstract: The molecular structure of coal is complex, and the existing research methods are limited, so it is difficult to clarify its influence mechanism on the spontaneous-combustion characteristics of coal. In this paper, the previous extraction, FTIR, TPR, TG-DSC and other experimental results are combined to analyze the extraction weakening effect and the correlation analysis of the spontaneous-combustion characteristic parameters of raffinate coal. The results show that extraction can destroy the connection bond of coal molecules, change the content of dominant active groups in the coal spontaneous-combustion reaction, increase the lower limit of the key temperature nodes of coal spontaneous-combustion or extend the temperature range, resulting in an increase in the ignition-point temperature of coal and a decrease in coal quality. This paper will provide a theoretical basis for the study of the microscopic mechanism of coal spontaneous-combustion and then provide new ideas for the development of an active prevention and control technology for coal spontaneous-combustion.

Keywords: coal spontaneous-combustion; solvent extraction; key active structures; correlation analysis; weakening mechanism

1. Introduction

As the most abundant fossil fuel in the world [1,2], coal resources are accompanied by many safety problems in the process of mining. The harm of coal spontaneous-combustion exists in the whole process of coal mining, storage and transportation [3,4]. The accidents caused by coal spontaneous-combustion account for more than 90% of mine fire accidents [5,6], which cause a lot of waste of coal resources and hinders the safe, green and sustainable development of China's coal industry [7,8].

Coal is a complex mixture of organic matter and inorganic matter, and its physical and chemical properties are affected by its molecular structure and active groups [9–12]. Qiu et al. [13] found that the content of oxygen-containing functional groups in the material is positively correlated with the specific surface area. When the content of oxygen-containing functional groups is too high, the specific surface area will shrink due to the blockage of micropores and the collapse of pore walls. Dhakate et al. [14] found that the proportion of surface reactive functional group content will affect the strength of carbon fiber in the process of studying carbon fiber materials. In the process of evaluating Russian needle coke carbon materials, Rudko et al. [15] found that materials with different hydrocarbon compositions and sulfur content would lead to changes in the structure and properties of the products. Coal spontaneous combustion is the result of heat storage and combustion after the coal–oxygen composite reaction between the coal molecular active structure and oxygen [16–18]. Previous studies have confirmed that the microstructures of the aliphatic hydrocarbon structure, aromatic hydrocarbon structure and oxygen-containing functional groups in coal molecules have different influence mechanisms and influence degrees in coal spontaneous combustion [19]. Therefore, controlling or eliminating the key active structure of coal molecules is an effective means to prevent and control coal spontaneous combustion. Solvent extraction technology can effectively strip the specific active structure of coal molecules and achieve the purpose of inhibiting coal spontaneous combustion [20]. Some scholars have successfully stripped the microscopic active structure of coal molecules by solvent extraction technology. Zhao et al. [21] found that the content of oxygen-containing functional groups such as phenols in coal samples dissolved by non-volatile ionic liquids was significantly reduced. Jin et al. [22] used diphenylamine to treat coal samples and found that the N-H in coal molecules was destroyed. Yao et al. [23] successfully stripped the aliphatic hydrocarbon structure in coal molecules using tetrahydrofuran combined with ultrasonic-assisted extraction technology. Barman et al. [24] reviewed the application of ultrasonic-assisted extraction technology in coal extraction and found that the content of functional groups on the surface of coal changed after a solvent combined with ultrasonic-assisted extraction.

On this basis, some scholars have studied the weakening mechanism of coal spontaneous combustion with changes in microstructure. Taraba et al. [25] used urea and other reagents to treat coal samples, and combined with a pulse flow calorimetry test, it was found that urea reduced the oxidation heat of coal at the chemical level, thus weakening the spontaneous-combustion activity of coal. Wang et al. [26] significantly delayed the ignition temperature of a test coal sample by adding dimethyl methylphosphonate (DMMP) material and determined by kinetic analysis that DMMP can also physically weaken the coal–oxygen composite reaction activity intensity in the low-temperature stage by blocking oxygen at the physical level. Onifade et al. [27] used gypsum powder to make antioxidant materials to treat the sponge coal of a South African mine. Through testing, gypsum materials can effectively reduce the oxygen consumption of coal and greatly reduce the spontaneous-combustion of coal. Deng et al. [28] found that when testing [BMIM][BF_4] materials the temperature threshold of the spontaneous-combustion reaction process interval of coal samples pretreated with ionic liquids was expanded, that is, the spontaneous-combustion of coal was weakened by delaying the characteristic temperature node of coal spontaneous combustion reaction acceleration. Zhang et al. [29] found that by reducing the content of the -OH structure, the quality and heat release of coal entering the combustion stage can be significantly reduced, and then the spontaneous-combustion of coal can be weakened. Slovák et al. [30] tested bituminous coal by using $CaCl_2$ and urea as a composite inhibitor and found that $CaCl_2$ and urea in the range of 100~250 °C will lead to a decrease in the heat released by coal oxidation. At the same time, $CaCl_2$ can inhibit the spontaneous-combustion of coal by increasing the activation energy in the process of coal spontaneous combustion.

The quantitative state (content and state) of the active structure of coal molecules is a key factor affecting the spontaneous-combustion characteristics of coal [31]. The above scholars have carried out research on the weakening mechanism of solvent extraction residue on the reactivity of coal spontaneous combustion and have achieved relevant results, but the influence of the key active structures on the characteristics of coal spontaneous combustion needs further study. Therefore, based on extraction technology, combined with an infrared test, temperature-programmed experiment and other means, this paper will test and analyze extracted coal and use the gray correlation method to analyze the correlation between the active group structure and the characteristic parameters of coal spontaneous combustion, to determine the weakening mechanism of the key active structure of coal molecules on the spontaneous-combustion characteristics. The research results provide research methods and methods for clarifying the microscopic reaction mechanism of coal spontaneous combustion and provide a theoretical basis for promoting the development of the active prevention and control of coal spontaneous combustion disasters in mines.

2. Experiment and Theoretical Analysis

2.1. Experiments of Extraction

In the experiment, 10 groups of extractants with 20% solute density were selected, and each group was configured with 5 L of solvent [32]. The extractants were as follows: cyclohexane + sodium dioctyl succinate sulfonate + anhydrous ethanol (CYC + AOT + AE), sodium dioctyl succinate sulfonate (AOT), cyclohexane (CYC), anhydrous ethanol (AE), cellulase (EG), tea polyphenols (TP), cellulase + tea polyphenols (EG + TP), tetrahydrofuran (THF), methanol (MT), n-hexane (CYH) and the raw coal control group (YM). During the extraction process, 200 g of samples with five particle sizes of 0~0.9, 0.9~3, 3~5, 5~7 and 7~10 mm was mixed into 1 kg of experimental coal samples, and the extraction solvent was added and fully mixed, followed by static extraction. After the extraction, the extracted coal samples were collected for drying treatment, and the coal samples were sealed and stored for later use.

2.2. Extraction of Residual Coal Group Content Test

A Vertex 70 V Fourier transform infrared spectrometer was used to test and determine the extraction effect of the group structure of the residual coal sample in each reagent group [32]. The experimental sample and dried potassium bromide powder (KBr) were mixed at a mass ratio of 1:150. After full grinding, 200 mg of the ground coal sample and KBr powder mixture was filled in a tableting metal mold. The mixed sample was placed on a tableting machine. Under a pressure of 15 Mpa, the mixed sample was made into an observation sheet and placed in a Fourier transform infrared spectrometer for experimental testing. The setting parameters were resolution 4 cm^{-1}, band 400^{-4}~4000 cm^{-1}, scan 32 times and export test analysis data preservation.

2.3. Determination of the Macroscopic Characteristic Parameters of Coal Oxidation

2.3.1. Temperature-Programmed Experiment Test

The experimental system included five parts, an adiabatic oxidation device, gas supply system, temperature-tracking detection and control system, oxidation gas product analysis system and mass change detection system, and was connected with a ZDC 7 mine fire multi-parameter intelligent monitoring device [33]. During the test, 800 g of dry raffinate coal sample was placed in the coal sample tank for sealing, and the air pump was adjusted for preventilation. The initial temperature of the box was set at 40 °C, and the heating rate was 0.8 °C/min to start the test. During the heating process, the ZDC 7 monitoring device was used to monitor the central temperature of the coal sample and the outlet oxygen volume fraction at the key node temperature in real time, and the temperature of the box was collected at the same time. The intersection point of the temperature– time curve of the coal center and the temperature–time curve of the box is the temperature of the intersection point. The data index can indirectly characterize the spontaneous-combustion strength of the test coal sample.

2.3.2. Isothermal Temperature Difference Leading Test

The main structure of the experiment included five parts: an adiabatic oxidation device, gas flow detection device, flow control system, temperature detection system and parameter control system [33]. In the experiment, 80 g of dry extracted coal sample was placed in the coal sample tank for heating up. The leading temperature was set to 10 °C, and the heating rate was 0.8 °C/min. The gas generated in the coal sample tank was introduced into the gas chromatography analyzer every 10 °C to detect the gas composition. The gas type and volume fraction data at different temperatures were recorded until the test was completed at 180 °C. Through experiments, the formation law of gaseous products in the oxidation process of raffinate coal samples can be obtained, and then the spontaneous-combustion reactivity intensity of each group of raffinate coal samples can be reflected.

2.3.3. Thermal Analysis Combined Test (TG-DSC)

An STA 8000-Spectrum thermal analyzer was used in the experiment [34]. During the experiment, 10 mg of dry extracted coal samples was placed in the sample room. The preloading temperature was 30 °C, the termination temperature was 800 °C, the heating rate was 10 °C/min, the preloaded gas flow rate was 100 mL/min, nitrogen was used as the protective gas and 21% oxygen was used as the purging gas during the test. At the same time, the temperature of each instrument interface and transmission network pipeline was maintained at 200 °C during the heating process of the control program. The changes in the mass and airflow heat data of the test samples during the heating process were recorded. Through a data fitting processing analysis, the differences in the thermal stability and thermal effect characterization parameters in the process of the oxidation and spontaneous-combustion of each group of raffinate coal were determined.

2.4. The Principle of Gray Correlation Analysis

Gray correlation analysis is an analysis method based on the existing data to calculate the correlation degree between the reference sequence and the comparison sequence [29]. The correlation degree between the research object and the characteristic parameters can be determined by gray correlation analysis. The larger the correlation degree is, the greater the correlation between the research object and the characteristic parameters. The specific calculation steps are as follows:

1. Determine the reference sequence (1) and the comparison sequence (2):

$$Y = Y(k)|k = 1,2\ldots n, \tag{1}$$

$$X_i = X_i(k)|k = 1,2\ldots n, i = 1,2\ldots m, \tag{2}$$

2. The initial value method is used to make the data dimensionless.
3. Calculate the correlation coefficient; the calculation formula is as follows (3):

$$\xi_i(k) = \frac{\min_i\min_k|y(k) - x_i(k)| + \rho\max_i\max_k|y(k) - x_i(k)|}{|y(k) - x_i(k)| + \rho\max_i\max_k|y(k) - x_i(k)|}, \tag{3}$$

4. Calculate the correlation degree, that is, calculate the average value of the obtained correlation coefficient to obtain the correlation degree, as shown in Equation (4):

$$r_i = \frac{1}{n}\sum_{k=1}^{n} \xi_i(k), k = 1,2,\ldots n, \tag{4}$$

In the formula, Y is the reference sequence, X_i is the comparison sequence, ξ_i is the correlation coefficient, ρ is the resolution coefficient, usually taken as $\rho = 0.5$, and r_i is the correlation degree.

2.5. Analysis Process of the Influence Mechanism of the Active Structure on Coal Spontaneous-Combustion Based on Macro–Micro Correlation

In this paper, the extraction experiment of coal samples was carried out based on extraction technology. In order to determine the influence mechanism of the active group structure on the reactivity of coal spontaneous combustion, a Fourier transform infrared spectroscopy test, thermal analysis combined test, temperature-programming experiment and isothermal temperature difference leading experiment test were carried out. By analyzing the differences in the characteristic data of the test and analysis results of each group of raffinate coal samples, the influence of different group structures on the actual coal spontaneous-combustion process is determined, and then the influence mechanism of the key active structure of coal molecules on the spontaneous-combustion characteristics is theoretically deduced. The specific technical roadmap is shown in Figure 1.

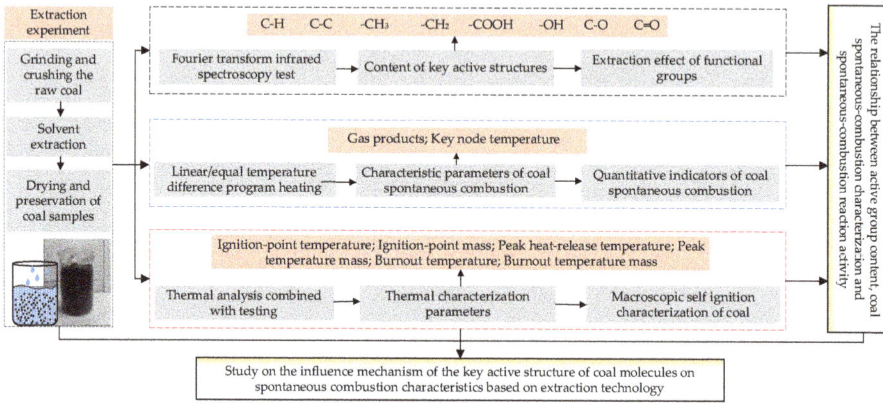

Figure 1. Technical roadmap.

3. Results and Analysis

In the previous study, the team carried out extraction experiments on Caojiatan coal samples and combined them with Fourier transform infrared spectroscopy to optimize the extraction solvent. Then, a temperature-programmed experiment was carried out on the extracted coal. By calculating the quantitative index of the spontaneous-combustion of extracted coal, the spontaneous-combustion weakening effect of coal samples treated with different extractants was obtained [32]. In addition, combined with the formation law of gas products and the change law of thermal physical parameters during the heating process of extracted coal, it was found that the temperature node of the spontaneous-combustion reaction interval of the coal sample treated by the extractant shifted, and the ignition-point temperature of the coal was delayed, resulting in a weakening of the spontaneous-combustion reaction activity of the coal [33,34].

However, the influence mechanism of the active structure on coal spontaneous combustion characteristics is not clear, and the quantitative characteristics between the active structure and coal spontaneous combustion characteristic parameters, gas products and thermophysical parameters are difficult to clarify. Therefore, this paper combines the experimental results and data of the previous team research to deeply analyze the problem.

3.1. Change in Active Group Content of Extracted Residual Coal

Based on the test and analysis process and calculation results of the previous raffinate coal experiment [32,33], the micro-group structure types of each group of raffinate coal samples were fitted according to the aromatic hydrocarbon structure, aliphatic hydrocarbon structure and oxygen-containing functional group structure. The absorption peaks were summarized, and the extraction effect of each extractant was analyzed. The infrared fitting peak area of the active structure of each raffinate coal group is shown in Table 1 and Figure 2 below.

Oxygen-containing functional groups are the most abundant in coal molecular groups and play an important role in the process of coal spontaneous combustion. The oxygen-containing functional groups represented by-OH play a leading role in the low-temperature oxidation stage of coal spontaneous combustion, and their content is positively correlated with the strength of coal spontaneous combustion. It can be seen from Figure 2 that regarding the correlation fitting peak area of each group of raffinate coal samples, the larger the fitting peak area, the worse the extraction effect. Therefore, the extraction effect of the oxygen-containing group structure from strong to weak is as follows: MT > EG > THF > AE > TP > CYC > EG + TP > CYC + AOT + AE > AOT > CYH > YM. Among them, the reagents with the best extraction effect of the oxygen-containing group structure were the

THF group, EG group and MT group, and the extraction rates were 13.9%, 14.7% and 19.1%, respectively.

Table 1. Infrared peak fitting results (dimensionless).

Group	Active Structure Category	YM	THF	MT	CYC	CYH	C + A + A	AE	AOT	EG	TP	EG + TP
Aromatic hydrocarbon	C-H	108	77	91	56	42	46	54	51	73	45	40
	C=C	185	199	157	190	204	201	166	211	173	179	172
Aliphatic hydrocarbon	$-CH_3$	52	43	15	49	29	40	54	130	58	65	45
	$-CH_2$	117	73	101	44	28	63	96	145	103	78	111
Oxygen-containing functional groups	-COOH	187	161	154	176	176	181	184	174	185	183	168
	-OH	435	388	310	389	463	428	357	480	357	367	425
	C-O	177	137	147	144	134	150	145	153	146	154	160
	C=O	66	59	89	75	93	75	63	56	50	50	50

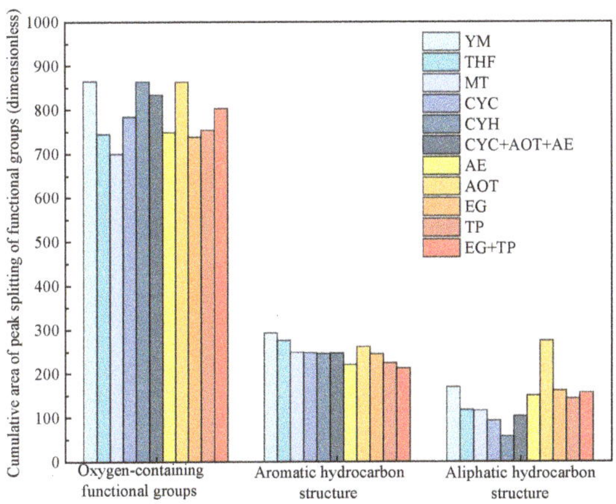

Figure 2. Comparison of the difference in group content under different extractants.

As the skeleton of coal molecules, the aromatic hydrocarbon structure is composed of a benzene ring, aromatic ring and so on. This kind of structure is very stable in coal molecules, and it is usually difficult to react with oxygen. It can be seen from Figure 2 that the order of the aromatic structure extraction effect from strong to weak is as follows: EG + TP > AE > TP > EG > CYH > CYC + AOT + AE > CYC > MT > AOT > THF > YM. Among them, the extraction residual coal group with the best extraction effect was the TP group, AE group and EG + TP compound group, and the extraction rate reached 23.5%, 24.9% and 27.6%, respectively.

The aliphatic hydrocarbon structure generally exists in coal molecules as aromatic ring side chains, and it is easy to react with oxygen during the coal spontaneous-combustion reaction. It can be seen from Figure 2 that the extraction effect of the aliphatic hydrocarbon structure from strong to weak is as follows: CYH > CYC > CYC + AOT + AE > MT > THF > TP > AE > EG + TP > EG > YM > AOT. Among them, the extraction residual coal group with the best extraction effect was the CYC + AOT + AE compound group, CYC group and CYH group, and the extraction rate reached 39.1%, 45.0% and 66.3%, respectively. The absorption peak area of aliphatic hydrocarbons in the AOT group increased. It is speculated that the AOT material contains a large amount of methyl structure, and the material is retained inside the coal pores during the extraction process, resulting in an increase in the area of infrared absorption peaks here. In addition, due to the physical and chemical properties of the AOT material itself, it is easy to form a reverse micelle layer wrapped on the surface of the coal body during the extraction process, which blocks the dissolution of aliphatic hydrocarbon structures such as methyl and methylene, resulting in the partial stripping of aliphatic hydrocarbon structures. It is also retained in the coal pores, resulting in an increase in the absorption peak area of the aliphatic hydrocarbon structure of the AOT group.

3.2. *Effect of Active Group Content on Coal Spontaneous Combustion*

The results of previous studies have found that [33] the spontaneous-combustion characteristics of extracted coal samples have changed to varying degrees. The determination method of coal oxidation kinetics [35] was used to calculate the spontaneous-combustion characteristics of each group of extracted coal samples based on the measured center temperature of each group of extracted coal samples and the volume fraction of O_2 at key points. It was found that except for the enhancement of the spontaneous-combustion of coal in the AE group, the spontaneous-combustion reactivity of coal in other groups was weakened to varying degrees. Another characterization of the change in coal spontaneous-combustion is the change in the coal spontaneous-combustion process interval. Figure 3 shows the change in the spontaneous-combustion process interval of each extraction residual coal group. It can be seen from the figure that the characteristic temperature interval of the coal spontaneous-combustion process of each extraction residual coal group has been extended. At the same time, there is no phenomenon of a compound self-heating stage in the extraction residual coal of the AE group.

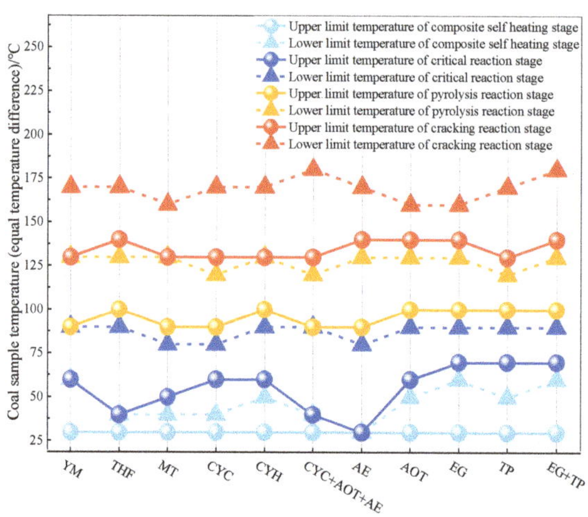

Figure 3. Coal spontaneous-combustion process interval stage change situation [33].

3.3. *Effect of Active Group Content on Gas Products*

According to the isothermal temperature difference leading test, the volume fraction of gas products in each group was drawn in a scatter plot, and the gas product data of different extraction residual coal groups were fitted by Logistic using Origin software to obtain the gas product fitting curve (as shown in Figure 4). The absolute production of each gas product was obtained by integrating the fitting curve, and then the correlation between the content of active groups and the concentration of gas products was analyzed to determine the effect of each active group on the concentration of gas products.

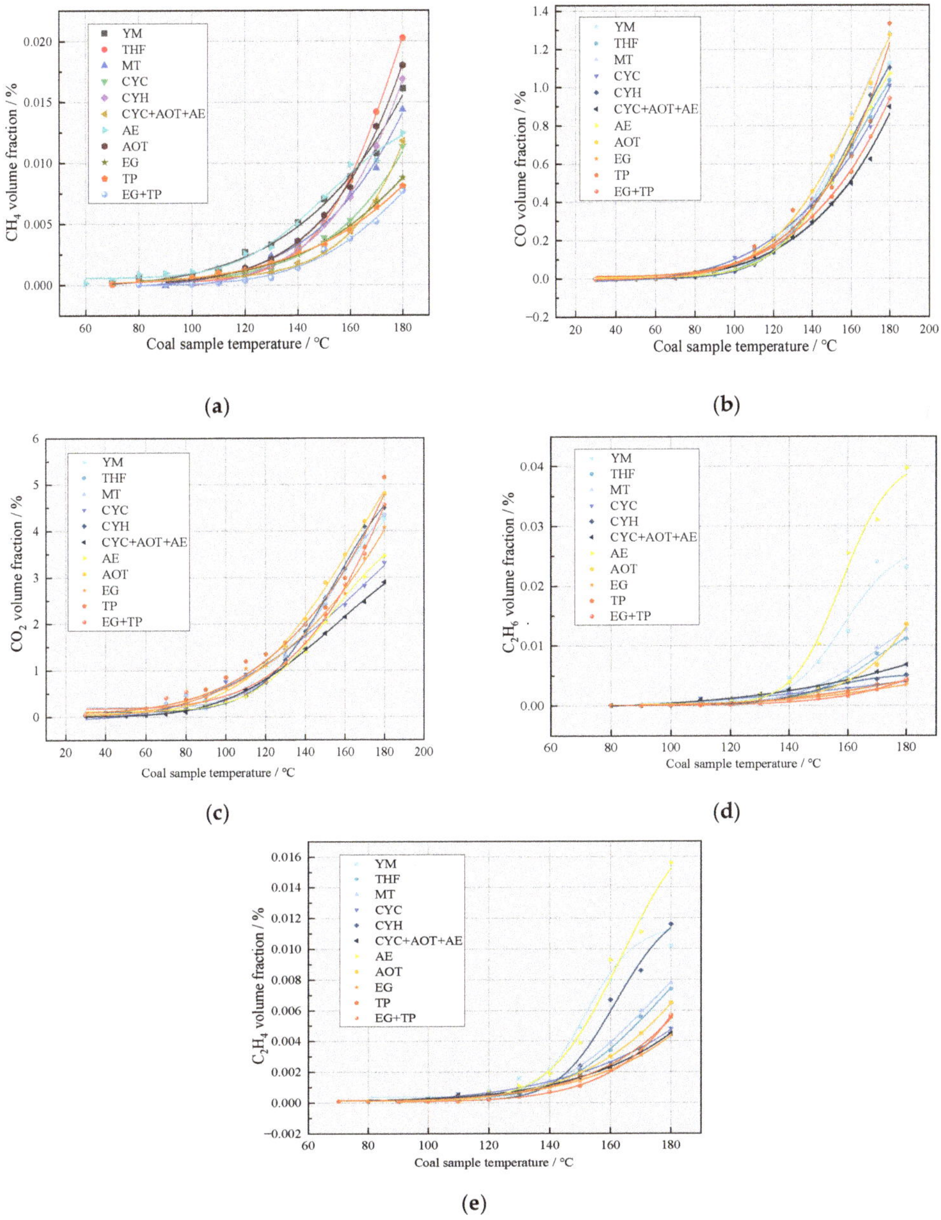

Figure 4. Fitting curves of the gas products of different raffinate coal groups [33]. (**a**) Fitting curve of CH_4 gas product; (**b**) Fitting curve of CO gas product; (**c**) Fitting curve of CO_2 gas product; (**d**) Fitting curve of C_2H_6 gas product; (**e**) Fitting curve of C_2H_4 gas product.

As shown in Figure 4, the fitting curves of five gas products of CH_4, CO, CO_2, C_2H_6 and C_2H_4 were obtained, respectively. Origin software was used to integrate the fitting curves, and the absolute production results of each gas product are shown in Table 2.

Table 2. The amount of gas products produced by different extraction residual coal groups.

Group	CH_4	CO	CO_2	C_2H_6	C_2H_4
YM	0.48549	38.33111	177.61669	0.62251	0.34461
THF	0.43321	35.06035	178.00860	0.2425	0.17216
MT	0.37073	40.99266	178.00860	0.28744	0.18847
CYC	0.27937	37.15417	158.69036	0.16944	0.14322
CYH	0.37136	36.34579	174.96191	0.1641	0.25814
CYC + AOT + AE	0.24421	28.80616	127.02185	0.23388	0.12843
AE	0.48858	36.17914	145.47469	0.9406	0.36273
AOT	0.43511	43.82697	201.89459	0.23651	0.15241
EG	0.26614	32.19166	173.87919	0.10831	0.11619
TP	0.26061	39.30838	200.26572	0.12891	0.13512
EG + TP	0.18521	32.19166	169.43187	0.09223	0.10690

Gray Correlation Analysis of Active Groups and Gas Products

The gas product can be used as a sign to divide the process of coal spontaneous combustion, and the change in its content can reflect the activity intensity of coal spontaneous combustion to a certain extent. To explore the relationship between the structure of active groups and the characteristics of coal spontaneous combustion, the content of each active group and the absolute production of different gas products were obtained by infrared peak fitting. The relationship between gas products and the structure of active groups was analyzed by gray correlation analysis, and then the influence of active groups on the characteristics of coal spontaneous combustion was determined [36]. The greater the correlation degree calculated, the greater the influence of the active group structure on the formation of gas products in the process of coal spontaneous combustion. In the process of this analysis, the amount of gas product generation is the reference sequence, and the content of different active groups is the comparison sequence. The results are shown in Figure 5.

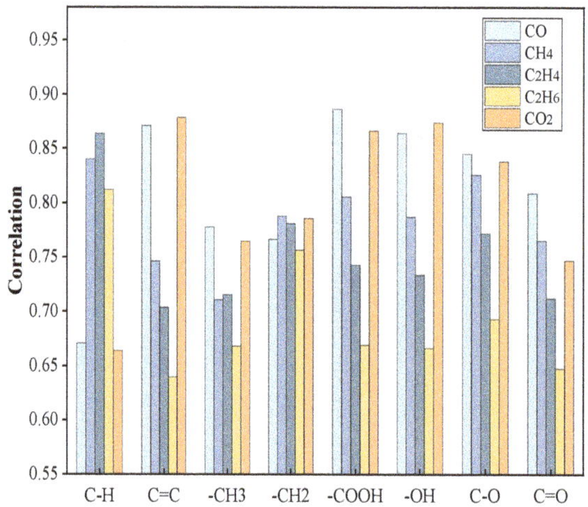

Figure 5. Gray correlation degree between active group structure and gas product.

(1) The correlation between CO, CO_2 and different active group structures

It can be seen from Figure 5 that the correlation between -COOH, C=C, -OH and the production of CO and CO_2 is high, indicating that these three types of active group structures have a great influence on the production of CO and CO_2. The correlation degree with CO is 0.8863, 0.8710 and 0.8641, and the correlation degree with CO_2 is 0.8661, 0.8783 and 0.8737, respectively. In the early stage of the coal spontaneous-combustion reaction,

-OH as the dominant group captures H and oxidizes to form an aldehyde radical, which further reacts with oxygen to form CO and CO_2, resulting in an increase in the production of CO and CO_2 gas. In the middle stage of the coal spontaneous-combustion reaction, -COOH combines with O_2 to generate CO and CO_2 gas, resulting in a continuous increase in CO and CO_2 generation. In this stage, -COOH has a great influence on gas production. In the later stage of the coal spontaneous-combustion reaction, the macromolecular benzene ring skeleton (C=C) as the dominant reaction group begins to break and reacts with free oxygen atoms and oxygen, resulting in a continuous increase in CO and CO_2 gas production. It can be inferred that a decrease in -COOH, C = C and-OH content will lead to a decrease in CO and CO_2 gas production, which will lead to a weakening of coal spontaneous combustion.

(2) The correlation between CH_4, C_2H_4, C_2H_6 and different active group structures

The results of the Origin fitting show that the production of CH_4, C_2H_4 and C_2H_6 gases is after the middle stage of the coal spontaneous-combustion reaction (60 °C). Therefore, the analysis of the correlation between the active structure and these three gases can reflect the influence of the active group structure on the spontaneous-combustion characteristics of coal after the middle stage of the oxidation reaction. It can be seen from Figure 5 that the groups with a high correlation with CH_4, C_2H_4 and C_2H_6 gases are C-H, -CH_2, -COOH and C-O structures. Therefore, it is believed that after the middle stage of coal spontaneous combustion, the oxygen-containing functional groups participate in the reaction, the aliphatic hydrocarbon structure reacts with the oxygen-containing functional groups to generate carbon–oxygen free radicals and further carbonylation will produce hydrocarbon gases such as CH_4 and C_2H_4. Among them, the active groups with the highest correlation with CH_4, C_2H_4 and C_2H_6 are all C-H structures, and their correlation degrees are 0.8400, 0.8638 and 0.8122, respectively. The C-H structure has the greatest influence on the production of the three gases. At the same time, -CH_2, -COOH and C-O all have a greater impact on the production of the three gases, indicating that after the coal spontaneous-combustion reaction enters the medium-term stage, the aromatic structure represented by the C-H structure and the aliphatic hydrocarbon structure represented by -CH_2 and some oxygen-containing functional groups have a greater impact on the coal spontaneous-combustion reaction activity. Therefore, it is considered that a decrease in aliphatic hydrocarbon structures such as C-H, -CH_2 and -COOH will lead to a decrease in hydrocarbon gas generation such as CH_4 and C_2H_4, thus weakening the spontaneous-combustion of coal.

3.4. Gray Correlation Analysis of Active Groups and Thermal Properties of Coal

The most intuitive characterization of the oxidation activity intensity of coal spontaneous combustion is the weight loss rate and exothermic intensity. Therefore, by determining the ignition-point temperature, heat-release peak temperature and burnout temperature characteristic points in the TG, DTG and DSC data curves, the difference in the coal combustion characteristics under the different spontaneous-combustion intensities of each group of raffinate coal samples is intuitively characterized. The characteristic temperature and corresponding TG data values at each feature point can be determined by processing the thermophysical characterization data of the extracted coal sample. All the determined data are summarized in Table 3.

In the process of this analysis, according to the data obtained in Table 3, the thermal physical parameters of different coals are taken as the reference sequence, and the structure of active groups is taken as the comparison sequence. The gray correlation analysis method is used to calculate the correlation between the structure of different active groups and the thermal properties of coal. The influence of different active group structures on the thermal properties of coal is determined, and then the influence of different active group structures on the spontaneous-combustion characteristics of coal is obtained. The calculation results are shown in Figure 6.

Table 3. The characteristic point temperature and corresponding mass value of each group of raffinate coal samples [34].

Group	Ignition Temperature/°C	Ignition-Point Quality/%	Peak Temperature of Heat Release/°C	Peak Temperature Quality/%	Burning Temperature/°C	Burnout Temperature Quality/%	Spontaneous-Combustion Index Value
YM	402.68	93.76	459.33	68.44	543.16	23.63	628.01
THF	415.09	82.86	464.42	61.76	542.79	24.13	773.34
MT	411.51	84.76	458.10	64.63	546.38	22.88	704.32
CYC	404.66	87.17	443.06	70.81	550.84	21.43	687.42
CYH	406.93	87.31	452.94	69.26	552.29	21.95	698.24
C + A + A	417.94	75.94	453.72	62.02	543.11	21.99	790.78
AE	402.88	92.18	457.74	68.13	540.10	22.58	625.89
AOT	410.32	85.30	452.58	68.91	540.76	23.40	692.23
EG	403.72	90.14	448.07	73.16	555.70	23.90	698.63
TP	400.65	89.73	454.21	72.42	545.80	23.48	712.48
EG + TP	416.52	81.35	449.09	72.51	557.23	23.27	843.20

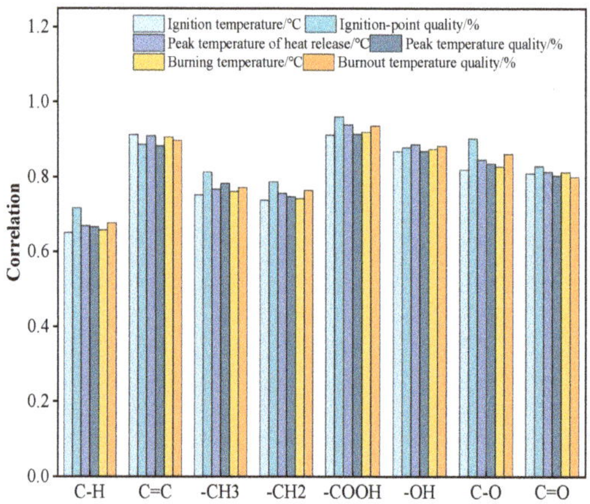

Figure 6. Correlation between active group structure and thermophysical parameters.

The ignition-point temperature of coal is the minimum temperature required for the continuous combustion of coal in air, and its size can reflect the strength of coal spontaneous combustion. The calculation results shown in Figure 6 by gray correlation analysis show that C=C, -COOH and -OH have a great correlation with the ignition-point temperature, and the correlation degrees are 0.9132, 0.9116 and 0.8678, respectively. In the process of coal spontaneous combustion, the active group structure reacts with oxygen to release heat. As the dominant groups participating in the reaction, i.e., -COOH, -OH and C=C, the heat released by their oxidation also has a great influence on the ignition temperature. The lower the mass remaining in the combustion stage of coal means the lower the energy intensity released by complete combustion. The analysis results show that the group structures with a great correlation with the quality of the ignition-point are -COOH, C-O and C=C, and their correlation degrees are 0.9604, 0.9026 and 0.8869, respectively. After extraction, the content of -COOH, C-O and C = C groups in coal decreases, resulting in a decrease in the intensity of the coal–oxygen composite reaction and a decrease in the amount of heat released, which is manifested in a decrease in coal quality. In addition, the peak temperature of heat release, the mass of peak temperature, and the mass of burnout temperature reflect the heat-release power of coal after entering the severe combustion stage. It can be seen from Figure 6 that the peak temperature of heat release, the mass of peak temperature, and the mass of burnout temperature have a great correlation with the structure of -COOH, C=C and -OH, indicating that the maximum heat-release power of coal is greatly affected by the structure of -COOH, C=C and -OH.

4. Discussion

The influence mechanism of the key active structure of coal molecules based on extraction technology on spontaneous-combustion is summarized by comparing and analyzing the test data by the gray correlation method, as shown in Figure 7.

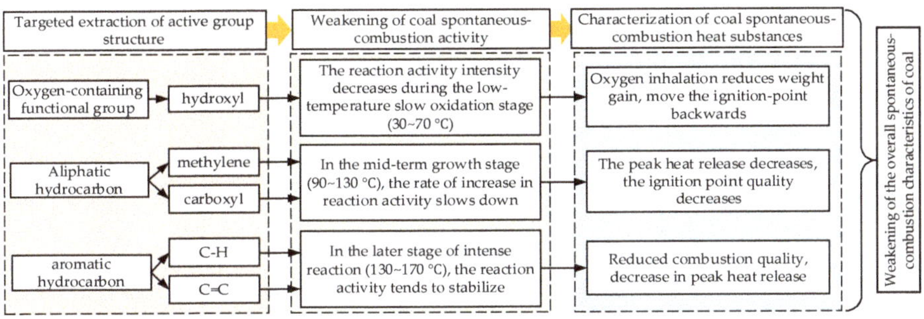

Figure 7. The influence mechanism of the key active structure of coal molecules on spontaneous combustion.

It can be seen from Figure 7 that the influence mechanism of the key active structure of coal molecules on spontaneous combustion characteristics is mainly divided into three major links.

(1) Extraction of the coal molecular active group structure

The most important role in this link is to strip the active group structure of coal molecules through extraction reagents. The essence is to destroy the connection structure between molecules through the chemical action of reagents or the dominant group structure itself in the process of the coal spontaneous-combustion reaction. For example, through various chemical energy modes of action or the characteristic components carried by the reagent, the related connection structure inside the coal molecule is destroyed, thereby stripping some functional group structures from the aromatic main chain of the coal molecules or destroying the macromolecular chain structure. THF reagent materials capture hydrogen bonds through a strong hydrogen bond acceptance ability, while other reagent materials such as pyridine [37], NMP [38] and ionic liquid [39] also have a similar group structure destruction ability.

(2) Weakening the reactivity intensity of coal spontaneous-combustion

By controlling the structure of the dominant group of the coal spontaneous-combustion reaction from the coal body, the lower limit of the key temperature node of the coal spontaneous-combustion is increased or the temperature range is prolonged, thereby weakening the reactivity of the coal spontaneous combustion in each process stage. For example, the oxygen-containing groups dominated by the -OH structure mainly affect the reactivity of coal spontaneous combustion at low temperature (30~70 °C). The aliphatic hydrocarbon structures represented by -COOH and -CH_2 structures mainly affect the reactivity of coal spontaneous combustion in the middle stage (90~130 °C). The aromatic hydrocarbon structures represented by C=C and C-H structures mainly act on the reactivity of coal spontaneous combustion in the later stage (130~170 °C). The effect of the above active group structures on the reactivity of coal spontaneous combustion is manifested in prolonging the temperature range of coal spontaneous combustion or increasing the lower limit of coal spontaneous combustion temperature. After extraction, the content of active structures such as oxygen-containing functional groups, aliphatic hydrocarbons and C=C participating in the oxidation reaction to generate gas is reduced, which leads to a decrease in the production of symbolic gases, and ultimately weakens the spontaneous-combustion of coal.

(3) Weakening the characterization strength of coal thermophysical properties

The microscopic mechanism of coal spontaneous combustion confirms that the essence of coal spontaneous combustion is the process of an oxygen capture reaction and the interaction of active groups in coal molecules, and the ultimate goal of coal spontaneous combustion weakening is to improve the thermophysical characterization strength of the coal oxidation spontaneous-combustion reaction. As the characterization parameters of coal spontaneous combustion activity, ignition-point temperature and ignition-point mass can reflect the size of coal spontaneous combustion reaction activity. By stripping the dominant group of the coal spontaneous-combustion reaction, the ignition temperature of coal will be delayed, and the quality of coal entering the combustion stage will be reduced. It is speculated that the stripping of the active structure will lead to a decrease in heat release and heat-release efficiency in the process of coal oxidation, thus weakening the effect of coal spontaneous combustion.

5. Conclusions

According to the previous research results and data of the team, combined with the quantitative indicators of the spontaneous-combustion of each group of raffinate coal, the migration law of spontaneous combustion process nodes and the characterization of thermal properties, the influence of various active group structures on the reaction of the coal spontaneous-combustion process was determined. The main conclusions are as follows:

(1) The influence of the active structure on the process of coal spontaneous combustion is different. The oxygen-containing structure represented by -OH mainly affects the reactivity in the low-temperature oxidation process (30~90 °C). The aliphatic structure represented by -COOH mainly affects the reactivity in the middle stage of coal spontaneous combustion (90~130 °C). The aromatic ring structure determines the reactivity intensity of coal spontaneous combustion in the high-temperature stage (130~170 °C).

(2) The gray correlation method was used to calculate the correlation degree between the active structure of coal molecules and gas products. It was found that the gas products produced in the process of coal spontaneous combustion had a large correlation degree with -OH, -COOH, $-CH_2$, C = C and other structures. With the development of the coal spontaneous-combustion process, various active structures have participated in the reaction, thus generating oxygen-containing gases and alkanes and other iconic gases. The stripping extraction of the active structure leads to a decrease in the content of the groups involved in the reaction, resulting in a decrease in gas production, thereby weakening the spontaneous-combustion of coal.

(3) Through gray correlation analysis, it is found that the main active groups affecting the thermal properties of coal are C=C in the molecular skeleton of coal, -COOH of aliphatic hydrocarbons and oxygen-containing functional groups represented by -OH. In the process of coal spontaneous combustion, the active structure reacts with oxygen to generate heat. The stripping of the active structure leads to a decrease in the intensity of the coal–oxygen composite reaction and a decrease in the amount of heat released, resulting in a delay in the ignition-point temperature and a decrease in the quality of the coal involved in the combustion. Finally, the coal spontaneous-combustion is weakened.

(4) The influence mechanism of the key active structure of coal molecules based on extraction technology on spontaneous combustion characteristics is divided into three parts. (1) Extraction destroys the connection bond between coal molecules or the dominant group structure of the coal spontaneous-combustion reaction. (2) Stripping the dominant group structure of the coal spontaneous-combustion reaction weakens the reaction activity intensity in each process stage of coal spontaneous combustion, prolongs the temperature range of the heat storage reaction or increases the lower limit of the rapid reaction temperature threshold. (3) As a result, the thermal physical

property characterization strength of coal spontaneous combustion is weakened, which is manifested as delaying the ignition-point temperature of coal or reducing the quality of coal involved in combustion.

Author Contributions: Conceptualization, supervision, investigation, writing—original draft, J.G.; writing—original draft, writing—review and editing, Y.W.; formal analysis, Y.L.; methodology, investigation, G.C.; methodology, formal analysis, D.L.; software, article proofreading Y.J. All authors have read and agreed to the published version of the manuscript.

Funding: This research was supported by the National Natural Science Foundation of China (grant nos. 52004209, 52304251 and 52174198), the Shaanxi Postdoctoral Science Foundation (grant no. 2023BSHEDZZ286) and the Shaanxi Science and Technology Association Young Talents Lifting Project (grant no. 20240205).

Institutional Review Board Statement: Not applicable.

Informed Consent Statement: Not applicable.

Data Availability Statement: Data available on request due to restrictions legal.

Conflicts of Interest: Author Yan Jin was employed by the company Taizhou Petroleum Branch, Sinopec Sales Co., Ltd. The remaining authors declare that the research was conducted in the absence of any commercial or financial relationships that could be construed as a potential conflict of interest.

References

1. Životić, D.; Jovanovski, G.; Simić, V.; Boev, I.; Cvetkov, V.; Makreski, P.; Polomčić, D.; Ristić, V.V. Coal: Exploration, reserves, and utilization. *ChemTexts* **2024**, *10*, 1. [CrossRef]
2. Coal reserves. *Energy Energy Conserv.* **2016**, *12*, 171.
3. Ulyanova, E.V.; Malinnikova, O.N.; Dokuchaeva, A.I.; Pashichev, B.N. Effect of structural nonuniformity on spontaneous combustion liability of coal. *Solid Fuel Chem.* **2022**, *56*, 426–431. [CrossRef]
4. Somekawa, T.; Manago, N.; Kurahashi, S.; Shiina, T.; Yogo, A.; Kuze, H. Hard-Target Reflection Laser Spectroscopy of Carbon Monoxide Gas Concentration for the Early Detection of Spontaneous Combustion of Coal. *Appl. Spectrosc.* **2024**, *78*, 398–402. [CrossRef] [PubMed]
5. Yetkin, M.E.; Özfırat, M.K.; Kun, M.; Pamukcu, C. The prediction of the risks of spontaneous combustion in underground coal mines using a fault tree analysis method. *MethodsX* **2024**, *13*, 102835. [CrossRef]
6. Onifade, M.; Genc, B. A review of research on spontaneous combustion of coal. *Int. J. Min. Sci. Technol.* **2020**, *30*, 303–311. [CrossRef]
7. Wang, D.M. *Mine Fire Science*; China University of Mining and Technology Press: Xuzhou, China, 2008.
8. Zhang, J.; Ren, T.; Liang, Y.T.; Wang, Z.W. A review on numerical solutions to self-heating of coal stockpile: Mechanism, theoretical basis, and variable study. *Fuel* **2016**, *182*, 80–109. [CrossRef]
9. Wang, F.R.; Tan, B.; Zan, X.Y.; Huang, J.L.; Fang, X.Y.; Fu, S.H.; Wang, H.Y.; Li, T.Z.; Qi, Q.J. Molecular simulation study of microstructural evolution during low-temperature oxidation of coal. *Energy* **2024**, *290*, 130054. [CrossRef]
10. Sharikov, F.Y.; Rudko, V.A.; Smyshlyaeva, K.I. Oxidation thermolysis kinetics of asphaltenes with various chemical prehistory. *Thermochim. Acta* **2023**, *726*, 179550. [CrossRef]
11. Gorlanov, E.S.; Mushihin, E.A.; Schneider, S.R.; Kovalskaya, K.V. Synthesis of Carbon–TiC/TiB Composites at the Electrolytic Reduction of Fused Salts. *J. Electrochem. Soc.* **2023**, *170*, 102501. [CrossRef]
12. Zhu, H.Q.; Li, B.R.; Xie, L.H.; Qu, B.L.; Hu, L.T.; Zhang, L.; Li, R. Comparative study on oxidation reaction characteristics of different forms of hydroxyl-containing coal molecules. *J. Mol. Struct.* **2024**, *1316*, 139046. [CrossRef]
13. Qiu, C.J.; Jiang, L.L.; Gao, Y.G.; Sheng, L.Z. Effects of oxygen-containing functional groups on carbon materials in supercapacitors: A review. *Mater. Des.* **2023**, *230*, 111952. [CrossRef]
14. Dhakate, S.R.; Bahl, O.P. Effect of carbon fiber surface functional groups on the mechanical properties of carbon–carbon composites with HTT. *Carbon* **2003**, *41*, 1193–1203. [CrossRef]
15. Rudko, V.A.; Gabdulkhakov, R.R.; Pyagai, I.N. Scientific and technical substantiation of the possibility for the organization of needle coke production in Russia. *J. Min. Inst.* **2023**, *263*, 795–809.
16. Larionov, K.B.; Gromov, A.A. Non-isothermal oxidation of coal with $Ce(NO_3)_3$ and $Cu(NO_3)_2$ additives. *Int. J. Coal Sci. Technol.* **2019**, *6*, 37–50. [CrossRef]
17. Choi, H.; Jo, W.; Kim, S.; Yoo, J.; Chun, D.; Rhim, Y.; Lim, J.; Lee, S. Comparison of spontaneous combustion susceptibility of coal dried by different processes from low-rank coal. *Korean J. Chem. Eng.* **2014**, *31*, 2151–2156. [CrossRef]
18. Taraba, B.; Pavelek, Z. Study of coal oxidation behaviour in re-opened sealed heating. *J. Loss Prev. Process Ind.* **2016**, *40*, 433–436. [CrossRef]

19. Xu, X.X.; Yuan, S.J.; Li, J.H.; Guo, S.L.; Yan, Z. Preparation of lignin-based intumescent nanogel and its mechanism of inhibiting coal spontaneous combustion. *Energy* **2023**, *275*, 127513. [CrossRef]
20. Stolarski, M.; Szczygiel, J.; Kulażyński, M. Extraction of brown coals with alcohols under supercritical conditions. *J. Energy Inst.* **2020**, *93*, 1993–1998. [CrossRef]
21. Zhao, H.; Franklin, M.S. Ionic liquids for coal dissolution, extraction and liquefaction. *J. Chem. Technol. Biotechnol.* **2020**, *95*, 2301–2310. [CrossRef]
22. Jin, J.Y.; Hao, C.Y.; Shen, W.M.; Yuan, Z.Q.; He, W.H.; Li, L. Study on the microcosmic mechanism of inhibition of coal spontaneous combustion by antioxidant diphenylamine. *J. Mol. Struct.* **2023**, *1294*, 136357. [CrossRef]
23. Yao, J.; Ji, H.J.; Lu, H.Z.; Gao, T.T. Effect of Tetrahydrofuran Extraction on Surface Functional Groups of Coking Coal and Its Wettability. *J. Anal. Methods Chem.* **2019**, *8*, 1285462. [CrossRef] [PubMed]
24. Barma, S.D. Ultrasonic-assisted coal beneficiation: A review. *Ultrason. Sonochem.* **2019**, *50*, 15–35. [CrossRef] [PubMed]
25. Taraba, B.; Peter, R.; Slovák, V. Calorimetric investigation of chemical additives affecting oxidation of coal at low temperatures. *Fuel Process. Tech.* **2011**, *92*, 712–715. [CrossRef]
26. Wang, F.S.; Sun, M.; Wang, J.T. Dimethyl Methylphosphonate for the Suppression of Coal Spontaneous Combustion. *Combust. Sci. Technol.* **2021**, *195*, 982–999. [CrossRef]
27. Onifade, M.; Genc, B.; Gbadamosi, A.R.; Morgan, A.; Ngoepe, T. Influence of antioxidants on spontaneous combustion and coal properties. *Process Saf. Environ.* **2021**, *148*, 1019–1032. [CrossRef]
28. Deng, J.; Lv, H.F.; Li, D.J.; Xiao, Y.; Wang, C.P.; Jiang, Z.G. Thermal behavior effects of 1-butyl-3-methylimidazole tetrafluoroborate on coals spontaneous combustion with different metamorphic levels. *J. China Coal Soc.* **2019**, *44*, 255–262.
29. Zhang, Y.T.; Yang, C.P.; Li, Y.Q.; Huang, Y.; Zhang, J.; Zhang, Y.B.; Li, Q.P. Ultrasonic extraction and oxidation characteristics of functional groups during coal spontaneous combustion. *Fuel* **2019**, *242*, 288–294. [CrossRef]
30. Slovák, V.; Taraba, B. Urea and $CaCl_2$ as inhibitors of coal low-temperature oxidation. *J. Therm. Anal. Calorim.* **2012**, *110*, 363–367. [CrossRef]
31. Jia, H.L.; Yang, Y.; Ren, W.X.; Kang, Z.H.; Shi, J.T. Experimental Study on the Characteristics of the Spontaneous Combustion of Coal at High Ground Temperatures. *Combust. Sci. Technol.* **2022**, *194*, 2880–2893. [CrossRef]
32. Guo, J.; Quan, Y.P.; Wen, H.; Zheng, X.Z.; Cai, G.B.; Jin, Y. The Effects of Solvent Extraction on the Functional Group Structure of Long-Flame Coal. *Fire* **2023**, *6*, 307. [CrossRef]
33. Liu, Y.; Guo, J.; Wu, Y.F.; Chen, C.M.; Zheng, X.Z.; Li, D.L.; Jin, Y. Response of Macroscopic Coal Oxidation and Pyrolysis Parameters to Different Extractants. *Combust. Sci. Technol.* **2024**, *56*, 1–22. [CrossRef]
34. Guo, J.; Gao, B.; Liu, Y.; Cai, G.B.; Jin, Y.; Chen, C.M.; Wang, L. Study on the influence of solvent extraction on thermal property parameters of coal oxidation. *J. Therm. Anal. Calorim.* **2024**, *56*, 1–10. [CrossRef]
35. *AQ/T 1068-2008*; The Test Method of Oxidation Kinetics for the Propensity of Coal to Spontaneous Combustion. State Administration of Work Safety: Beijing, China, 2008.
36. Rahui, A.; Allouch, M.; Alami, M. Optimization of the tungsten inert gas welding parameters of mild steel thin sheets through the gray relational analysis method. *Int. J. Adv. Manuf. Technol.* **2024**, *133*, 4437–4452. [CrossRef]
37. Fletcher, T.H.; Bai, S.; Pugmire, R.J.; Solum, M.S.; Wood, S.; Grant, D.M. Chemical structural features of pyridine extracts and residues of the Argonne Premium coals using solid-state C-13 NMR spectroscopy. *Energy Fuels* **1993**, *7*, 734–742. [CrossRef]
38. Coto, B.; Suárez, I.; Tenorio, M.J.; Huerga, I. Extraction of aromatic and polyaromatic compounds with NMP: Experimental and model description. *Fluid Phase Equilib.* **2022**, *554*, 113293. [CrossRef]
39. Olea, F.; Estay, H.; Jofré-Ulloa, P.P.; Cabezas, R.; Merlet, G.; Araya, A.; Quijada-Maldonado, E. A review towards selective cobalt recovery in solvent extraction: Potential of ionic liquids as diluents and extractants. *Sep. Purif. Technol.* **2024**, *348*, 127758. [CrossRef]

fire

MDPI

Article

Spatiotemporal Evolution and Frontier Focus Analysis Based on Coal Fire Control Body of Knowledge

Dandan Han [1,2], Guchen Niu [1,*], Bing Liu [1], Feiran Wang [1], Yongbo Ren [1], Chang Su [1], Yutong Yao [1] and Zining Zhao [1]

1 School of Emergency Management and Safety Engineering, China University of Mining and Technology (Beijing), Beijing 100083, China; handd.2006@tsinghua.org.cn (D.H.); bingliu2024@163.com (B.L.); wangfeiran2003@163.com (F.W.); ryb1092499545@163.com (Y.R.); suchang3203@163.com (C.S.); 2210150609@student.cumtb.edu.cn (Y.Y.); zhaozining2022@outlook.com (Z.Z.)

2 International Exchange and Cooperation Center, Ministry of Emergency Management, Beijing 100012, China

* Correspondence: nnniu188@163.com

Abstract: Mine fire accidents frequently constitute a major threat to mining safety, and their potential consequences are extremely severe, which highlights the urgency of fire prevention and control research. In this study, the CiteSpace software was used to conduct a metrological analysis of 717 relevant studies in the field of mine fire prevention and control (MFPC), aiming to reveal the research trends and trends in this field. This analysis found that the annual number of MFPC articles showed a significant upward trend, indicating that it is in rapid development during the active period. China, the United States, and Australia are the main contributors in this field, and the institutional contribution of China University of Mining and Technology is particularly outstanding, reflecting the regional concentration of research activities. The analysis of cooperation networks reveals the close cross-regional collaboration among European countries. The inhibition effect and evaluation criteria and the inhibition technology under different coal characteristics have become the focus of research. Activation energy, release, and quantum chemistry have become recent hot spots, reflecting the research on the mechanism of forward physicochemical synergistic inhibition and the in-depth exploration of the molecular level. It indicates that future research will focus on the development of temperature-responsive retardant materials, the application of quantum chemistry theory, and the exploration of the microscopic mechanism of coal spontaneous combustion through molecular simulation technology to further optimize the fire prevention strategy. In summary, the findings of this study not only provide a comprehensive picture of current research activities in the MFPC field but also indicate potential directions for future research and have important guiding significance for promoting the development of this field.

Keywords: knowledge graph; visualization; mine fire prevention and control; CiteSpace

1. Introduction

China's energy system is notably characterized by an abundance of coal resources and a relative scarcity of petroleum resources. Coal occupies a vital position as the cornerstone of national energy [1–4]. Due to the complex and changeable geological conditions of coal seams in our country, the safety production of coal mines is faced with severe challenges, so various disasters occur frequently. Among them, the spontaneous combustion of coal (SCC) is particularly concerning because of the seriousness of its potential hazards, the difficulty of prevention and control, and the far-reaching impact on resources and the environment [5,6]. Figure 1 shows the trend of the incidence and death toll of coal mine fire accidents in China from 2008 to 2023. Although, in recent years, China's supervision and management of coal mine safety production has been continuously strengthened, which has led to a significant decrease in the incidence rate and death rate of coal mine accidents. However, due to the huge coal production, wide distribution of mines, complex mining conditions, and differences in various levels, mine fire prevention and control (MFPC) work is still facing a severe situation [7].

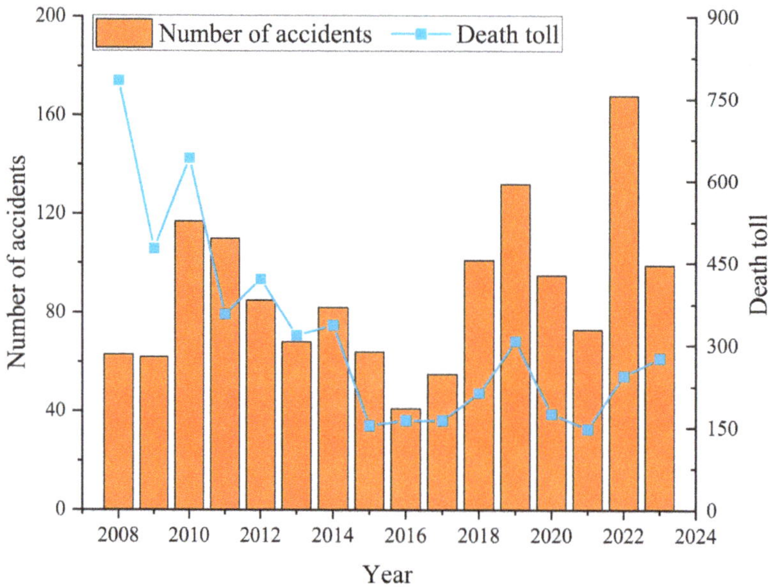

Figure 1. Number of coal mine accidents and death toll in the past 15 years.

Scholars from various countries have developed and researched a wide range of fire suppression materials for use in dealing with spontaneous coal combustion. These include slurries, foams, inert gases, colloids, and retardants [8–16]. Inhibition technology can be divided into the following three basic categories according to their principle of action: physical inhibition, chemical inhibition, and physicochemical synergistic inhibition with both characteristics [17–28]. The physical inhibition method is mainly to slow down the occurrence of coal oxygen reaction by endothermic cooling, isolating oxygen, and covering the surface active center of coal. The chemical inhibition method is to add appropriate inhibitors according to the reaction mechanism of coal and oxygen and slow down the oxidation rate of coal samples by reducing the reactivity of groups on the surface of coal, reducing the content of free radicals in the system, and changing the reaction path. These fire-fighting materials with different characteristics and concerns, in their respective research and evolution processes, have not only greatly enriched and improved the knowledge system in the MFPC field but also accumulated rich scientific research results and effectively promoted technological innovation and development in the field.

To date, many scholars and experts have invested extensive and in-depth research work in the direction of MFPC, but there is less organization of existing results [29–32]. The current literature review has significant limitations in revealing the network relationships among knowledge communities and their evolution and fails to clearly present the dynamic development of scholars' understanding of knowledge, the intrinsic connections among research themes, key documents and journals that have made a difference in the course of research, and clues that indicate the future research trends of scholars. In order to clearly and accurately explore these issues, it is necessary to use bibliometric tools and textual analysis techniques, with a view to extracting scholars from the voluminous literature and realizing in-depth insights and a scientific analysis of relevant research fields. Currently, the main tools commonly used in academia for visualization and analysis are CiteSpace, SPSS, Ucinet, and Vosviewer [33–37]. In order to explore the research status and development trend of coal porosity, Shao et al. [38] analyzed two visual analysis software (Vosviewer and

CiteSpace) and found and summarized the main knowledge framework of coal porosity research. The results show that there has been a significant paradigm shift in the study of coal pores, that is, from the traditional macroscopic description and analysis to the more detailed microscopic level. In addition, several current hot trends in coal pore research are pointed out in the analysis as follows: molecular simulation technology, pore structure, methane storage, and methane adsorption behavior. Hu et al. [39] conducted a detailed analysis of the literature on coal fire detection technology in the past decade using CiteSpace and Vosviewer software, revealing that although a solid basic theoretical system and research framework has been established in this field, prospective research is still insufficient. The focus of coal fire detection research is to continuously improve the level of refinement of data processing and the accuracy of detection devices, which is exactly the focus of current research frontiers in this field. Duan et al. [40] drew a knowledge domain map using Vosviewer and CiteSpace to study the development and trend of coal-related free radicals. The research results show that the main research in this direction includes the spontaneous combustion of coal, pyrolysis, control, coal and biomass, environmental hazards, and coal geochemistry. The depth and breadth of research in this field have been significantly expanded. The overall background and influencing factors of the coal free radical phenomenon have been discussed not only at the macro level but also at the micro level. The research by Wang et al. [41] reveals that in recent years, the research on functional groups in coal has been significantly deepened to the microscopic level, marking the maturity and perfection of the basic theoretical system and research means in this field. The core framework of the current research revolves around the following key pillars: gas adsorption in coal, functional group evolution analysis during heat treatment, and functional group classification and spatial distribution analysis. In the future, the research of functional groups in coal will focus on the application of quantum chemistry theory and molecular simulation experiments. In order to study the research status of underground coal mine fire scene, Wang et al. [42] found that the research on "underground coal mine fire detection technology" is a research hotspot through visual analysis. The advantages of detection methods currently used in this field are analyzed. Zhao et al. [43] systematically combed and analyzed the academic literature on the application of acoustic emission technology in coal published worldwide from 2010 to 2020 in-depth using the visual analysis software CiteSpace. The results show that the time domain parameters of AE, such as cycle count, energy, waveform, and signal intensity, have reached a mature level. However, further improvement of acoustic emission positioning technology will be an important focus of future research. At present, the fractal dimension of signal strength, acoustic emission count, and B-value are generally regarded as the mainstream means to identify the fracture phenomenon. It is worth paying attention to the obvious inadequacies in dealing with the complex challenges encountered in practical engineering projects regarding the activity laws of steep-dip coal seam and its surrounding rocks during deformation. Yang et al. [44] used CiteSpace to analyze the correlation among countries, authors, and keywords in the research field of SCC. The research shows that the low-temperature oxidation process, kinetic principle, internal mechanism construction and model development constitute the core issues of SCC research. Recently, the research focus is mainly on the research and development of three-phase foam technology, fire extinguishing strategy, and foam preparations.

Based on the above, in order to accurately grasp the research achievements and research directions in MFPC field, the Web of Science (WOS) core database is used as the data source, the CiteSpace analysis method is used to analyze the literature distribution and hot research achievements, and the hot issues in the MFPC field are summarized. It provides a reference for the research frontier, the determination of scientific problems, and the choice of research direction and provides a scientific basis for reducing the risk of coal mine fires.

2. Data and Methods

2.1. Data Sources

WOS is an influential professional and comprehensive scientific information service platform in the research community, which has been accepted by many scholars and personnel [45], contains more than 12,000 world-authoritative and high-impact scholarly journals in a variety of research fields, such as natural sciences, engineering, and technology, and is considered to be the most suitable database for bibliometric analysis. Therefore, this paper uses the core data set of WOS as the source data. The subject of the search is "mine fire prevention and control" or "prevention of spontaneous combustion of coal", which means that the title, abstract, and keywords of the journal are not required to be included in the search. This means that as long as there are terms related to mine fire prevention and control or prevention of spontaneous combustion of coal in the title, abstract, or keywords, we can find these documents. The search node time is December 2023. A total of 717 documents were obtained by manually removing irrelevant documents one by one. A total of 717 references were obtained by lifting and filtering in CiteSpace.

2.2. Methodology

This paper mainly uses the bibliometric analysis method to study the knowledge structure and evolution process in this field. Bibliometrics is a statistical method used to evaluate and quantify the emergence and development trend of a specific research topic [46]. CiteSpace is a multi-view information visualization software. Professor CM Chen developed the CiteSpace knowledge graph analysis software and introduced it to China. Its main function is to display evolution trends and knowledge associations of the scientific frontier through visual analysis, including keyword co-occurrence analysis, keyword clustering, institution distribution disclosure, author cooperation exploration, literature integration, and other visual means. Its unique feature is that the method can accurately translate the massive data contained in the WOS database into various knowledge graphs and reveal the deep architectural features and associated connotations of the scientific research field through visual methods. The data processing and analysis work adopted in this paper is accomplished with the help of CiteSpace version 6.3.R1, an advanced visual bibliometric tool.

3. Results and Discussion

3.1. Year-to-Year Trends in the Number of Articles Issued

As a quantitative index, the number of studies can reveal the degree of academic activity and its development in a certain subject field to some extent. It directly reflects the accumulation of research results of related disciplines and the active situation of knowledge innovation. Figure 2 shows the number of publications in the direction of MFPC from 2000 to 2023. According to the change trend of MFPC's annual publication volume, it can be divided into the following three stages: the initial period and the exploration period (2000–2008), and the average number of MFPC papers in this stage is only four per year. This may be due to the fact that the field was still in its infancy during this period, and researchers were exploring new methods and technologies to prevent and control mine fires [47–50]. In addition, technologies such as the Internet and social media were not yet widespread during this period, so the dissemination of articles was limited. The period of steady growth (2009–2019) was an important phase in the global MFPC research journey, which lasted for a decade, and the annual publication volume in the field showed a continuous and steady increase, with an average of about 25 academic articles published each year [51–55]. During the explosive growth period (2020 to 2023), the average number of papers published in the direction of MFPC in this stage is 107 per year, and the number of papers has surged [56–63]. The main reason is that the research in MFPC field continues to deepen, resulting in an increasing number of fruitful results accumulated by researchers, which not only build a solid foundation for subsequent research but also indicate the direction of research. The increasing number of publications each year clearly indicates that MFPC is receiving unprecedented attention, indicating that the field is about to enter a new stage of development.

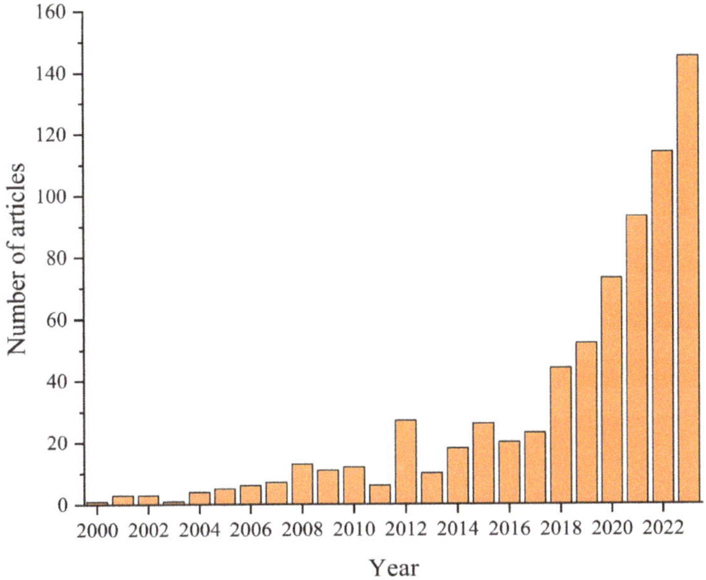

Figure 2. Annual production of articles in the field of mine fire control.

3.2. Analysis of National Publications and Cooperation

We used CiteSpace to create a knowledge map for country/region cooperation analysis to identify the major countries in the MFPC space and their partnerships, as shown in Figure 3. It was observed that as many as 36 countries participated in the MFPC study. The top three countries were China, with a total of 642 publications (89.54%), the United States, with 27 publications (3.77%), and Australia, with 21 publications (2.93%). Each node in Figure 3 represents a different country, and the number of articles for each country is represented by the size of the node. The colors of the nodes and the thicknesses of the lines carry specific meanings. The darker the center color of the node, the earlier the research in this field was carried out. The lighter the external color of the node, the more active the MFPC research has been in the country/region in recent years. These nodes are connected to each other by wires, which graphically show the cooperative links between different geographical areas. The thickness of the connection can directly reflect the closeness of cooperation and interaction between countries and regions. The thicker the line between the two nodes, the closer the connection between the two countries. On the other hand, the smaller the connection, the weaker the cooperation between the two countries. When a node is surrounded by a pink circle, it means that the node has a large centrality and plays an important role in the network. It is clear that European countries cooperate more closely with each other than in any other part of the world. China has close cooperation with Australia, Canada, and the United States. Among these, China, the United States, England, and other countries have an important role in the research of MFPC field in the whole national network. In order to further promote this field to a higher stage of development, China, Canada, and the United States, which are rich in global scientific research output, need to strengthen the deep cooperation and collaborative innovation among the three countries. The United States, China, and some European countries started MFPC research earlier, and in recent years, China, the United States, and Canada mainly have been committed to MFPC research.

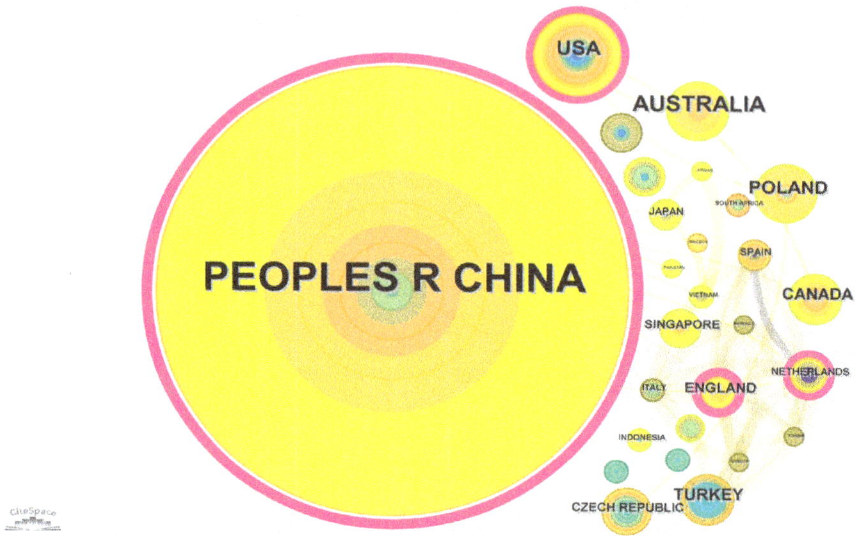

Figure 3. Knowledge map of cooperating countries.

3.3. Analysis of Research Institutions and Cooperation

In order to identify the most productive institutions in the MFPC field and the collaborations between them, an analysis of research institutions and inter-institutional collaborations is provided [38]. CiteSpace-related parameters are determined to generate a research institution collaboration graph with 247 nodes and 318 connectivity lines, as shown in Figure 4. The level of research output of the institutions is visually displayed by the node size, and the connections between the nodes depict the closeness of cooperation between the research institutions. As can be seen in Figure 4, in the MFPC field, six universities, including the China University of Mining and Technology (CUMT), Xi'an University of Science and Technology (XUST), Shandong University of Science and Technology (SDUST), Henan Polytechnic University (HPU), Anhui University of Science and Technology (AUST), and University of Science and Technology of Beijing (USTB), by virtue of their remarkable achievements in related scientific research output, are firmly in the forefront of all research institutions, which is specifically reflected in the number of papers ranked in the top six. CUMT, XUST, and SDUST play an important role in the whole network of research institutions. Among these, CUMT has established close cooperation with more than 20 research institutions. Examples include SDUST, USTB, HPU, the Chinese Academy of Sciences, Hunan University of Science and Technology, Pennsylvania State University-University Park, Curtin University, West Virginia University, the Commonwealth Scientific and Industrial Research Organisation (CSIRO), the Pennsylvania Commonwealth System of Higher Education (PCSHE), and the Czech Academy of Sciences. This research reveals that the cooperation between academic institutions presents obvious regional characteristics, that is, the cooperation behavior mostly occurs between institutions in similar geographical locations or in the same region, forming a relatively dense regional collaboration network. In contrast, cross-border or even intercontinental cooperation is rare. This phenomenon shows that despite the increasing attention paid to international scientific research cooperation under the trend of globalization, in practice, academic institutions still tend to give priority to geographical proximity and convenience of resource sharing within a region when choosing partners, resulting in transnational and trans-continental cooperation being less frequent than intra-regional cooperation. Of course, this does not mean that research cooperation between countries or continents cannot be achieved, but in the existing research cooperation models, it has not yet become the mainstream and needs to be further promoted and developed.

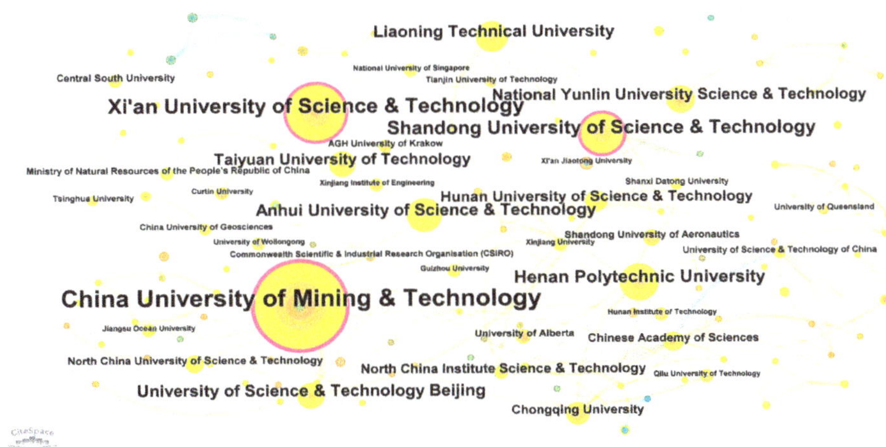

Figure 4. Analysis of research institution collaboration.

3.4. Author Analysis

3.4.1. Analysis of the Volume of Publications and Cooperation

Price's Law, which was developed by British physicist and historian of science Derek J. de Solla Price, is an important theory to describe the growth of scientific literature [64]. The specific formula is as follows:

$$M = 0.794 \times \sqrt{N_{max}} \quad (1)$$

M represents the minimum number of highly active, influential, and significant contributors in a certain research direction. N_{max} represents the maximum number of papers published by a single author in their field of expertise within a particular research direction. Through systematic statistics and in-depth mining of the literature information data, we can comprehensively obtain a series of key information about relevant authors, including, but not limited to, scientific research output, the average citation of a single paper, the influence distribution of journal-published papers, the coverage and depth of research topics, the composition and stability of cooperation networks, and other multidimensional indicators. These detailed data not only help us objectively evaluate the efficiency and influence of the author's academic output but are also crucial to identifying the author's status and role in the academic world. In the in-depth analysis of author groups, the high-yield authors and the core authors have undoubtedly become the focus of in-depth analysis of researchers because of their unique status and significant contributions. Table 1 lists the top 10 authors and their related information. In Table 1, the N_{max} is 39, that is, the M is 4.96, so core authors in the MFPC field need to publish at least five articles.

Table 1 details the list of the top ten authors in the number of published papers, as well as their respective countries, academic institutions, specific paper output, and corresponding output ratio. Among them, Deng Jun of XUST published 34 papers and is ranked first, accounting for 4.7 percent. Shu Chimin of the National Yunlin University of Science and Technology published 27 articles (3.8%), ranking second. Hu Xiangming from SDUST (22 articles) and Xiao Yang from XUST (22 articles) tied for third place, accounting for 3.1% each.

Table 1. Institutions to which the main authors belong, the number of publications, and their percentage share of the total number of publications.

Rank	Author	Organization	Volume of Publication	Proportions	Rank	Author	Organization	Volume of Publication	Proportions
1	Deng Jun	Xi'an University of Science and Technology	34	4.7%	6	Wen Hu	Xi'an University of Science and Technology	19	2.6%
2	Shu Chimin	National Yunlin University of Science and Technology	27	3.8%	7	Wang Wei	Tsinghua University	18	2.5%
3	Hu Xiangming	Shandong University of Science and Technology	22	3.1%	8	Lu Wei	Shandong University of Science and Technology	16	2.2%
4	Xiao Yang	Xi'an University of Science and Technology	22	3.1%	9	Liang Yuntao	Shandong University of Science and Technology	15	2.1%
5	Yang Shengqiang	China University of Mining and Technology	20	2.8%	10	Qin Botao	China University of Mining and Technology	15	2.1%

The WOS data set was imported into the CiteSpace platform for analysis. The research period was set to 2000 to 2023, and the interval was set to be every two years. When performing the "node type" analysis, the focus was placed on the "author" category. The "Pathfinder" path algorithm is used, supplemented by "Pruning networks" and "pruning the merge network" pruning network technologies to conduct an in-depth analysis of the occurrence frequency of keywords and thus generate a graph showing the author cooperation mode.

As can be seen in Figure 5, in the lower left is a cooperative network composed of Deng Jun, Shu Chimin, and Bai Zujin [65–69]. One of them, Deng Jun, is also closely associated with Yang Shengqiang of CUMT. In the middle section is a collaborative network consisting of Hu Xiangming, Lu Wei, Huang Zhian, Xue Di, and Liang Yuntao [70–77]. In the bottom right is a collaborative network consisting of Wang Deming, Qin Botao, Shi Quanlin, and Wang Wei [78–82]. Among them, Wang Deming first proposed a new type of prevention and control technology, three-phase foam, in 2004. In the top left, is a collaborative network consisting of Yang Shengqiang and Zhou Buzhuang, among others [83–87]. In the upper right center, is a collaborative network consisting of Tan Bo, Wang Haiyan, Niu Huiyong, and Shao Zhuangzhuang [88–96]. On the right is a collaborative network consisting of Li Zenghua and Kong Biao et al. [97–99]. According to the author cooperation map, most of the MFPC researchers are from CUMT, XUST, HPU, SDUST, and the China University of Mining and Technology (Beijing). In addition, the author's institutions show significant geographical proximity and particularly close cooperation with each other, which indirectly confirms the large number of scholars in the MFPC field, but the research activities show obvious fragmentation characteristics, lacking large-scale and organized collaborative research and knowledge exchange. In the development of the MFPC field, in-depth academic exchanges and cooperation among leading scholars have played a decisive role in promoting the progress of the entire discipline. In view of this, it is urgent to take measures to strengthen the cooperation and communication among researchers in the MFPC field to promote resource integration and knowledge sharing.

3.4.2. Author Co-Citation Analysis

In-depth author co-citation analysis of the academic literature is a key research method to reveal and understand the law of scientific knowledge dissemination and disciplinary interaction. Its core lies in the detailed tracking and analysis of a specific phenomenon, that is, when two or more authors' independent works become the common objects of other researchers' papers at the same time, which is called the "co-citation phenomenon". Co-citation is not only an intuitive manifestation of academic influence but also an important indicator to measure the potential connection and cooperation tendency between authors and even among their subject fields.

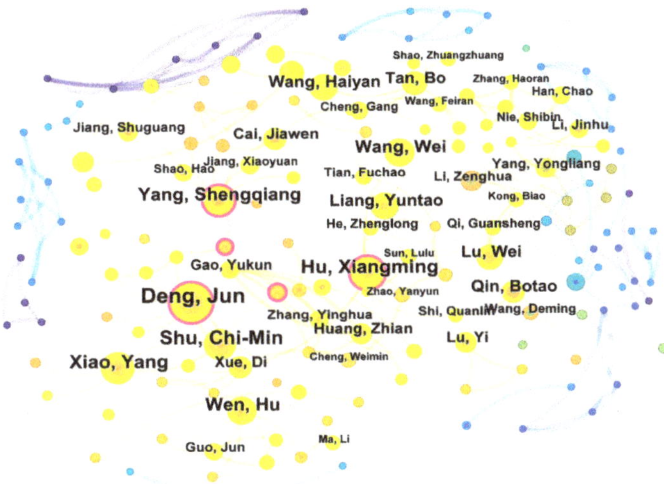

Figure 5. Author collaboration analysis.

The highly cited authors, such as Deng Jun, Qin Botao, Cheng Weimin, Song Zeyang, and Wang Deming, revealed in Figure 6, represent active and influential academic forces in this research field. Among them, Professor Deng Jun not only performs well in academic output but also has had these achievements widely recognized and highly cited in academic circles. Professor Qin Botao and Professor Wang Deming have made remarkable achievements in applying three-phase foam to SCC. Carras JN from CSIRO has done a remarkable job studying SCC phenomena and the use of carbon dioxide injection to prevent coal seam fires. On the other hand, Song Zeyang from XUST focuses on in-depth research in areas such as energy conversion and coal field fires.

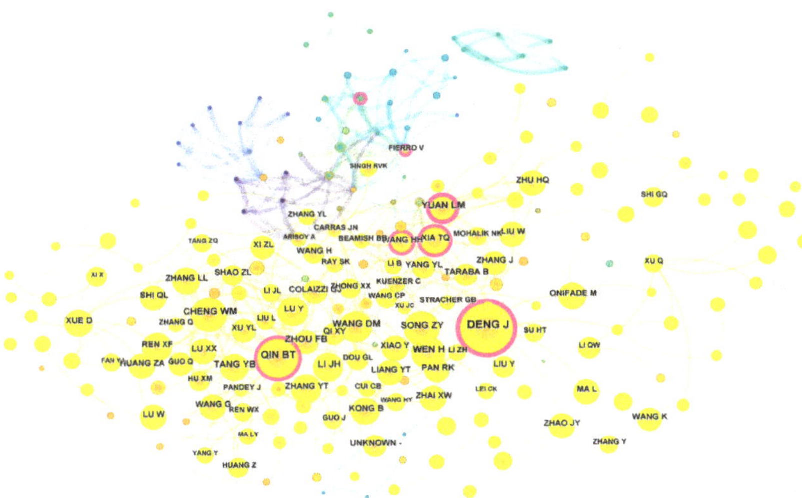

Figure 6. Mapping of co-cited authors.

3.5. Journal Analysis

3.5.1. Analysis of Major Journals

Academic journals play a central and irreplaceable role in the exchange and diffusion of scientific research [35]. Based on the detailed data provided by the WOS database, we conducted an in-depth review of MFPC papers published in the field and selected the top journals from them; the specific information is shown in Table 2. Among them, *Fuel* magazine, with its 90 published papers, topped the list, accounting for 12.55%; *Combustion Science and Technology*, with 72 papers (10.04%), and *Energy*, with 33 papers (4.60%), followed. *Process Safety and Environmental Protection* ranked first, with an average citation of 35.8, and *Fuel* ranked second, with an average citation of 30.9888. *Fuel* and *Process Safety and Environmental Protection* are at the top of the list in terms of the number of publications (first and fourth), impact factor (7.4 and 7.8), and average citations (second and first). The top ten journals focus on energy science, chemical engineering, and geochemistry, which together constitute a key platform for knowledge production and interaction in related research fields.

Table 2. Top ten journals in terms of number of publications.

Rank	Journal	Volume of Publications	Proportions	Impact Factor	Total Citations	Average Citation	Citation Index
1	Fuel	90	12.55%	7.4	2789	30.9888	SCI
2	Combustion Science and Technology	72	10.04%	1.9	522	7.2500	SCI
3	Energy	33	4.60%	9.0	658	19.9394	SCI
4	Process Safety and Environmental Protection	30	4.18%	7.8	1074	35.8000	SCI
5	ACS Omega	29	4.05%	4.1	202	6.9655	SCI
6	Journal of Thermal Analysis and Calorimetry	19	2.65%	4.4	229	12.0526	SCI
7	Environmental Science and Pollution Research	18	2.51%	5.8	379	21.0556	SCI
8	Energies	17	2.37%	3.2	113	6.6471	SCI
9	Energy Sources Part A-Recovery Utilization and Environmental Effects	17	2.37%	2.9	116	6.8235	SCI
10	Journal of Loss Prevention in the Process Industries	15	2.09%	3.5	155	10.3333	SCI

3.5.2. Journal Co-Citation Analysis

Select the node type as "Cited Journal" and the path algorithm as "Pathfinder" to obtain the co-citation graph of journals, as shown in Figure 7. Table 3 provides details of the top 10 journals with high citations.

Table 3. Top ten highly cited journals.

Rank	Journal	Co-Citation	Rank	Journal	Co-Citation
1	Fuel	519	6	Energy and Fuels	248
2	International Journal of Coal Geology	358	7	International Journal of Mining Science and Technology	240
3	Process Safety and Environmental Protection	349	8	Journal of China Coal Society	234
4	Journal of Loss Prevention in the Process Industries	279	9	Combustion Science and Technology	230
5	Fuel Processing Technology	271	10	Energy	228

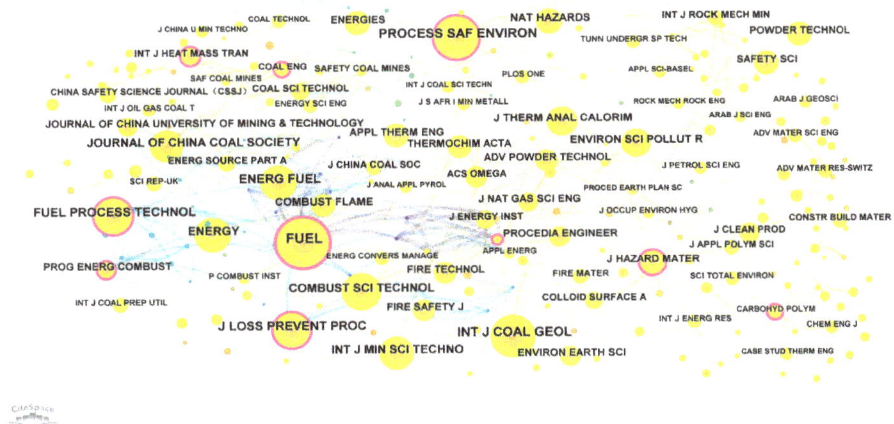

Figure 7. Journal co-citation knowledge map.

Among journal co-citations, *Fuel*, the *International Journal of Coal Geology*, *Process Safety and Environmental Protection*, the *Journal of Loss Prevention in the Process Industries*, and *Fuel Processing Technology* are frequently cited. In terms of co-citation, *Fuel* ranked first, with 519, the *International Journal of Coal Geology* ranked second, with 358, and *Process Safety and Environmental Protection* ranked third, with 349. Among these, *Fuel* and *Process Safety and Environmental Protection* play a key role in both average citation and co-citation. In general, the ten journals listed in the table are all prestigious academic publishing platforms around the world, and they have high profiles and influence in the MFPC research field. These include *Fuel*, *Energy*, *Process Safety and Environmental Protection*, and *Combustion Science and Technology*. These journals play a key role in mine fire control research.

3.6. Core Literature Citation Analysis

For an in-depth analysis of the most influential literature in the MFPC field, we selected the ten most frequently cited studies as research objects, including one review article and nine monographs. As shown in Table 4, the publication time distribution of these highly cited studies is as follows: one of them was published before 2008, which constituted the early representative achievement; six papers were published between 2009 and 2019, showing the active research trend in this field during the decade; and the remaining three papers, published after 2020, reflect the latest research in the field in recent years. Of these ten papers, four are related to colloid control technology (#1, #2, #6, #8), two are related to foam control technology (#3, #7), two are related to injection inerting control technology (#9, #10), one is related to mine fire suppressant retardant (#5), and one paper summarizes the existing technology in the MFPC field (#4). Cheng et al. [100] published an article entitled "An intelligent gel designed to control the spontaneous combustion of coal: Fire prevention and extinguishing properties" in *Fuel* in 2017, which was cited 354 times; Ren et al. [101] published an article entitled "Novel sodium silicate/polymer composite gels for the prevention of spontaneous combustion of coal" in the *Journal of Hazardous Materials* in 2019, which was cited 154 times. The two most cited articles are research articles. The reason is that the MFPC field entered a period of stable development around 2009, during which the number of published studies is relatively limited, and it is not enough to support a comprehensive and systematic review or summational research. Since 2019, the MFPC field has entered a stage of rapid development, especially in foam fire prevention technology and colloidal fire prevention technology, showing significant innovative vitality and technological progress.

In terms of colloidal fire prevention technology, researchers have actively absorbed the mature experience and advanced technology of medical, food, and other related industries and achieved a substantial jump in technology and innovation. After two years of rapid development, the MFPC field finally gave birth to the first comprehensive and systematic review and analysis of various current fire prevention technologies, marking the integration of theoretical research and practical experience in this field to a new height.

Table 4. Top ten highly cited studies.

Rank	Title	Types	Journal	Author	Year	Citation
#1	An intelligent gel designed to control the spontaneous combustion of coal: Fire prevention and extinguishing properties	Article	Fuel	Cheng et al. [100]	2017	354
#2	Novel sodium silicate/polymer composite gels for the prevention of spontaneous combustion of coal	Article	Journal of Hazardous Materials	Ren et al. [101]	2019	154
#3	Prevention, control and/or extinguishment of coal seam fires using cellular grout	Article	International Journal of Coal Geology	Colaizzi, GJ [102]	2004	131
#4	Overview of commonly used materials for coal spontaneous combustion prevention	Review	Fuel	Li et al. [103]	2020	112
#5	Inhibiting effects of three commercial inhibitors in spontaneous coal combustion	Article	Energy	Deng et al. [104]	2018	104
#6	Synthesis and characteristics of fire extinguishing gel with high water absorption for coal mines	Article	Process Safety and Environmental Protection	Li et al. [105]	2019	102
#7	Aqueous three-phase foam supported by fly ash for coal spontaneous combustion prevention and control	Article	Advanced Powder Technology	Qin et al. [106]	2014	98
#8	Fire prevention and control using gel-stabilization foam to inhibit spontaneous combustion of coal: Characteristics and engineering applications	Article	Fuel	Xue et al. [107]	2020	96
#9	Coal spontaneous combustion and N2 suppression in triple goafs: A numerical simulation and experimental study	Article	Fuel	Liu et al. [108]	2020	91
#10	A New Approach to Control a Serious Mine Fire with Using Liquid Nitrogen as Extinguishing Media	Article	Fire Technology	Zhou et al. [109]	2015	84

3.7. Hot Spots and Frontiers of Mine Fire Prevention and Control Research

3.7.1. Keyword Co-Occurrence Analysis

Keywords play an irreplaceable role in the structure of academic papers. In essence, they are highly condensed and precise refinements of the thesis theme, aiming to convey the core issues and ideological essence of the article in the most concise form. They are like "labels" for the content of the article, conveying the core issues and research directions in a concise and clear manner. Therefore, keyword analysis is actually a process of systematic testing and in-depth interpretation of the key points of the paper [34]. The WOS data were formatted and converted using CiteSpace, the time was 2000–2023, the time slice was 2 years, and the "keyword" was set to "Node type" to analyze the content. The keyword co-occurrence network that was generated according to the above settings has been visually displayed in Figure 8. The meaning of each element in the figure is provided by keywords that appear in the form of nodes. The line reveals the association law of keywords as follows: if a pair of keywords appear in the same study, there must be a line between them. In the network diagram, there is a direct proportional relationship between the size design of nodes and the actual occurrence times of the keywords they represent in the literature data set. Specifically, the size of the node area becomes an intuitive scale to measure the actual use of the keyword in the research literature, following the principle of "the larger the area, the more mentions". This means that when a node presents a larger area in the graph, it actually reveals that the corresponding keyword has been frequently used and emphasized in the literature collection.

Figure 8. Keyword co-occurrence mapping.

As can be seen in the figure, spontaneous combustion and low-temperature oxidation are the core keywords; their significant network center status and extensive connectivity highlight their critical role in the field of mine fire research. This not only reflects the deep concern of the scientific research community for SCC processes, especially oxidation reactions occurring at low temperatures, but also reveals the seriousness of these processes as potential causes of mine fire accidents, so it has become the focus of research by researchers. At the same time, the high frequency of the keywords prevention and behavior and their close correlation with many other keywords further reveal the focus of the research. The strong correlation of prevention with words such as fire agent, inhibition mechanism, and design indicates how to effectively prevent the development of fire sources through scientific design and the selection of appropriate fire extinguishing agents and inhibition mechanisms in MFPC research. It became the key technology strategy of MFPC. This highlights the central role of preventive measures in reducing mine fire risk. On the other hand, the close connection between the keywords behavior and oxidation, temperature, and model predictions reveals researchers' great concern for the characteristics of mine fire behavior and how to use models to predict their dynamic changes. This indicates that researchers are committed to studying the interaction between the physical and chemical behavior of fire and environmental variables in order to more accurately predict the development trend of fire, and then formulate more targeted prevention and control strategies. In addition, the keywords 3 phase foam, gas, ionic liquid, and gel in the map are closely related to the above important concepts, reflecting the cutting-edge trend of MFPC field in technical applications. These advanced technologies not only play an important role in physical, chemical, or physicochemical inhibition strategies but they are directly related to the solution of core problems such as spontaneous combustion, low-temperature oxidation, prevention, oxidation, and behavior. This shows the deep integration of theory and practice. In summary, this analysis not only reveals the focus of mine fire prevention and control research, such as spontaneous combustion and low-temperature oxidation mechanism, but also emphasizes the necessity of prevention measures and fire behavior research and shows the development of key technologies in the MFPC field, providing clear directional guidance and potential innovation points for subsequent research.

3.7.2. Keyword Clustering

Thirty-three clusters were obtained from the 290 keyword extractions, and the clusters with less than ten clusters were filtered by CiteSpace's built-in filtering function. Figure 9 shows the main clusters in this paper, according to # 0, # 1. . . . . .# 7 sort. In the MFPC field, the CiteSpace analysis map reveals the research on the inhibition effect of different control technologies on various types of coal under different storage methods, states, and external environments. According to the unique properties of different coal, especially the characteristic temperature and the formation of signature gas during spontaneous combustion, the effects of various inhibition techniques are evaluated. In recent years, blocking mechanism and gel foam have become hot topics and have widely drawn the attention of academia and industry, and many scholars have focused their research on the blocking mechanism of anti-fire gels [110–114]. The rapid development of computer technology and software technology is promoting the progress of scientific research with unprecedented efforts, especially in the field of MFPC; the numerical simulation of various inhibition technologies has gradually emerged and become the focus of academic and industrial attention [115–122]. This kind of research mainly discusses how various materials affect the active site of SCC at the molecular level. In short, the research in the MFPC field is deepening, and new prevention and control technologies and evaluation systems are gradually taking shape, which will further promote scientific and technological progress and development in this field.

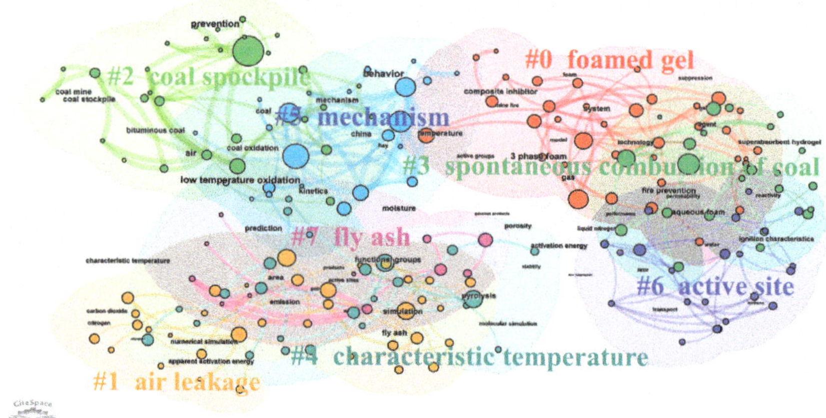

Figure 9. Knowledge map for keyword clustering.

3.7.3. Keyword Bursting

CiteSpace, a scientific knowledge graph analysis tool, is used to dig deeply into the literature data in MFPC field in order to reveal its frontier fields and future development trends, and the mutation word analysis function built into the software is adopted. The purpose of this function is to identify keywords with significant changes in frequency over a period of time, that is, abrupt changes through a specific algorithm. These keywords are often closely related to emerging trends and hot issues in the research field and are important clues to reveal the development context and future direction of the discipline. A specific value ($\gamma = 0.6$) was assigned to the key parameter γ in the mutation word analysis, and 33 keywords meeting the mutation word criteria were successfully identified from the literature data in the MFPC field. In order to focus more on the most representative and influential mutation words, the top 25 keywords were further selected from the 33 keywords. These keywords were more prominent in the degree of mutation, and their Burst value ranked first among all mutation words. These keywords are sorted by Burst value size, and the sorting results are visually shown in Figure 10. Year indicates the year in which the keyword first appears; begin and end indicate the start and end year in which the keyword is used as the frontier; and strength indicates the emergence strength. The red line represents the specific period in which the keyword has become a hot topic of academic research, light blue represents that the node has not yet appeared, and dark blue indicates that the node has begun to appear.

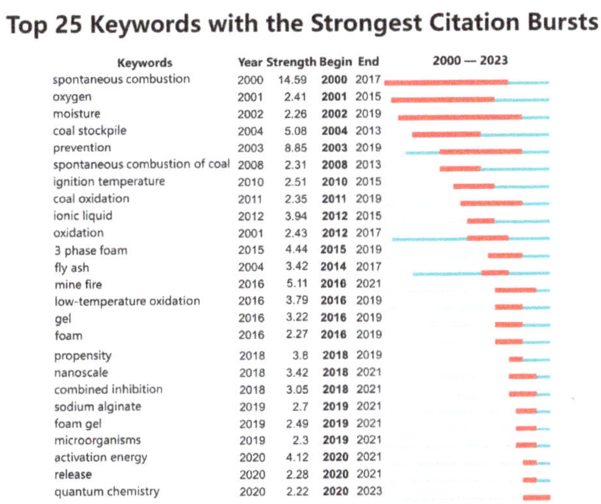

Figure 10. Sorting the top 25 emergent keywords.

As can be seen in the figure, spontaneous combustion ranks first, with a significant outburst intensity value of 14.59, which indicates the deep concern about the spontaneous combustion mechanism of coal and its prevention and control strategies in this field. Prevention followed, with a strength of 8.85, underscoring the centrality of prevention measures in mine fire prevention and the emphasis placed on highly effective inhibitors and suppression mechanisms. The intensity values of mine fire and coal stockpile are 5.11 and 5.08, respectively, which suggests that researchers should consider fire risk control in specific application scenarios, especially the safety management of open coal stockpile, a fire-prone area. The high breakout strength of three-phase foam and release (4.44 and 4.12, respectively) reflects the rise of innovative retarding technologies, particularly those that can release retarding ingredients in a timely manner to enhance flame retardancy. It is worth noting that the discovery of activation energy and quantum chemistry marks a change in research perspective, that is, from the traditional macroscopic analysis to the in-depth exploration of the microscopic mechanism of coal spontaneous combustion. This is closely related to the popularity of molecular simulation software, which enables researchers to explore in more detail how different inhibitors affect the oxidation reaction path and activation site of coal at the molecular level, thus promoting the development of physicochemical synergistic inhibition methods. The decline of moisture and ignition temperature means that the research on the basic physical properties of coal has become mature, and the rise of new keywords, such as release and quantum chemistry, indicates that MFPC research is entering a new stage of development. It focuses on deeper theoretical understanding and technological innovation. In summary, the emergence and change in these keywords not only reveal the shift of research hotspots in the MFPC field but also provide clues for future research directions, namely, strengthening the exploration of the inhibition mechanism at the microscopic level, developing intelligent responsive inhibition materials, and optimizing the design and application of inhibitors using advanced computational techniques. This map provides valuable visual evidence for understanding the evolution of knowledge structures in the MFPC field, facilitating cross-disciplinary exchanges and collaboration, and providing new ideas for addressing complex scientific challenges.

4. Conclusions

In this study, the CiteSpace software is used to conduct an in-depth metrological analysis of the MFPC literature, which provides a strong basis for the research trend and future development direction of this field. The analysis results show the following:

(1) The annual volume of articles in the MFPC research field has grown significantly, especially between 2020 and 2023, indicating that the field is experiencing an unprecedented period of active development. China's research contributions in the MFPC field are the most prominent, followed by the United States and Australia, showing the dominance of these countries in international research. CUMT ranks among the best research institutions. *Fuel* and *Process Safety and Environmental Protection* play a key role in both average citation and co-citation. The top 10 cited articles are one review and nine papers. One of these is entitled "An intelligent gel designed to control the spontaneous combustion of coal". The research paper "Fire prevention and extinguishing properties" topped the list for citations;

(2) The high occurrence rate of prevention and behavior keywords emphasizes the importance of prevention measures and coal spontaneous combustion behavior research. In terms of technology, the reference to retarding technologies, such as three-phase foams, gels, and ionic liquids, highlights the key role of materials and technological advances in the MFPC field. A keyword cluster analysis reveals the in-depth discussion of MFPC research in the evaluation of inhibition effect, adaptability of different coal characteristics, influence of external environmental factors, etc., and further confirms the status of inhibition mechanism as a current research hotspot;

(3) The emergence of activation energy, release, quantum chemistry, and other recent hot words indicates that the research focus is shifting to the mechanism of physicochemical synergistic inhibition, energy release control, and the application of quantum chemistry theory in the study of coal spontaneous combustion. It indicates that the field is developing in the direction of more micro and theoretical depth;

(4) The CiteSpace analysis reveals the dynamic changes in MFPC research, including the diversification of research topics, the iterative updating of technologies and the frontier transformation of research hotspots. Future research should continue to focus on the in-depth exploration of the coal spontaneous combustion mechanism, strengthen the development of physicochemical synergistic inhibition materials, and apply advanced computational methods, such as quantum chemical simulation, to improve the theoretical guidance and practical application level of MFPC. Exploring new chemical inhibitors, simplifying the preparation and application process of materials, improving the thermal stability of materials, extending the action time, and improving the environmental safety of materials are the main problems facing the development of fireproof materials. These findings not only provide a comprehensive overview of the current MFPC research but also point the way for future research direction and have important scientific value for reducing mine fire risk and ensuring mining safety.

Author Contributions: Conceptualization, D.H. and G.N.; data curation, B.L.; investigation, Y.R., C.S., Y.Y. and Z.Z.; methodology, D.H. and F.W.; writing—original draft, D.H. and G.N.; writing—

review and editing, G.N., B.L., F.W. and Y.R. All authors have read and agreed to the published version of the manuscript.

Funding: This work was financially sponsored by the China University of Mining and Technology (Beijing) College Student Innovation Training Program (202412032) and supported by the Fundamental Research Funds for the Central Universities.

Institutional Review Board Statement: Not applicable.

Informed Consent Statement: Not applicable.

Data Availability Statement: The original contributions presented in the study are included in the article, further inquiries can be directed to the corresponding author.

Conflicts of Interest: The authors declare no conflicts of interest.

Nomenclature

MFPC	mine fire prevention and control
SCC	spontaneous combustion of coal
WOS	Web of Science
CUMT	China University of Mining and Technology
XUST	Xi'an University of Science and Technology
SDUST	Shandong University of Science and Technology
HPU	Henan Polytechnic University
AUST	Anhui University of Science and Technology
USTB	University of Science and Technology of Beijing
CSIRO	Commonwealth Scientific and Industrial Research Organisation
PCSHE	Pennsylvania Commonwealth System of Higher Education

References

1. Yan, D.; Su, F.; Wang, Z.; Chen, Z.; Lei, Y.; Ye, B. Achieving High-Quality Development in China's Coal-Based Cities: How Heterogeneities Green Innovation Promote Carbon Emission Performance. *Environ. Dev. Sustain.* **2024**, *26*, 13921–13941. [CrossRef]
2. Chen, J.; Lv, Y.; Fan, T. Research on the Evolution and Driving Factors of Digitalization of Energy in China—A New Perspective Based on Coupling Coordination. *Heliyon* **2023**, *9*, e14138. [CrossRef] [PubMed]
3. Huang, A.-C.; Huang, C.-F.; Shu, C.-M. A Case Study for an Assessment of Fire Station Selection in the Central Urban Area. *Safety* **2023**, *9*, 84. [CrossRef]
4. Huang, A.-C.; Cao, F.-C.; Ma, X.-Y. A Comparative Approach Study on the Thermal and Calorimetric Analysis of Fire-Extinguishing Powders. *Safety* **2024**, *10*, 31. [CrossRef]
5. Pei, Q.; Jia, Z.; Liu, J.; Wang, Y.; Wang, J.; Zhang, Y. Prediction of Coal Spontaneous Combustion Hazard Grades Based on Fuzzy Clustered Case-Based Reasoning. *Fire* **2024**, *7*, 107. [CrossRef]
6. Yan, L.; Qin, Y.; Chu, C.; Sun, B.; Xu, Y.; Jiang, Q.; Song, Y.; Tang, F.; Xu, H. Experimental Study on Inhibition of Coal Spontaneous Combustion by Carbon Dioxide and Nitrogen. Energy Sources Part A-Recovery Util. *Environ. Eff.* **2024**, *46*, 6444–6460.
7. Zhang, J.; Xu, K.; Reniers, G.; You, G. Statistical Analysis the Characteristics of Extraordinarily Severe Coal Mine Accidents (ESCMAs) in China from 1950 to 2018. *Process Saf. Environ. Prot.* **2020**, *133*, 332–340. [CrossRef]
8. Yang, F.; Lu, Y.; Yan, Z.; Wang, G.G.X.; Hu, X.; Gu, W. Colloidal Particle-Stabilized Foam To Control the Coal Spontaneous Combustion: Stability Mechanism Analysis and Extinguishing Properties. *Energy Fuels* **2020**, *34*, 14822–14831. [CrossRef]
9. Zhang, Y.; Wu, B.; Liu, S.-H.; Lei, B.; Zhao, J.; Zhao, Y. Thermal Kinetics of Nitrogen Inhibiting Spontaneous Combustion of Secondary Oxidation Coal and Extinguishing Effects. *Fuel* **2020**, *278*, 118223. [CrossRef]
10. Dong, K.; Wang, J.; Zhang, Y.; Liang, Z.; Shi, Q. Performance of Fire Extinguishing Gel with Strong Stability for Coal Mine. *Combust. Sci. Technol.* **2022**, *194*, 1661–1677. [CrossRef]
11. Ge, L.; Shao, Y.; Wang, Y.; Zhang, G.; Zhang, Z.; Liu, L. Experimental Research on Inerting Characteristics of Carbon Dioxide Used for Fire Extinguishment in a Large Sealed Space. *Process Saf. Environ. Prot.* **2020**, *142*, 174–190. [CrossRef]
12. Tang, Y.; Guo, Q.; Yerman, L. Experimental Investigation on Using Chloride/Hydroxide Aerosol to Control Spontaneous Combustion of Lignite in Underground Coal Mines. *Energy Fuels* **2020**, *34*, 10607–10618. [CrossRef]
13. Zhai, X.-W.; Pan, W.-J.; Xiao, Y.; Wang, S.; Ouyang, L. Inhibition of Coal Spontaneous Combustion by an Environment-Friendly, Water-Based Fire Extinguishing Agent. *J. Therm. Anal. Calorim.* **2021**, *144*, 325–334. [CrossRef]
14. Xi, X.; Shi, Q.; Jiang, S.; Zhang, W.; Wang, K.; Zhengyan, W. Study on the Effect of Ionic Liquids on Coal Spontaneous Combustion Characteristic by Microstructure and Thermodynamic. *Process Saf. Environ. Protect.* **2020**, *140*, 190–198. [CrossRef]

15. Lu, Y.; Liu, Y.; Shi, S.; Wang, G.G.X.; Li, H.; Wang, T. Micro-Particles Stabilized Aqueous Foam for Coal Spontaneous Combustion Control and Its Flow Characteristics. *Process Saf. Environ. Protect.* **2020**, *139*, 262–272. [CrossRef]
16. Zhang, X.; Yang, J.; Xie, P.; Liu, H.; Deng, Q.; Dai, F. Experimental Study on Controlled-Release Inhibitor Foam for Restraining Spontaneous Combustion of Coal. *Energy Explor. Exploit.* **2020**, *38*, 1159–1177. [CrossRef]
17. Han, C.; Nie, S.; Liu, Z.; Liu, S.; Zhang, H.; Li, J.; Zhang, H.; Wang, Z. A Novel Biomass Sodium Alginate Gel Foam to Inhibit the Spontaneous Combustion of Coal. *Fuel* **2022**, *314*, 122779. [CrossRef]
18. Lu, W.; Sun, X.; Gao, L.; Hu, X.; Song, H.; Kong, B. Study on the Characteristics and Mechanism of DL-Malic Acid in Inhibiting Spontaneous Combustion of Lignite and Bituminous Coal. *Fuel* **2022**, *308*, 122012. [CrossRef]
19. Xue, D.; Hu, X.; Cheng, W.; Yu, X.; Wu, M.; Zhao, Y.; Lu, Y.; Pan, R.; Niu, H.; Hu, S. Development of a Novel Composite Inhibitor Modified with Proanthocyanidins and Mixed with Ammonium Polyphosphate. *Energy* **2020**, *213*, 118901. [CrossRef]
20. Geng, H.-J.; Zhao, Y.-M.; Liu, X.-L.; Tian, F.-C. The Application of Nitrogen Curtain Technology to Longwall Goaf to Prevent the Spontaneous Combustion of Coal: A Case Study in Shajihai Coalmine, China. *Fire* **2023**, *6*, 363. [CrossRef]
21. Guo, J.; Cai, G.; Jin, Y.; Zheng, X.; Liu, Y. An Improved Composite Fly Ash Gel to Extinguish Underground Coal Fire in Close Distance Coal Seams: A Case Study. *Adv. Mater. Sci. Eng.* **2020**, *2020*, 5695471. [CrossRef]
22. Zhang, F.; Li, Z.; Cheng, J.; Wang, Y.; Chen, J.; Jiang, Z. Preparation and Performance Study of OPC-MF-PEG Microcapsules Based on Strong Antioxidant Groups against Spontaneous Coal Combustion. *Constr. Build. Mater.* **2024**, *425*, 135808. [CrossRef]
23. Du, Q.; Mu, X.; Deng, C.; Duan, H.; Jing, P.; Qiao, L.; Guo, Y. Preparation of Complex Physicochemical Synergistic Inhibitors and Study on Inhibition of Spontaneous Combustion of Coal. *Int. J. Coal Prep. Util.* **2024**. [CrossRef]
24. Xi, X.; Xia, C.; Yin, C.; Jiang, S. Preparation and Fire Extinguishing Characteristics of Inorganic Solidified Foam with Large Amount of Slag. Energy Sources Part A-Recovery Util. *Environ. Eff.* **2024**, *46*, 5522–5535.
25. Li, L.; Liu, T.; Li, Z.; Chen, X.; Wang, L.; Feng, S. Different Prevention Effects of Ventilation Dilution on Methane Accumulation at High Temperature Zone in Coal Mine Goafs. *Energies* **2023**, *16*, 3168. [CrossRef]
26. Wang, Y.; Zheng, Q.; Su, H.; Huang, Z.; Wang, G. Synthesis and Characteristics of a pH-Sensitive Sol-Gel Transition Colloid for Coal Fire Extinguishing. *Gels* **2023**, *9*, 69. [CrossRef] [PubMed]
27. Zhang, X.; Pan, Y. Preparation, Properties and Application of Gel Materials for Coal Gangue Control. *Energies* **2022**, *15*, 557. [CrossRef]
28. Zhang, Y.; Xu, J.; Wang, D. Experimental Study on the Inhibition Effects of Nitrogen and Carbon Dioxide on Coal Spontaneous Combustion. *Energies* **2020**, *13*, 5256. [CrossRef]
29. Yan, Q.; Lu, Y.; Wang, Q.; Gu, W.; Wu, F. Research Progress and Development Trend of Coal Spontaneous Combustion Prevention Technology. *Combust. Sci. Technol.* **2024**. [CrossRef]
30. Liu, S.; Shi, G.; Ren, H.; Cao, J.; Wang, D. Key Parameters of Separated Nitrogen Injection in Goaf Based on Orthogonal Test. *Combust. Sci. Technol.* **2023**, *9*, 15428–15438. [CrossRef]
31. Nádudvari, Á.; Abramowicz, A.; Fabiańska, M.; Misz-Kennan, M.; Ciesielczuk, J. Classification of Fires in Coal Waste Dumps Based on Landsat, Aster Thermal Bands and Thermal Camera in Polish and Ukrainian Mining Regions. *Int. J. Coal Sci. Technol.* **2021**, *8*, 441–456. [CrossRef]
32. Alipour, M.; Alizadeh, B.; Jahangard, A.; GandomiSani, A. Wildfire Events at the Triassic–Jurassic Boundary of the Tabas Basin, Central Iran. *Int. J. Coal Sci. Technol.* **2021**, *8*, 897–907. [CrossRef]
33. Ding, X.; Yang, Z. Knowledge Mapping of Platform Research: A Visual Analysis Using VOSviewer and CiteSpace. *Electron. Commer. Res.* **2022**, *22*, 787–809. [CrossRef]
34. Yue, T.; Liu, H.; Long, R.; Chen, H.; Gan, X.; Liu, J. Research Trends and Hotspots Related to Global Carbon Footprint Based on Bibliometric Analysis: 2007–2018. *Environ. Sci. Pollut. Res.* **2020**, *27*, 17671–17691. [CrossRef] [PubMed]
35. Liu, H.; Hong, R.; Xiang, C.; Lv, C.; Li, H. Visualization and Analysis of Mapping Knowledge Domains for Spontaneous Combustion Studies. *Fuel* **2020**, *262*, 116598. [CrossRef]
36. Qiu, H.-H.; Liu, L.-G. A Study on the Evolution of Carbon Capture and Storage Technology Based on Knowledge Mapping. *Energies* **2018**, *11*, 1103. [CrossRef]
37. Hernandez-Torrano, D.; Ibrayeva, L. Creativity and Education: A Bibliometric Mapping of the Research Literature (1975–2019). *Think. Skills Creat.* **2020**, *35*, 100625. [CrossRef]
38. Shao, Z.; Tan, B.; Guo, Y.; Li, T.; Li, X.; Fang, X.; Wang, F.; Zhang, Q.; Wang, H. Visualization and Analysis of Mapping Knowledge Domains for Coal Pores Studies. *Fuel* **2022**, *320*, 123761. [CrossRef]
39. Hu, L.; Zhu, H.; Liao, Q.; Qu, B.; Gao, R.; Tao, R.; Fu, M. Visual Analysis of Coal Fire Detection Research Based on Bibliometrics. *Int. J. Remote Sens.* **2023**, *44*, 5976–6011. [CrossRef]
40. Duan, Z.; Zhang, Y.; Deng, J.; Shu, P.; Yao, D. A Systematic Exploration of Mapping Knowledge Domains for Free Radical Research Related to Coal. *Energy* **2023**, *282*, 128914. [CrossRef]
41. Wang, T.; Wang, H.; Liu, Y.; Zhou, Z.; Wang, G.; Wang, Y.; Fang, X.; Chen, Z.; Qi, Q. Visualization and Analysis of Mapping Knowledge Domains for Coal Functional Groups. *Combust. Sci. Technol.* **2024**. [CrossRef]
42. Wang, T.; Wang, H.; Fang, X.; Wang, G.; Chen, Y.; Xu, Z.; Qi, Q. Research Progress and Visualization of Underground Coal Fire Detection Methods. *Environ. Sci. Pollut. Res.* **2023**, *30*, 74671–74690. [CrossRef] [PubMed]
43. Zhao, S.; Chao, Q.; Yang, L.; Qin, K.; Zuo, J. A Review on Application of Acoustic Emission in Coal-Analysis Based on CiteSpace Knowledge Network. *Processes* **2022**, *10*, 2397. [CrossRef]

44. Yang, F.; Qiu, D. Exploring Coal Spontaneous Combustion by Bibliometric Analysis. *Process Saf. Environ. Protect.* **2019**, *132*, 1–10. [CrossRef]
45. Yan, C.; Li, H.; Pu, R.; Deeprasert, J.; Jotikasthira, N. Knowledge Mapping of Research Data in China: A Bibliometric Study Using Visual Analysis. *Libr. Hi Tech* **2024**, *42*, 331–349. [CrossRef]
46. Kokol, P.; Blazun Vosner, H.; Zavrsnik, J. Application of Bibliometrics in Medicine: A Historical Bibliometrics Analysis. *Health Inf. Libr. J.* **2021**, *38*, 125–138. [CrossRef] [PubMed]
47. Singh, R.V.K.; Tripathi, D.D.; Singh, V.K. Evaluation of Suitable Technology for Prevention and Control of Spontaneous Heating/Fire in Coal Mines. *Arch. Min. Sci.* **2008**, *53*, 555–564.
48. Wu, C.; Li, Z.; Yang, F.; Hu, H.; Gu, D. Risk Forecast of Spontaneous Combustion of Sulfide Ore Dump in a Stope and Controlling Approaches of the Fire. *Arch. Min. Sci.* **2008**, *53*, 565–579.
49. Xie, Z.; Li, R. Study on Prediction Technology of Coal Seam Spontaneous Combustion of Renlou Coal Mine. In Proceedings of the Progress in Safety Science and Technology, Vol VII, PTS A and B, Bejing, China, 24 September 2008; Li, S.C., Wang, Y.J., An, Y., Sun, X.Y., Li, X., Eds.; Science Press Beijing: Beijing, China, 2008; Volume 7, pp. 1419–1423.
50. Wang, J.; Fan, Z.; Zhang, J. Prevention and Control for Coal Spontaneous Combustionin by Fully Mechanized Top-Coal Caving of Extremely Thick Coal Seam. In Proceedings of the Mine Hazards Prevention and Control Technology, Brisbane, Australia, 19–21 November 2013; Wang, C., Guo, W.J., Cheng, J.L., Eds.; Science Press Beijing: Qingdao, China, 2007; pp. 186–191.
51. Deng, J.; Lue, H.-F.; Xiao, Y.; Wang, C.-P.; Shu, C.-M.; Jiang, Z.-G. Thermal Effect of Ionic Liquids on Coal Spontaneous Combustion. *J. Therm. Anal. Calorim.* **2019**, *138*, 3415–3424. [CrossRef]
52. Xi, Z.; Wang, X.; Wang, X.; Wang, L.; Li, D.; Guo, X.; Jin, L. Self-Hardening Thermoplastic Foam for the Inhibition of Coal Oxidation at Low Temperatures. *Combust. Sci. Technol.* **2019**, *191*, 1942–1959. [CrossRef]
53. Wang, L.; Tawiah, B.; Shi, Y.; Cai, S.; Rao, X.; Liu, C.; Yang, Y.; Yang, F.; Yu, B.; Liang, Y.; et al. Highly Effective Flame-Retardant Rigid Polyurethane Foams: Fabrication and Applications in Inhibition of Coal Combustion. *Polymers* **2019**, *11*, 1776. [CrossRef] [PubMed]
54. Bin, L.; Tsai, Y.-T.; Liu, S.-H.; Deng, J.; Xiao, Y.; Wang, Q.-H.; Shu, C.-M. Effects of 1-Butyl-3-Methylimidazolium Nitrate on the Thermal Hazardous Properties of Lignitous and Long Flame Coal through a Green Approach and Thermokinetic Models. *Process Saf. Environ. Protect.* **2019**, *131*, 127–134.
55. Lu, H.-F.; Xiao, Y.; Deng, J.; Li, D.; Yin, L.; Shu, C.-M. Inhibiting Effects of 1-Butyl-3-Methyl Imidazole Tetrafluoroborate on Coal Spontaneous Combustion under Different Oxygen Concentrations. *Energy* **2019**, *186*, 115907. [CrossRef]
56. Wang, K.; Fan, H.; Gao, P.; He, Y.; Shu, P. Spontaneous Combustion Characteristics of Wetting Coal under Different Prepyrolysis Temperatures. *ACS Omega* **2020**, *5*, 33347–33356. [CrossRef] [PubMed]
57. Wang, C.-P.; Bai, Z.-J.; Xiao, Y.; Deng, J.; Shu, C.-M. Effects of FeS_2 on the Process of Coal Spontaneous Combustion at Low Temperatures. *Process Saf. Environ. Protect.* **2020**, *142*, 165–173. [CrossRef]
58. Lei, B.; He, B.; Xiao, B.; Du, P.; Wu, B. Effects of N_2 and 1,1,1,3,3,3-Hexafluoropropane ($C_3H_2F_6$) on Inhibition of Coal Flames. *J. Energy Resour. Technol.-Trans. ASME* **2020**, *142*, 102304. [CrossRef]
59. Cheng, J.; Wu, Y.; Dong, Z.; Zhang, R.; Wang, W.; Wei, G.; Chu, T.; Yu, Z.; Qin, Y.; Liu, G.; et al. A Novel Composite Inorganic Retarding Gel for Preventing Coal Spontaneous Combustion. *Case Stud. Therm. Eng.* **2021**, *28*, 101648. [CrossRef]
60. Xue, D.; Hu, X.; Dong, H.; Cheng, W.; Wang, W.; Liang, Y. Examination of Characteristics of Anti-Oxidation Compound Inhibitor for Preventing the Spontaneous Combustion of Coal. *Fuel* **2022**, *310*, 122160. [CrossRef]
61. Dou, G.; Liu, J.; Jiang, Z.; Jian, H.; Zhong, X. Preparation and Characterization of a Lignin Based Hydrogel Inhibitor on Coal Spontaneous Combustion. *Fuel* **2022**, *308*, 122074. [CrossRef]
62. Wang, F.; Sun, M.; Wang, J. Dimethyl Methylphosphonate for the Suppression of Coal Spontaneous Combustion. *Combust. Sci. Technol.* **2023**, *195*, 982–999. [CrossRef]
63. Wang, L.; Liu, Z.; Yang, H.; Li, H.; Yu, M.; He, T.; Luo, Z.; Liu, F. A Novel Biomass Thermoresponsive Konjac Glucomannan Composite Gel Developed to Control the Coal Spontaneous Combustion: Fire Prevention and Extinguishing Properties. *Fuel* **2021**, *306*, 121757. [CrossRef]
64. Cardella, G.M.; Hernández-Sánchez, B.R.; Sánchez-García, J.C. Women Entrepreneurship: A Systematic Review to Outline the Boundaries of Scientific Literature. *Front. Psychol.* **2020**, *11*, 1557. [CrossRef] [PubMed]
65. Dou, Z.; Li, L.-L.; Liu, Z.; Min, Y.-M.; Guo, S.-J.; Chen, L.-C.; Wang, Q.-L.; Li, J.-H.; Yang, J.-F.; Zhang, J.-W.; et al. Oxidation to Spontaneous Combustion of Pyrophoric Iron Sulphides in the Process Industries: A Review. *J. Loss Prev. Process Ind.* **2023**, *85*, 105171. [CrossRef]
66. Wang, C.; Duan, X.; Deng, Y.; Deng, J.; Bai, Z.; Yang, N.; Qu, G. Study of Inflatable Temperature-Sensitive Hydrogel to Prevent the Coal Spontaneous Combustion. *Fuel* **2023**, *354*, 129333. [CrossRef]
67. He, Y.; Deng, J.; Yi, X.; Xiao, Y.; Deng, Y.; Chen, W. Effect of Rare-Earth-Containing Inhibitors on the Low-Temperature Oxidation Characteristics and Thermodynamic Properties of Coal. *Energy* **2023**, *281*, 128316. [CrossRef]
68. Bai, Z.; Deng, J.; Wang, C.; Hou, Y.; Zhang, Y.; Kang, F.; Ramakrishna, S. Study on the Mechanism of Lignite Oxidation Inhibition by Antioxidant Resveratrol. *Energy* **2023**, *273*, 127235. [CrossRef]
69. Bai, Z.; Deng, J.; Wang, C.; Zhang, Y.; Shu, C.-M.; Ramakrishna, S. Effect of Anions in Ionic Liquids on Microstructure and Oxidation Characteristics of Lignite. *Fuel* **2023**, *339*, 127446. [CrossRef]

70. Zhao, Y.; Sun, G.; Hu, X.; Xue, D.; Wang, K. Study on the Preparation of Inorganic Solidified Foam with High Stability and Its Prevention and Treatment of Coal Spontaneous Combustion. *Energy Fuels* **2023**, *37*, 14268–14279. [CrossRef]
71. Chen, J.; Jia, B.; Fu, S.; Wen, Y.; Liang, Y.; Tian, F. Novel PFA-Based Inorganic Three-Phase Foam for Inhibiting Coal Spontaneous Combustion. *ACS Omega* **2023**, *8*, 24615–24623. [CrossRef]
72. Xue, D.; Hu, X.; Sun, G.; Wang, K.; Liu, T.; Wang, J.; Wang, F. A Study on a Janus-Type Composite Solidified Foam and Its Characteristics for Preventing and Controlling Spontaneous Combustion of Coal. *Energy* **2023**, *275*, 127433. [CrossRef]
73. Dong, H.; Hu, X.; Yu, A.; Wang, W.; Zhao, Q.; Wei, H.; Yang, Z.; Wang, X.; Luo, C. Study on the Mechanism of an Enteromorpha-Based Compound Inhibitor for Inhibiting the Spontaneous Combustion of Coal Using in Situ Infrared Spectroscopy and Thermal Analysis Kinetics. *J. Environ. Chem. Eng.* **2023**, *11*, 109577. [CrossRef]
74. Yin, Y.; Zhang, Y.; Huang, Z.; Ding, H.; Hu, X.; Gao, Y.; Yang, Y. Study on the Synergistic Antioxidant Effect of Coal Inhibitors and the DFT Calculation. *Combust. Sci. Technol.* **2023**. [CrossRef]
75. Ma, W.; Ma, Q.; Zhao, Y.; Hu, X.; He, Z.; Chi, L.; Zhao, X.; Zhao, Y.; Sun, F.; Chu, C. Preparation and Resource Utilization of Sludge-Based Fire-Preventive and Extinguishing Composite Gel for Coal Mine. *J. Environ. Chem. Eng.* **2023**, *11*, 109465. [CrossRef]
76. Cao, H.; Lu, W.; Yin, Z.; Fu, H.; Mou, X.; Pan, X.; Kong, B. Research on the Tracing and Fire Prevention Performance of New Mining Composite Gel. *Combust. Sci. Technol.* **2023**. [CrossRef]
77. Zhou, G.; Li, S.; Zhang, Q.; Liang, Y.; Song, S.; Sun, B.; Wang, Q. Preparation and Properties of Bio-Based Self-Healing Fire Prevention Gel Reinforced with Expanded Graphite Ligand. *J. Build. Eng.* **2023**, *66*, 105845. [CrossRef]
78. Shi, Q.; Qin, B.; Hao, Y.; Li, H. Experimental Investigation of the Flow and Extinguishment Characteristics of Gel-Stabilized Foam Used to Control Coal Fire. *Energy* **2022**, *247*, 123484. [CrossRef]
79. Shi, Q.; Qin, B.; Xu, Y.; Hao, M.; Shao, X.; Zhuo, H. Experimental Investigation of the Drainage Characteristic and Stability Mechanism of Gel-Stabilized Foam Used to Extinguish Coal Fire. *Fuel* **2022**, *313*, 122685. [CrossRef]
80. Lu, X.-X.; Han, Y.; Xue, X.; Wang, D.-M. Research on a Noble Extinguish Material for the Underground Fire Prevention. *Fire Mater.* **2020**, *44*, 230–241. [CrossRef]
81. Shi, Q.; Qin, B. Experimental Research on Gel-Stabilized Foam Designed to Prevent and Control Spontaneous Combustion of Coal. *Fuel* **2019**, *254*, 115558. [CrossRef]
82. Shi, Q.; Qin, B.; Bi, Q.; Qu, B. Fly Ash Suspensions Stabilized by Hydroxypropyl Guar Gum and Xanthan Gum for Retarding Spontaneous Combustion of Coal. *Combust. Sci. Technol.* **2018**, *190*, 2097–2110. [CrossRef]
83. Miao, J.; Yang, S.; Jiang, X.; Hou, Z.; Shao, H. Effect of the Sudden Change of Ambient Atmosphere on Free Radicals in Coal Body by CO_2 Fire Prevention Gas Injection. *Energy Sources Part A-Recovery Util. Environ. Eff.* **2023**, *45*, 829–840. [CrossRef]
84. Tang, Z.; Zhang, H.; Deng, J.; Yang, S.; Xu, G.; Zheng, W.; Chang, P. Synergistic Inhibiting and Sealing Material for Preventing the Compound Disaster of Gas and Coal Spontaneous Combustion: Ratio Optimization and Inhibition Test. *Combust. Sci. Technol.* 2022. [CrossRef]
85. Zhou, B.; Yang, S.; Yang, W.; Jiang, X.; Song, W.; Cai, J.; Xu, Q.; Tang, Z. Variation Characteristics of Active Groups and Macroscopic Gas Products during Low-Temperature Oxidation of Coal under the Action of Inert Gases N_2 and CO_2. *Fuel* **2022**, *307*, 121893. [CrossRef]
86. Cai, J.; Yang, S.; Zhong, Y.; Song, W.; Zheng, W. A Physical-Chemical Synergetic Inhibitor for Coal Spontaneous Combustion and Its Fire Prevention Performance. *Combust. Sci. Technol.* **2022**, *194*, 1155–1167. [CrossRef]
87. Zhao, T.; Yang, S.; Hu, X.; Song, W.; Cai, J.; Xu, Q. Restraining Effect of Nitrogen on Coal Oxidation in Different Stages: Non-Isothermal TG-DSC and EPR Research. *Int. J. Min. Sci. Technol.* **2020**, *30*, 387–395. [CrossRef]
88. Gao, L.; Tan, B.; Fan, L.; Wang, H.; Li, X.; Lu, W.; Jiang, Y. Comparison and Analysis of Spontaneous Combustion Control between Coal Storage Silos and Biomass Silos. *Energy* **2024**, *286*, 129623. [CrossRef]
89. Sun, Q.; Niu, H.; Yang, Y.; Sun, S. Study on the Inhibitory Effect of DL-Malic Acid on the Phase Structure and Combustion Characteristics of Coal. *Fire Mater.* **2024**, *48*, 238–247. [CrossRef]
90. Cheng, G.; Wang, H.; Tan, B.; Fu, S. Carbon Dioxide Prevents Oxygen Adsorption at Low-Temperature Oxidation Stage of Low-Rank Coal: Laboratory Study and Molecular Simulation. *Processes* **2023**, *11*, 2504. [CrossRef]
91. Niu, H.; Liu, Y.; Wang, H.; Tan, B.; Li, S. Effect of Functional Groups on Spontaneous Combustion Characteristics of High Water-Rich Coal after Long-Term Soaking and Drying. *Energy Sources Part A-Recovery Util. Environ. Eff.* **2023**, *45*, 9325–9340. [CrossRef]
92. Niu, H.; Sun, Q.; Bu, Y.; Yang, Y.; Sun, S.; Li, S.; Tao, M.; Mao, Z. Review and Prospects of Research on Materials to Prevent and Extinguish Mine Fires. *Fire Mater.* **2023**, *47*, 739–757. [CrossRef]
93. Tan, B.; Song, X.; Zhang, B.; Shao, Z.; Li, Z.; Liu, S. Study on the Effect of Different Seawater Mass Ratio on Coal Spontaneous Combustion Characteristics. *Thermochim. Acta* **2022**, *716*, 179328. [CrossRef]
94. Fang, X.; Wang, H.; Tan, B.; Wang, F.; Shao, Z.Z.; Cheng, G.; Yao, H. Experimental Comparison Study of CO_2 and N_2 Inerted Loose Coal Based on Atmospheric Pressure Gas Replacement. *Fuel* **2022**, *328*, 125347. [CrossRef]
95. Tan, B.; Li, X.; Zhang, X.; Zhang, Z.; Zhang, H. Research on Initial Prevention of Spontaneous Combustion in Coal Bunkers Based on Fire-Extinguishing and Fireproof Inerting. *ACS Omega* **2022**, *7*, 3359–3368. [CrossRef] [PubMed]
96. Wang, H.; Tan, B.; Shao, Z.; Guo, Y.; Zhang, Z.; Xu, C. Influence of Different Content of FeS_2 on Spontaneous Combustion Characteristics of Coal. *Fuel* **2021**, *288*, 119582. [CrossRef]

97. Li, J.; Li, Z.; Yang, Y.; Zhang, X.; Yan, D.; Liu, L. Inhibitive Effects of Antioxidants on Coal Spontaneous Combustion. *Energy Fuels* **2017**, *31*, 14180–14190. [CrossRef]
98. Yang, Y.; Li, Z.; Si, L.; Hou, S.; Zhou, Y.; Qi, Q. Consolidation Grouting Technology for Fire Prevention in Mined-out Areas of Working Face with Large Inclined Angle and Its Application. *Fire Mater.* **2017**, *41*, 700–715. [CrossRef]
99. Yang, Y.; Li, Z.; Tang, Y.; Liu, Z.; Ji, H. Fine Coal Covering for Preventing Spontaneous Combustion of Coal Pile. *Nat. Hazards* **2014**, *74*, 603–622. [CrossRef]
100. Cheng, W.; Hu, X.; Xie, J.; Zhao, Y. An Intelligent Gel Designed to Control the Spontaneous Combustion of Coal: Fire Prevention and Extinguishing Properties. *Fuel* **2017**, *210*, 826–835. [CrossRef]
101. Ren, X.; Hu, X.; Xue, D.; Li, Y.; Shao, Z.; Dong, H.; Cheng, W.; Zhao, Y.; Xin, L.; Lu, W. Novel Sodium Silicate/Polymer Composite Gels for the Prevention of Spontaneous Combustion of Coal. *J. Hazard. Mater.* **2019**, *371*, 643–654. [CrossRef]
102. Colaizzi, G.J. Prevention, Control and/or Extinguishment of Coal Seam Fires Using Cellular Grout. *Int. J. Coal Geol.* **2004**, *59*, 75–81. [CrossRef]
103. Li, Q.-W.; Xiao, Y.; Zhong, K.-Q.; Shu, C.-M.; Lu, H.-F.; Deng, J.; Wu, S. Overview of Commonly Used Materials for Coal Spontaneous Combustion Prevention. *Fuel* **2020**, *275*, 117981. [CrossRef]
104. Deng, J.; Yang, Y.; Zhang, Y.-N.; Liu, B.; Shu, C.-M. Inhibiting Effects of Three Commercial Inhibitors in Spontaneous Coal Combustion. *Energy* **2018**, *160*, 1174–1185. [CrossRef]
105. Synthesis and Characteristics of Fire Extinguishing Gel with High Water Absorption for Coal Mines (Topic)—1—All Databases. Available online: https://webofscience-clarivate-cn-s.vpn.cumtb.edu.cn:8118/wos/alldb/summary/519b1f54-a5c7-481a-9d4c-13169e237368-bc4da01d/relevance/1 (accessed on 11 December 2023).
106. Qin, B.; Lu, Y.; Li, Y.; Wang, D. Aqueous Three-Phase Foam Supported by Fly Ash for Coal Spontaneous Combustion Prevention and Control. *Adv. Powder Technol.* **2014**, *25*, 1527–1533. [CrossRef]
107. Xue, D.; Hu, X.; Cheng, W.; Wei, J.; Zhao, Y.; Shen, L. Fire Prevention and Control Using Gel-Stabilization Foam to Inhibit Spontaneous Combustion of Coal: Characteristics and Engineering Applications. *Fuel* **2020**, *264*, 116903. [CrossRef]
108. Liu, Y.; Wen, H.; Guo, J.; Jin, Y.; Wei, G.; Yang, Z. Coal Spontaneous Combustion and N_2 Suppression in Triple Goafs: A Numerical Simulation and Experimental Study. *Fuel* **2020**, *271*, 117625. [CrossRef]
109. Zhou, F.; Shi, B.; Cheng, J.; Ma, L. A New Approach to Control a Serious Mine Fire with Using Liquid Nitrogen as Extinguishing Media. *Fire Technol.* **2015**, *51*, 325–334.
110. Xue, D.; Hu, X.; Cheng, W.; Wu, M.; Shao, Z.; Li, Y.; Zhao, Y.; Zhang, K. Carbon Dioxide Sealing-Based Inhibition of Coal Spontaneous Combustion: A Temperature-Sensitive Micro-Encapsulated Fire-Retardant Foamed Gel. *Fuel* **2020**, *266*, 117036. [CrossRef]
111. Fan, Y.-J.; Zhao, Y.-Y.; Hu, X.-M.; Wu, M.-Y.; Xue, D. A Novel Fire Prevention and Control Plastogel to Inhibit Spontaneous Combustion of Coal: Its Characteristics and Engineering Applications. *Fuel* **2020**, *263*, 116693. [CrossRef]
112. Cao, Q.; Shen, J.; Guo, W. Effects of Foaming Agent on Properties of Foamed Gel. *Adv. Compos. Lett.* **2020**, *29*, 0963693519881649. [CrossRef]
113. Li, Y.; Hu, X.; Cheng, W.; Shao, Z.; Xue, D.; Zhao, Y.; Lu, W. A Novel High-Toughness, Organic/Inorganic Double-Network Fire-Retardant Gel for Coal-Seam with High Ground Temperature. *Fuel* **2020**, *263*, 116779. [CrossRef]
114. Gao, J.; Chu, R.; Meng, X.; Yang, J.; Yang, D.; Li, X.; Lou, W. Synergistic Mechanism of CO_2 and Active Functional Groups during Low Temperature Oxidation of Lignite. *Fuel* **2020**, *278*, 118407. [CrossRef]
115. Jin, J.; Hao, C.; Shen, W.; Yuan, Z.; He, W.; Li, L. Study on the Microcosmic Mechanism of Inhibition of Coal Spontaneous Combustion by Antioxidant Diphenylamine. *J. Mol. Struct.* **2023**, *1294*, 136357. [CrossRef]
116. Li, H.; Shen, X.; Lu, J.; Lu, Y.; Shi, S.; Shao, S. Experimental and Quantum Chemical Investigation on the Inhibitory Effects of Resveratrol on Coal Spontaneous Combustion. *Fuel* **2023**, *354*, 129297. [CrossRef]
117. Gao, F.; Jia, Z.; Bai, Q.; Dong, J. Characteristics and Mechanism of Glutathione in Inhibiting Coal Spontaneous Combustion. *Fuel* **2023**, *352*, 129006. [CrossRef]
118. Zhang, X.; Lu, B.; Zhang, J.; Fu, X.; Deng, H.; Qiao, L.; Ding, C.; Gao, F. Experimental and Simulation Study on Hydroxyl Group Promoting Low-Temperature Oxidation of Active Groups in Coal. *Fuel* **2023**, *340*, 127501. [CrossRef]
119. Zhang, X.; Yu, C.; Lu, B.; Gao, F.; Shan, C.; Zou, J. Study on the Inhibitory Mechanism of Dehydrogenated Antioxidants on Coal Spontaneous Combustion. *Sci. Rep.* **2022**, *12*, 21237. [CrossRef] [PubMed]
120. Li, L.; Hao, C.; Zhai, R.; He, W.; Deng, C. Study on the Mechanism of Free Radical Scavenger TEMPO Blocking in Coal Oxidation Chain Reaction. *Fuel* **2023**, *331*, 125853. [CrossRef]
121. Dai, F.; Qin, G.; Huang, G.; Deng, H.; Zhang, X. A Comparison of the Inhibition Effect of Zn^{2+} or Fe^{2+} in the Oxidation of a-Thiophenol Groups in Coal. *Chem. Eng. Sci.* **2022**, *260*, 117863. [CrossRef]
122. Zhang, L.; Han, Y.; Xu, D.; Jiang, Q.; Xin, H.; Fu, C.; He, W. Study on the Reaction Path of -CH_3 and -CHO Functional Groups during Coal Spontaneous Combustion: Quantum Chemistry and Experimental Research. *Energies* **2022**, *15*, 4891. [CrossRef]

fire

MDPI

Article

Mitigating Coal Spontaneous Combustion Risk within Goaf of Gob-Side Entry Retaining by Roof Cutting: Investigation of Air Leakage Characteristics and Effective Plugging Techniques

Zhipeng Zhang [1,2], Xiaokun Chen [1,2], Zhijin Yu [1,2,*], Hao Sun [3], Dewei Huang [1,2], Jiangle Wu [1,2] and Hao Zhang [1,2]

1 School of Safety Science and Engineering, Xi'an University of Science & Technology, Xi'an 710054, China; 23120089020@stu.xust.edu.cn (Z.Z.); chenxk@xust.edu.cn (X.C.); 22220226118@stu.xust.edu.cn (D.H.); 23120089019@stu.xust.edu.cn (J.W.); 23220226178@stu.xust.edu.cn (H.Z.)
2 Shaanxi Key Laboratory of Prevention and Control of Coal Fire, Xi'an 710054, China
3 Heilongjiang Longmei Shuangyashan Mining Co., Ltd., Shuangyashan 155100, China; 15193515517@163.com
* Correspondence: yuzhijin@xust.edu.cn

Abstract: Relative to conventional coal pillar retention mining technology (the 121 mining method), gob-side entry retaining by cutting roof (the 110 mining method), a non-pillar mining technique, efficiently addresses issues like poor coal resource recovery and significant rock burst damage. Nonetheless, the open-type goaf created by 110 mining techniques suffers from complex and significant air leaks, increasing the likelihood of coal spontaneous combustion (CSC) within the gob area. To address the CSC problem caused by complex air leakage within the goaf of gob-side entry retaining by roof cutting, this study takes the 17202 working face of Dongrong Second Coal Mine as the object of study. Field tests and simulation calculations are conducted to research the features of air leakage and the distribution of the oxidation zone within the goaf. Subsequently, plugging technology with varying plugging lengths is proposed and implemented. The tests and simulations reveal that the airflow migration within the goaf follows an L-shaped pattern, while air leakage primarily originates from gaps found in the gob-side entry retaining wall. The amount of air leaking into the gob-side entry retaining section is 171.59 m^3/min, which represents 7.3% of the overall airflow. The maximum oxidation zone within the goaf ranges from 58.7 m to 151.8 m. After the air leakage is blocked, the airflow migration route within the goaf is transformed into a U-shaped distribution, and the maximum oxidation zone range changes from 42.8 m to 80.7 m. Engineering practice demonstrates that after air leakage plugging, the total air leakage volume within the gob-side entry retaining section significantly reduces to 20.59 m^3/min, representing only 0.78% of the total airflow volume. This research provides reference on how to prevent the occurrence of CSC in similar mine goafs.

Keywords: gob-side entry retaining by roof cutting; goaf; airflow leakage; plugging technology; coal spontaneous combustion

1. Introduction

Mine fire is one of the five main disasters in the process of coal mining, most of which are caused by CSC in the goaf [1–6]. The mining technology, ventilation type, and air leakage degree of the working face are the primary factors affecting CSC in the goaf [7–9]. Over the past 50 years, China's mines have usually used a longwall to a stay-coal pillar method from a coal mining method called the "121 mining method" (employing a U-shaped ventilation pattern, as depicted in Figure 1a). Mining a working face requires pre-excavation of two roadways; at the same time, to balance the roof pressure brought by the previous working face, a specific width of coal pillars must be maintained between them [10–12]. However, several coal pillars with a width of tens of meters are discarded when the mine working face is mined using the "121 mining method", leading to a serious coal resource loss of up to 40% of the mine's recoverable reserves. In recent years, to enhance the coal extraction rate and minimize discarded resources, many mines in China have realized coal pillar-free mining by gob-side roof cutting, known as the "110 mining method" (consider the Y-type ventilation as a case in point, as shown in Figure 1b). The technology uses advanced blasting to pre-split the coal seam roof and uses the stope periodic pressure to cut the roof along the goaf. The caving rock mass forms a support structure for the overlying basic roof rock beam, which controls the rotary sinking deformation of the main roof. At the same time, the roof cutting forms the roadway side to cut off the goaf, automatically forms the roadway to be used for the next working face mining, and finally realizes that a working face only needs to excavate one roadway, which means that a mining working face only needs to excavate one roadway, leaving zero coal pillar mining modes [13–18]. Implementing the 110 mining method minimizes roadway excavation, mitigates the risk of rock bursts, and significantly enhances the coal resource recovery rate. Concurrently, it alters the permeability of the goaf, increases the complexity of internal airflow migration, and expands the peril of CSC in the goaf.

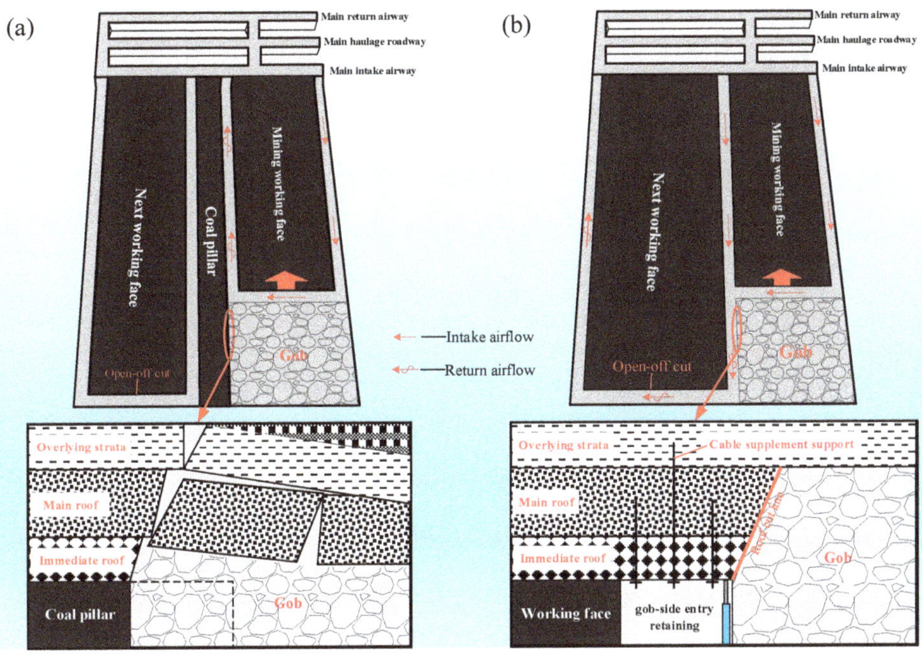

Figure 1. Process of the "121 mining method" and the "110 mining method". (**a**) The 121 mining method. (**b**) The 110 mining method.

Compared with the traditional "121 mining method", the "110 mining method" creates an open gob area with serious air leakage potential [19,20]. The gob-side entry retaining wall automatically formed by roof cutting is directly connected to the caving zone in the goaf, leading to a significant reduction in the one-way local resistance of the air leakage pathway within the goaf, thus expanding the air leakage area and promoting air leakage within the goaf [21,22]. Under the "121 mining method", most of the working faces adopt the "U"-shaped ventilation patterns. Air enters the operational area through the intake airflow roadway and exits through the return airflow roadway. In this time frame, some of the airflow moves into the goaf from the working plane adjacent to the intake airflow roadway and then flows into the working face near the return airflow roadway after a while. Therefore, the gas that builds up in the goaf is released from the upper corner due to the airflow, leading to the challenge of easily surpassing the safety limit for disaster gas concentration in that area [23–26]. By implementing the "110 mining method", the ventilation pattern of the mine is altered from a "U-shaped" to a "Y-type", modifying the airflow in the working area and gas transportation in the gob region. This adjustment effectively addresses the issue of hazardous gas buildup within the upper corner of the working plane [27–29]. However, the open type of the gob area formed under this model has serious wind leakage and complex patterns, which have important impacts on CSC in the goaf. Consequently, studying the airflow migration pattern in the goaf is crucial, especially when implementing gob-side entry retaining by roof cutting. Developing specific blocking measures is essential for the prevention of CSC in the goaf region.

CSC in the goaf is intimately linked to the ventilation method of the working face and the degree of air leakage within the goaf. Many researchers have extensively studied to determine the air leakage features within the goaf with various ventilation methods [30–35]. Tian et al. [36] employed the SF_6 tracer gas to test air leakage in the U-shaped ventilation working face at Wenzhuang Coal Mine. The findings revealed numerous air leakage channels within the gob area of the working plane, with a significant increase in air leakage speed at the interface between the hydraulic support and the gob area. Zhai et al. [37] utilized field observation and computer simulation to research the distribution of oxygen within the gob area of the working face under the U+L ventilation pattern, assessing the level of air leakage in the gob at different locations. Li et al. [38] utilized Comsol Multiphysics software to examine the air leakage patterns and gas distribution properties within the gob area of the working plane site with the Y-shaped ventilation method and identified the key air leakage zone and its extent. Guo et al. [39] conducted comprehensive research on the air leakage level and gas concentration of a working face under W-shaped and U-type ventilation systems through field testing and theoretical analysis. The research revealed the impact of various ventilation patterns on gas migration and the concentration of harmful gases in the mined-out area. The above-mentioned research mainly analyzed the air leakage situation in the goaf under different ventilation modes through field testing and numerical simulation methods, providing reliable directives for the prevention of CSC in the gob area. However, the air leakage law of the open-type goaf formed under the technical condition of gob-side entry retaining by roof cutting has not been grasped at this stage, and implementing specific prevention and control measures to prevent CSC in the goaf is challenging.

Therefore, this study focuses on the 17202 comprehensive mining working plane of Dongrong Second Coal Mine to investigate air leakage patterns within the open-type goaf. By employing a combination of field testing and computer simulation methods, plugging technology is introduced for field application. This approach successfully minimizes the risk of CSC caused by air leakage in the gob area, offering valuable insights for enhancing safety measures in similar mining operations.

2. Engineering Background

The Dongrong Second Coal Mine is located in Heilongjiang Province, China. The 17202 working face is located in the No.17 coal seam of Dongrong Second Coal Mine. The working face applies the process of gob-side entry retaining by roof cutting and comprehensive mechanization longwall retreating coal mining method. The 17202 working plane measures 978 m in strike length and 180 m in tendency length and features an average coal seam thickness of 4.5 m. It also has a design mining height of 3.5 m and a coal seam inclination angle ranging from 19 to 21 degrees. This working face utilizes the "two-inlet-one return" Y-shaped ventilation system, where the belt lane serves as the main inlet with an airflow of 1680 m^3/min. The track lane functions as the auxiliary inlet with an airflow of 530 m^3/min, and the gob-side entry retaining by roof cutting serves as the return lane with an airflow of 2348 m^3/min. The mine's geographical location and the layout of the 17202 working plane are illustrated in Figure 2.

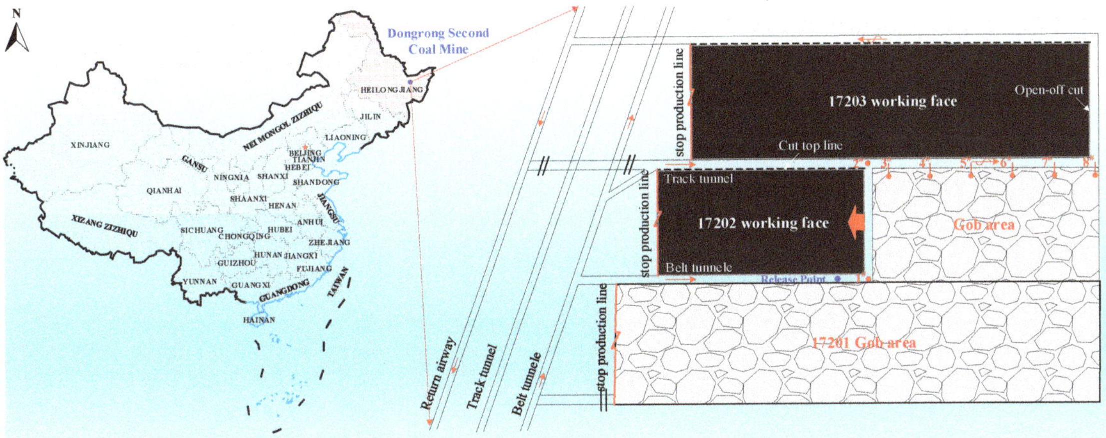

Figure 2. Schematic diagram of the geographic location of the mine and the layout of the working face.

3. Methods and Simulation

3.1. SF_6 Tracer Gas Test Air Leakage

3.1.1. Air Leakage Tests in Goaf

The SF_6 pulse release approach is utilized to measure the airflow migration pathway within the goaf of gob-side entry retaining through roof cutting and calculate the air leakage velocity.

The minimum air leakage speed within the goaf is calculated by the following formula [40]:

$$v_{\min} = L/t \tag{1}$$

where $v_{\min}$ is the minimum leakage wind speed, m/s; L is the linear distance from the tracer gas release point to the sampling point, m; t is the peak time from the tracer gas release to the detection of the tracer gas, s.

The formula for calculating the air leakage rate in the gob area is as follows:

$$k = \frac{m_{SF_6-\text{inflow}} - m_{SF_6-\text{outflow}}}{m_{SF_6-\text{inflow}}} \times 100\% \tag{2}$$

$$m_{SF_6} = V_{SF_6}\rho_{SF_6} \tag{3}$$

$$V_{SF_6} = \int_0^t \varphi Q dt = vA\int_0^t \varphi dt \tag{4}$$

$$Q = vA \tag{5}$$

where k is the air leakage rate; m_{SF_6} is the mass of the tracer gas, kg; V_{SF_6} is the volume of the tracer gas, m^3; ρ_{SF_6} is the density of the tracer gas, kg/m^3; φ is the volume fraction of the tracer gas detected, 10^{-6}; Q is the air volume, m^3/s; v is the wind speed, m/s; A is the cross-sectional area of the tunnel, m^2.

Following the current ventilation conditions in the mining working face, the release point of tracer gas is set 30 m from the inlet corner of the belt lane of the 17202 working face, sampling point $1^\#$ is set at the inlet corner of the belt lane of the 17202 working face, sampling point $2^\#$ is set at the junction of the railroad track lane and gob-side entry, sampling points $3^\#$, $4^\#$, $5^\#$, $6^\#$, $7^\#$, and $8^\#$ are arranged at a depth of 0.5 m and a spacing of 55 m inside the gob-side entry retaining wall, as shown in the diagram in Figure 2.

The SF_6 tracer gas cylinder is connected to the pressure-reducing valve and flow meter, the release flow rate of the SF_6 tracer gas is set to 109.49 L/min and is kept releasing for 30 min. From 0~5 min, each point is sampled every 1 min; from 5~30 min, each point is sampled every 5 min; from 30~180 min, each point is sampled every 10 min. A total of 200 gas samples are collected.

3.1.2. Air Leakage Tests in Gob-Side Entry Retaining Section

The continuous quantitative release method is employed to measure the extent of air leakage within the gob-side entry retaining section. The calculation method for continuous air leakage within the roadway is based on the estimated airflow migration route within the goaf of the 17202 working plane. The principle is to set sampling points $1^{\#}$ and $2^{\#}$ along the airflow direction in the roadway, and the tracer gas is continuously and quantitatively released at the release point. After the tracer gas reaches a stable state in the airflow in the roadway, the air volume of sampling point $1^{\#}$ is recorded as Q_1, and the detected SF_6 tracer gas concentration is recorded as C_1. The air volume of the sampling point $2^{\#}$ is recorded as Q_2, and the concentration of SF_6 tracer gas is recorded as C_2 [30], as shown in Figure 3.

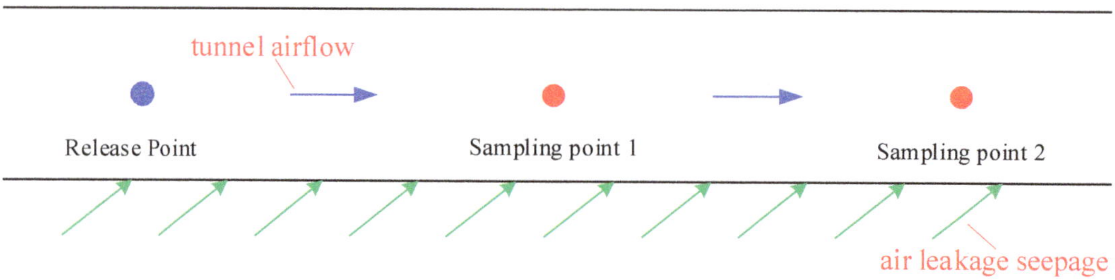

Figure 3. Leakage into the detection space detection schematic.

Let the amount of SF_6 gas released be q. Based on the law of conservation of mass, $q = Q_1 \cdot C_1 = Q_2 \cdot C_2$. If there is air leakage between two points, the amount of air leakage between the two points is $\Delta Q = Q_2 - Q_1$. Therefore, we can obtain:

$$\Delta Q = Q_2 - Q_1 = \frac{q}{C_1} - \frac{q}{C_2} \tag{6}$$

The calculation formula for the air leakage rate of the laneway:

$$\alpha_i = \frac{C_{i+1} - C_i}{C_i} \times 100\% \tag{7}$$

Based on the continuous constant release SF_6 flow formula:

$$q = KCQ \times 10^{-6} \tag{8}$$

where q is the SF_6 tracer gas release, ml/min; K is the error coefficient, taken as 0.05; Q is the air volume of the lane; C is the minimum concentration of SF_6 in the expected airflow, taken as 10^{-8}.

To determine the amount of air leakage within the gob-side entry retaining section, the release point of SF_6 tracer gas is established at measurement point $2^{\#}$ in Figure 2, while measurement points $3^{\#}$ to $8^{\#}$ are designated as the sampling points. The SF_6 release volume is set to 1200 mL/min according to the preset SF_6 tracer gas release flow rate. After 20 min of release, it is uniformly mixed with the airflow of the roadway to reach a stable state, and gas sampling is performed at each sampling detection point.

3.2. Air Leakage Law Simulation in the Gob

3.2.1. Geometric Model and Simulation Conditions

A three-dimensional physical model of the goaf under the Y-shaped ventilation pattern of "two intakes and one return" is established based on the real working conditions of the 17202 working plane, as illustrated in Figure 4. The goaf has a depth of 285 m, a working face tendency length of 180, a height of 30 m, and a floating coal thickness of 0.5 m, and the overlying strata in the goaf have a thickness of 29.5 m. To enhance calculation accuracy and decrease computation time, the model is segmented using free triangle and quadrilateral meshes. The area around the working plane and the gob-side entry retaining section is further refined, resulting in a total of 75,617 unit grids.

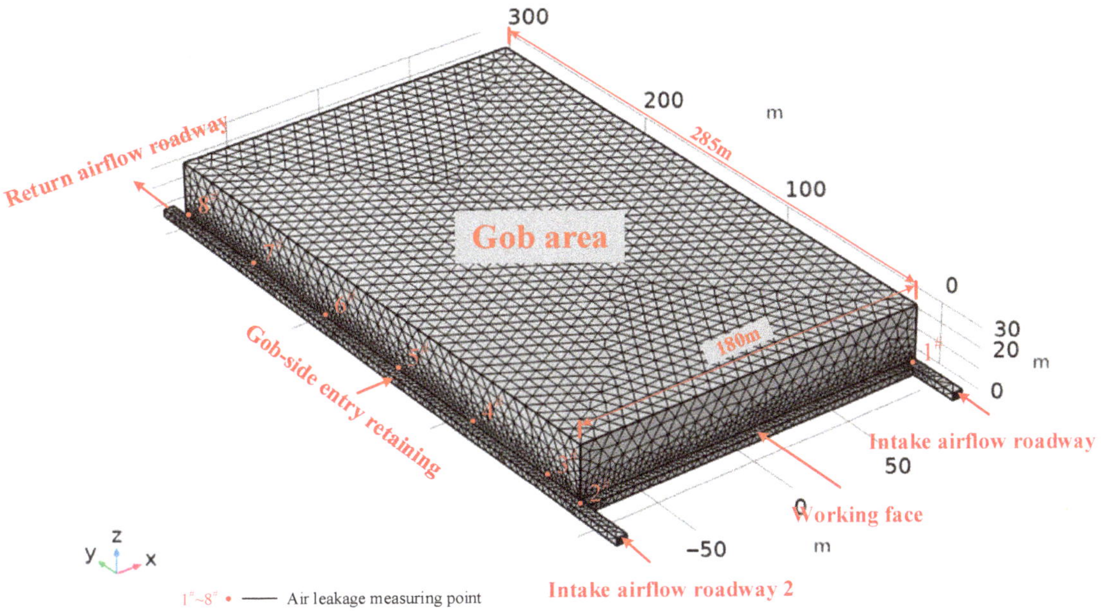

Figure 4. Geometric model and grid generation.

To facilitate modeling and calculation, the following assumptions are made: (1) no other chemical reactions are occurring in the stable internal flow field of the goaf; (2) the fragmented coal and rock material within the goaf forms an isotropic porous medium; (3) the gas in the whole flow field model is ideal incompressible gas; (4) there is no heat source in the goaf, and the model does not consider the energy equation. The temperature of the goaf, intake roadway, and return roadway is implanted to 298 K, the oxygen concentration of the working face in standard state is implanted to 21%, and the oxygen concentration of goaf is set to 0. The specific geometric dimensions, boundary conditions, and simulation parameters for the three-dimensional goaf physical model are displayed in Table 1.

Table 1. Geometric size and simulation conditions of the simulation model.

Category	Parameter	Value or Condition
Geometric size	Working face: x(m) × y(m) × z(m)	180 × 5 × 3
	Gob area: x(m) × y(m) × z(m)	180 × 285 × 30
	Intake airflow roadway 1: x(m) × y(m) × z(m)	5 × 25 × 3
	Intake airflow roadway 2: x(m) × y(m) × z(m)	5 × 25 × 3
	Return airflow roadway: x(m) × y(m) × z(m)	5 × 300 × 3

Table 1. *Cont.*

Category	Parameter	Value or Condition
Boundary conditions and simulation parameters	Inlet velocity 1: wind speed of belt roadway (m/s)	1.6
	Inlet velocity 2: wind speed in track lanes (m/s)	0.5
	Outlet	Outflow, $p = 0$
	Gob's internal state	Porous zone
	Other walls	Zero slip
	Porosity of porous media in goaf	User-defined function
	Thermal conductivity of coal ($W \cdot m^{-1} \cdot K^{-1}$)	0.2
	Porosity of floating coal	0.3
	Air diffusion coefficient ($m^2 \cdot s^{-1}$)	1.5×10^{-5}
	Activation energy ($J \cdot mol^{-1}$)	5×10^4
	Pre-exponential factor (s^{-1})	180
	Gas constant ($J \cdot mol^{-1} \cdot K^{-1}$)	8.314
	Oxygen consumption rate	$AC^{n}e^{(-E_a/RT)\cdot[(1-q)\cdot q]}$

Note: A is the pre-exponential factor; C is the oxygen concentration; E_a is the activation energy; R is a gas constant; T is the temperature of the coal sample; q is the porosity of floating coal.

3.2.2. Control Equations [27,40]

(1) Conservation of mass equation:

$$\frac{\partial \rho}{\partial t} + \nabla \cdot (\rho v) = S_m \tag{9}$$

where ρ is the density, kg/m^3; t is the time, s; v is the velocity, m/s; S_m is the mass, $kg/(m^3 \cdot s)$.

(2) Conservation of momentum equation:

$$\frac{\partial \rho}{\partial t}(\rho v) + \nabla \cdot (\rho v) = -\nabla p + \nabla \cdot (\tau) + \rho g + F \tag{10}$$

where t is the time, s; p is the hydrostatic pressure, Pa; τ is the stress tensor, N/m^2; ρg is the gravitational body force, N; F is the external body force, N.

(3) Ideal gas equation of state:

$$\rho = \frac{pM}{RT} \tag{11}$$

where p is the pressure, Pa; M is the molar mass of the gas, g/mol; R is the ideal gas constant, 8.314 J/(mol·K); T is the temperature, K.

(4) Gas component transport equation:

$$\frac{\partial}{\partial t}(C_i) = \nabla(D_i \nabla C_i) - \nabla(u C_i) + R_i \tag{12}$$

where C_i is the gas component inside the goaf, mol/m^3; D_i is the diffusion coefficient of the gas component, m^2/s; R_i is the source (sink) terms of gas components, $mol/(m^3 \cdot s)$; u is the velocity field.

4. Results Analysis

4.1. Analysis of Air Leakage Using SF_6 Tracer Gas Testing

4.1.1. Air Leakage Characteristics in Goaf

The information presented in Figure 5 indicates that the SF_6 tracer gas is released at the end of the sampling, and the whole detection process is 180 min. Among them, SF_6 tracer gas is detected at point 1# closest to the release point for the first time, and SF_6 concentration reaches a detection peak of 71.47 ppm at 15 min. The SF_6 concentration of sampling point 2# reaches a detection peak of 49.38 ppm at 25 min. The SF_6 concentration of sampling point 3# reaches a detection peak of 39.63 ppm at 30 min. The SF_6 concentration of sampling point 4# reaches a detection peak of 27.36 ppm at 40 min. The SF_6 concentration of sampling point $5^{\#}$ reaches a detection peak of 16.36 ppm at 50 min. The SF_6 concentration of sampling point $6^{\#}$ reaches a detection peak of 12.27 ppm at 70 min. The SF_6 concentration of sampling point $7^{\#}$ reaches a detection peak of 6.23 ppm at 90 min. The SF_6 concentration of sampling point $8^{\#}$ reaches a detection peak of 3.87 ppm at 110 min. This shows that the farther the distance from the SF6 tracer gas release point, the longer the peak time of the detected SF6 concentration and the smaller the peak concentration.

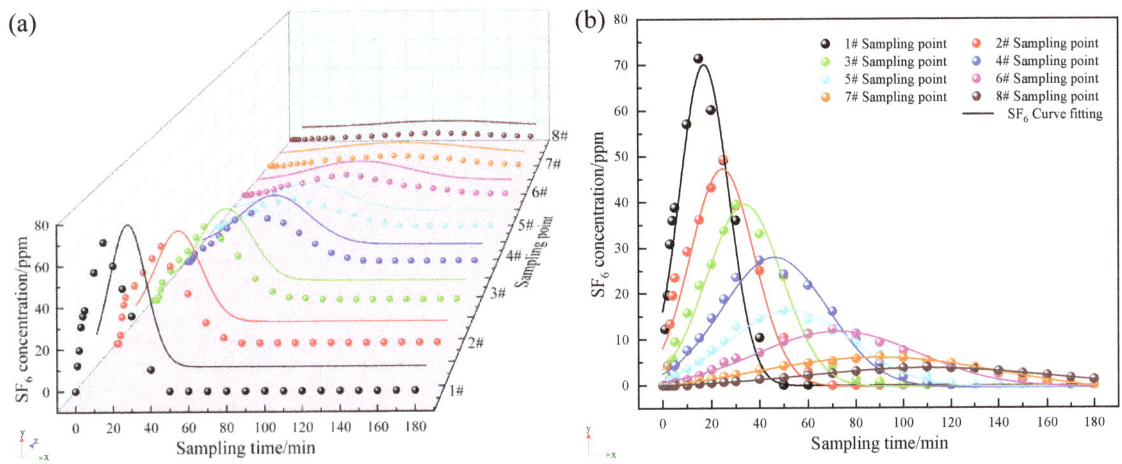

Figure 5. Prediction time history of SF_6 at each sampling point. (**a**) Three-dimensional perspective. (**b**) Two-dimensional perspective.

According to the formula for calculating the minimum air leakage within the goaf and the formula for calculating the air leakage rate within the goaf, it is calculated that the minimum air leakage speeds of sampling points $3^{\#}$~$8^{\#}$ within the gob area of the 17202 working face are 0.1013 m/s, 0.0838 m/s, 0.0781 m/s, 0.0658 m/s, 0.0601 m/s, and 0.0569 m/s in order, and the leakage rate in the gob area is 20.07%. Data fitting is used to analyze the air leakage velocity and peak concentration at each sampling point, with the specific fitting results displayed in Figures 6 and 7.

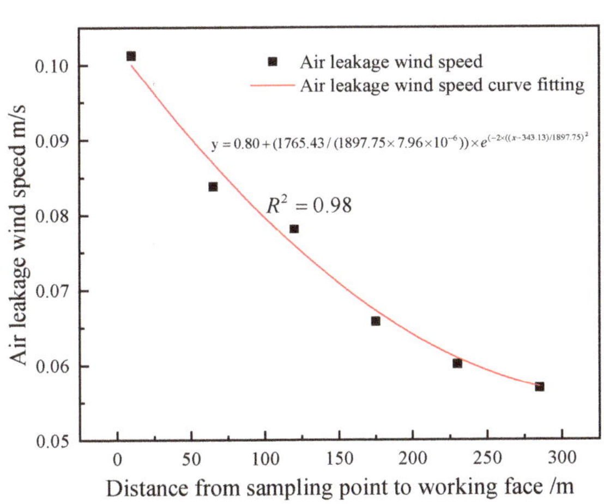

Figure 6. Prediction curve between leakage wind speed and working face distance.

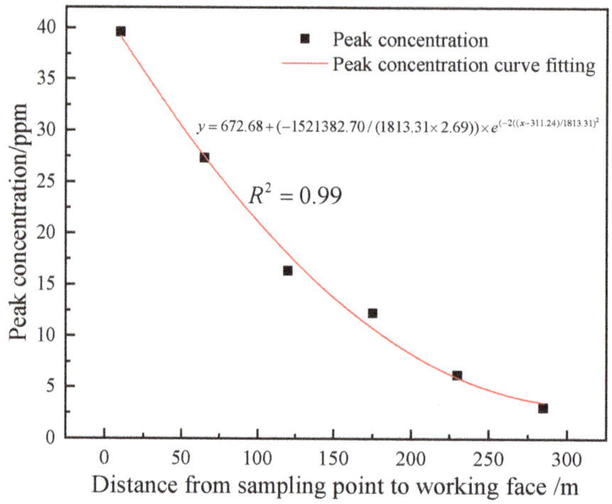

Figure 7. Prediction curve between peak concentration and working face distance.

As can be seen from Figures 6 and 7, with the depth in the buried depth of the goaf, the overlying rock activity inside the goaf tends to stabilize and the falling rock gradually compacts, which leads to a significant decrease in the leakage wind speed and peak concentration in goaf.

After conducting calculations, the approximate migration route of airflow in the goaf of 17202 working face is found. At first, the airflow from the primary main intake airflow roadway into the goaf seeps into the goaf region. Over time, the airflow escapes through the gob-side entry retaining wall.

4.1.2. Air Leakage Degree of Gob-Side Entry Retaining Section

After the SF_6 tracer gas is released for 20 min and the roadway air flow is uniformly mixed to a stable state, a 5 L sampling airbag is used to sample three times at sampling points $3^{\#}$~$8^{\#}$ set up within the gob-side entry retaining, and the sampling time and position are marked on the surface of the airbag and sent to the laboratory for chromatographic analysis. Finally, the SF_6 concentration at each sampling point is obtained by averaging. The findings are displayed in Table 2.

Table 2. Measurement results of air leakage volume within gob-side entry retaining section.

Release Point	Sampling Points	Release Amount (mL/min)	Detection Concentration (ppm)
The intersection of the 17202 working face and the gob-side entry retaining wall	$3^{\#}$	1200	62.21
	$4^{\#}$		59.03
	$5^{\#}$		58.11
	$6^{\#}$		57.54
	$7^{\#}$		57.23
	$8^{\#}$		57.12

Data from the detection show that the distance between the SF6 sampling point and release point increases, and the concentration of the SF6 tracer gas decreases.

According to Equations (6) and (7), the air leakage volume and air leakage rate of the section of gob-side entry retaining and the findings are displayed in Table 3.

As can be seen from Table 3, the degree of air leakage in sections $3^{\#}$~$4^{\#}$ is extremely serious, with the air leakage amount reaching 103.91 m^3/min. The air leakage rate is 5.38%, followed by the rest of the sections. As the waste material in the mined-out section becomes more compressed, the amount of air escaping from the mined-out area into the gob-side entry retaining gradually diminishes. In the 17202 working plane, the volume of air leakage total for the gob-side entry retaining section is 171.59 m^3/min, representing 7.3% of its overall air volume.

Table 3. Measurement results of air leakage volume and air leakage rate of unsealed gob-side entry retaining wall.

Measurement Section	Air Leakage Volume (m^3/min)	Air Leakage Rate (%)	Total Air Leakage Volume (m^3/min)
3#~4#	103.91	5.38	
4#~5#	32.19	1.58	
5#~5#	20.15	0.99	171.59
6#~7#	11.30	0.54	
7#~8#	4.04	0.19	

4.2. Analysis of Simulation Results Regarding Air Leakage in the Goaf

4.2.1. The Law of Airflow Migration in the Goaf

Simulation calculations are used to determine the airflow migration path and distribution of air leakage velocity within the goaf of the 17202 working plane, as depicted in Figures 8 and 9. Additionally, Figure 10 compares the actual measurements of air leakage within the goaf, with the results obtained from numerical simulations.

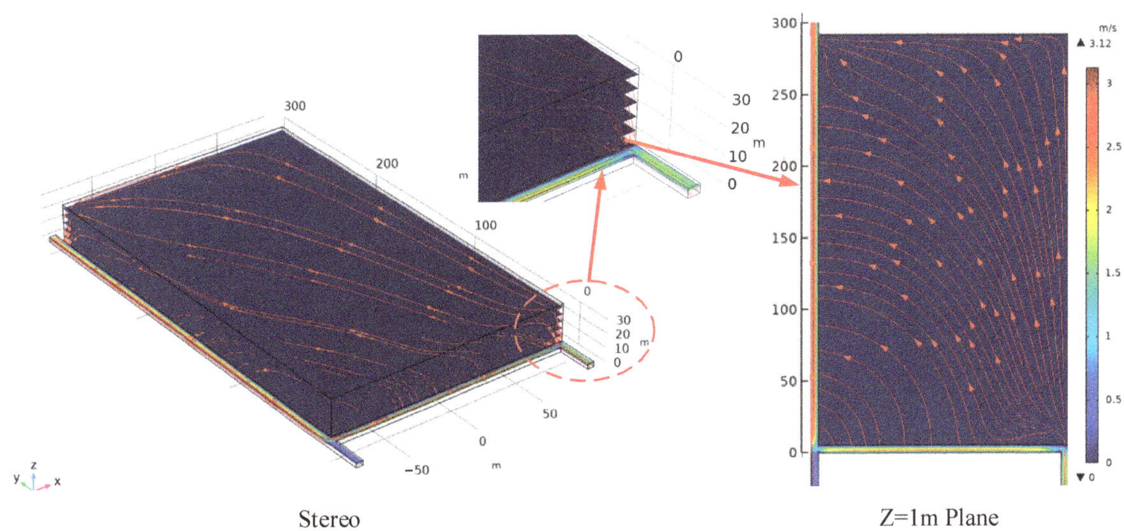

Figure 8. Prediction of airflow migration route in goaf.

As depicted in Figure 8, the airflow leaks into the goaf from the working plane and then seeps into the mined-out area. As it reaches a specific height, the airflow shifts horizontally towards the gob-side entry retaining side. Upon reaching the gob-side entry retaining area, the airflow descends to the lower part of the gob region and escapes from the gob-side entry retaining side. The airflow migration streamlines within the gob area of the 17202 working plane are shaped like an L when viewed on the z = 1 m plane. As airflow leaks from the working plane into the mined-out area and reaches a particular level, it is redirected approximately 90° towards the gob-side entry retaining wall and finally escapes from the gob-side entry retaining wall.

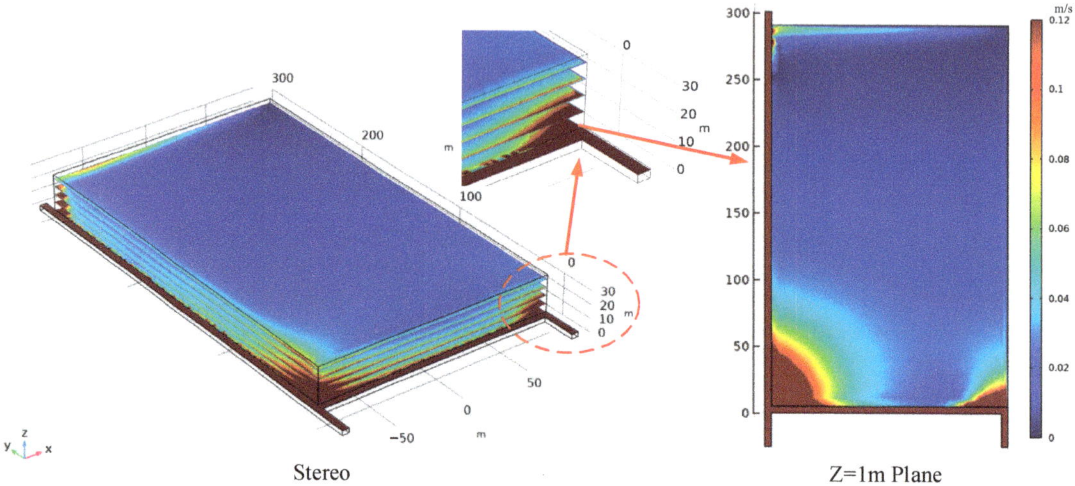

Figure 9. Diagram of the distribution of air leakage velocity within the goaf area.

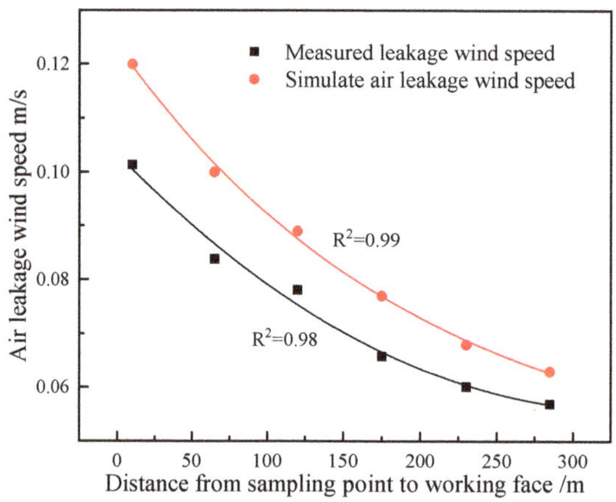

Figure 10. Comparison involving the measurable air leakage velocity within the goaf with the results obtained from numerical simulations.

The larger air leakage wind speed within the goaf of the 17202 working plane can be observed in Figure 9, located at the corners on both sides of the working face. The significant air leakage area at the corner of the main air intake roadway is attributed to the high wind speed within the main intake airflow roadway, the low degree of roof collapse and compaction in the goaf, and the low-pressure zone on the side of the gob-side entry retaining wall. The large air leakage wind speed in the corner of the gob-side entry retaining is caused by the delay in the collapse of the wall and the loose roof connection, resulting in a direct connection between the gob-side entry retaining and the goaf.

The results from the numerical simulation in Figure 10 indicate that the air leakage within the goaf is slightly higher than what is observed in the field measurements. The reason is that the real measurement value is slightly lower than the numerical simulation value because of the impact of field conditions and sampling techniques, detection equipment, and other factors in the actual test process, which is in line with the actual situation within the allowable range of error. Therefore, the numerical simulation value is accurate and effective.

4.2.2. Oxygen Concentration Distribution Characteristics in Goaf

The result in Figure 11 shows a U-shaped distribution of three-dimensional oxygen concentration within the gob region of the 17202 working face. This pattern indicates that oxygen levels are high on the edges of the gob area and low in the center. The distribution of oxygen concentration within the mined-out area forms an L-shape at the z = 1 m plane. A spontaneous combustion zone emerges within a certain distance from the working plane, with a delineated banded spontaneous combustion zone visible alongside the gob area retaining roadway. The oxygen concentration index serves as the primary criterion for categorizing the "three zones" of CSC. These zones include the heat dissipation zone (oxygen concentration > 18%), the oxidation zone (18% $\leq$ oxygen concentration $\leq$ 8%), and the suffocation zone (oxygen concentration < 8%) in goaf areas. Based on this, the distribution scope of the heat dissipation zone, oxidation heating zone, and suffocation zone within the mined-out area for the 17202 working face is defined, as detailed in Table 4.

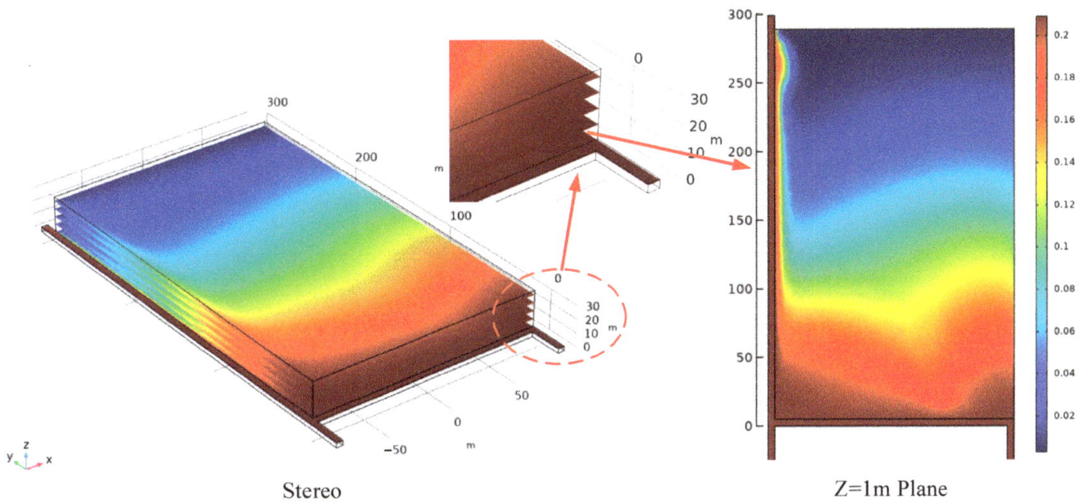

Figure 11. Forecasting the distribution of oxygen concentration in the goaf.

Table 4. 17202 working face gob area spontaneous combustion "three zones" distribution range.

Location within the Goaf	Heat Dissipation Zone	Oxidation Zone	Suffocation Zone
Main air inlet tunnel side	0~58.7 m	58.7 m~151.8 m	>151.8 m
Middle	0~36.8 m	63.8 m~139 m	>139 m
Gob-side entry retaining side	0~62.6 m	62.6 m~125.3 m	>125.3 m

4.3. Analysis of Air Leakage Prevention in Goaf

The research findings above indicate a significant issue of air leakage within the gob-side entry retaining wall in the goaf. In this case, the peril of CSC within the goaf is high. Therefore, specific preventive measures must be implemented to avoid CSC within the mined-out area. In this regard, for the serious air leakage problem within the gob-side entry retaining wall in the goaf, an air leakage blocking technology is proposed (spraying and plugging the gob-side entry retaining wall in the goaf since the open-off cut of the working plane), and the flow field change rule of the goaf under the blocking distances of 40 m, 80 m, 120 m, 160 m, 200 m, 240 m, and 285 m is investigated; the placement of the spraying and plugging is presented in Figure 12.

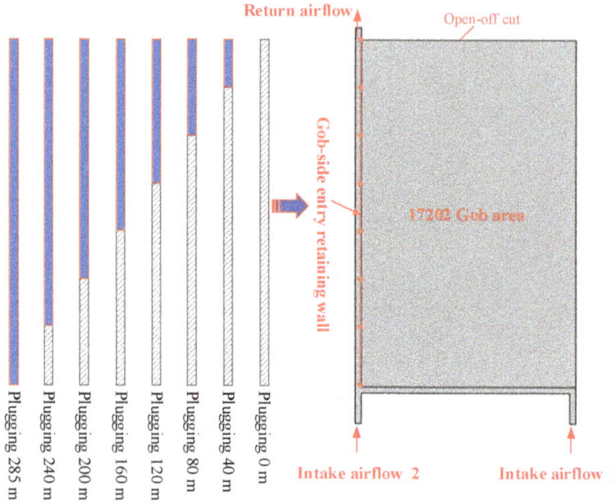

Figure 12. Schematic diagram of phased plugging of air leakage.

4.3.1. Characteristics of Flow Field Distribution in the Goaf

The graph in Figure 13 illustrates that, as the spraying distance on the side of gob-side entry retention increases, there is a noticeable shift in the airflow pattern from the working plane into the goaf, transitioning from an L-shaped distribution to a more distinct U-shaped distribution. Once the gob-side entry retaining wall is fully sealed off, the airflow flows into the mined-out area from the working plane on the side of the belt roadway. After a while, most of the airflow returns to the working plane on the side of the gob-side entry retaining. Only a small part of the airflow leaks into the goaf, which greatly reduces the air leakage within the mined-out area, thus reducing the risk of CSC within the goaf.

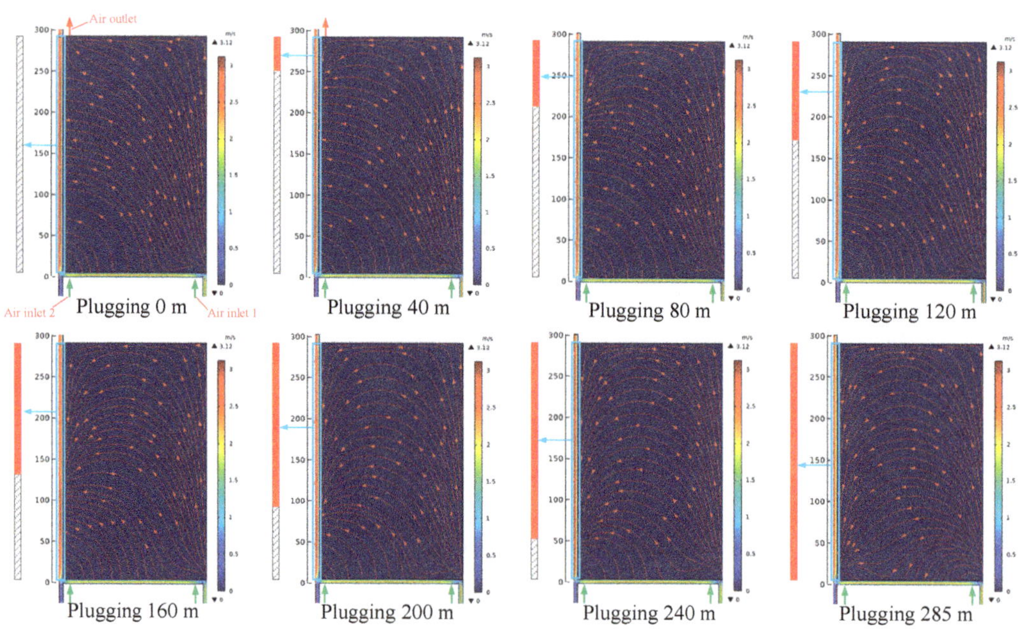

Figure 13. Prediction of airflow migration route in the goaf during staged plugging process.

4.3.2. Characteristics of Oxygen Concentration Distribution in the Goaf

The results in Figures 14–16 illustrate that addressing spraying and plugging leakage on the gob-side entry retaining wall can alter the extent of the oxidation zone distribution within the mined-out area. As the distance of spraying sealant increases, the spread of the oxidation zone within the goaf reduces. Once the spraying sealing distance extends to 285 m, the zone of oxidation within the gob area on the main air inlet side reduces from 58.7 m to 151.8 m to 42.8 m to 80.7 m, while the zone of oxidation in the middle of the gob area decreases from 63.8 m to 139 m to 6.8 m to 10.4 m. The oxidation zone in the mined-out area at the gob-side entry retaining walls decreases from 62.8 m~125.3 m to 13.3 m~20.1 m. It can be shown that spraying and sealing on the gob area side can effectively reduce the migration and diffusion range of oxygen within the gob area, decrease the extent of the oxidation zone distribution in the gob region, and decrease the likelihood of CSC in the gob area.

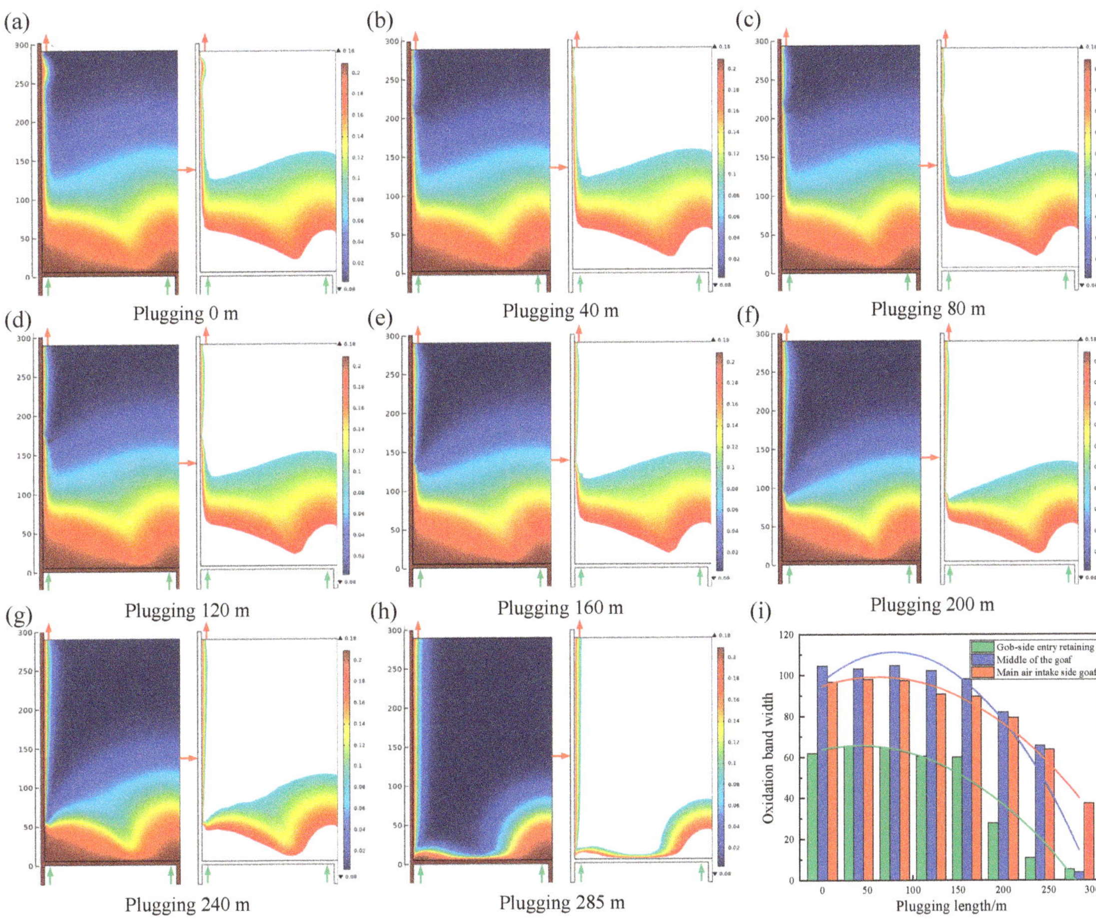

Figure 14. Prediction of oxygen concentration distribution in the goaf during staged plugging process. (**a**) plugging 0 m. (**b**) plugging 40 m. (**c**) plugging 80 m. (**d**) plugging 120 m. (**e**) plugging 160 m. (**f**) plugging 200 m. (**g**) plugging 240 m. (**h**) plugging 285 m. (**i**) The distribution of oxidation zone in goaf under different plugging distance.

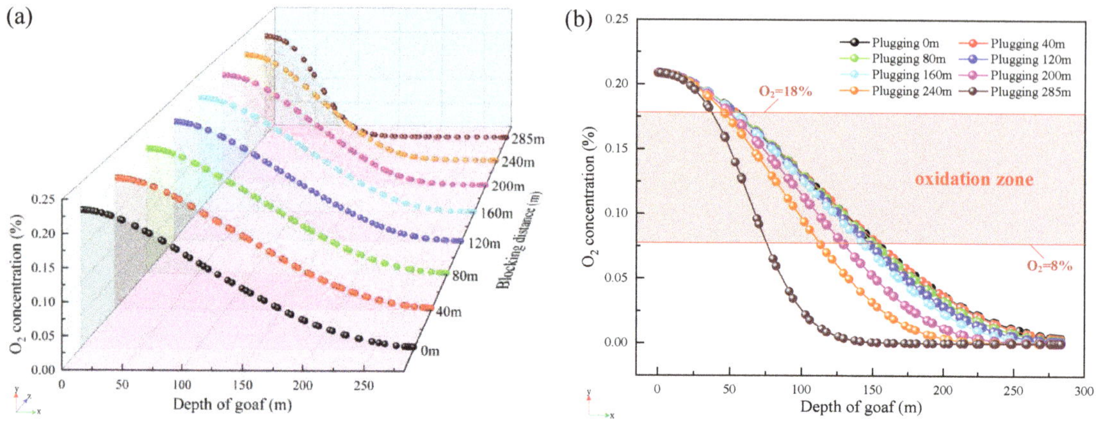

Figure 15. Changes of oxygen concentration within the goaf of the main air inlet side under different plugging distances. (**a**) Three-dimensional perspective. (**b**) Two-dimensional perspective.

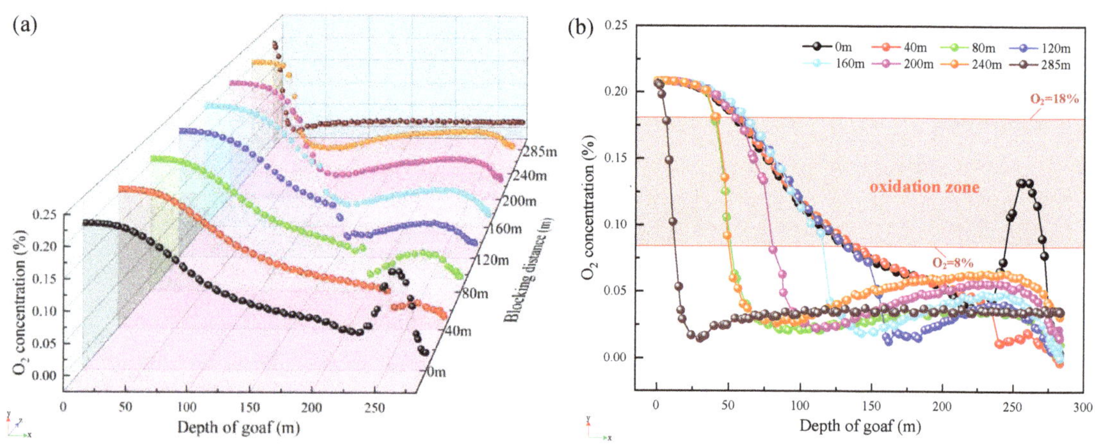

Figure 16. Changes of oxygen concentration in the goaf of gob-side entry retaining side under different plugging distances. (**a**) Three-dimensional perspective. (**b**) Two-dimensional perspective.

5. Applications

5.1. Plugging Construction

On the surface of a coal mine roadway, sprayed concrete or mortar is usually used to plug the air leakage, but, due to the lack of toughness of cement mortar after curing, the roadway is easily prone to cracks and falls away when it is deformed under pressure, thus losing the role of plugging the air leakage. To solve the problems existing in the process of plugging the roadway by traditional concrete spraying, after much market research, a two-component composite modified polyurethane roadway spraying and plugging material is found to be able to quickly and effectively seal the air leakage channels existing on the roadway surface of underground coal and rock bodies [41–45].

The material is tested according to the test method required by AQ 1116-2020 "General Safety Technical Specification for Polymer Materials for Reinforcement, Water Plugging, Filling and Spraying in Coal Mines" [46], and the measurement results are shown in Table 5.

Table 5. Performance test results of two-component composite modified polyurethane spray plugging materials.

Serial Number	Inspection Items			Technical Specifications	Test Results
1	Maximum reaction temperature,°C			≤140	94.7
2	Expansion multiplier			≥25	31
3	Stability (70 °C ± 2 °C, 48 h), %			≤0.1	0.03
4	Compressive strength		Pressure strain 10%, kPa	≥10	65
			Pressure strain 30%, kPa	≥10	26
			Pressure strain 70%, kPa	≥40	43
5	Oxygen index, %			≥28	28.3
6	Air permeability, m^2			≤0.05	≤0.005
7	Flame retardant performance	Alcohol torch burning test	Flame burning time, s	≤3	0.3
			Flameless burning time, s	≤10	1.8
			Flame extension length, mm	≤280	110
		Alcohol lamp burning test	Flame burning time, s	≤6	0.3
			Flameless burning time, s	≤20	2.2
			Flame extension length, mm	≤250	83
8	Surface resistance, Ω			$\leq 3 \times 10^8$	5.6×10^7

According to the air leakage plugging effect of numerical simulation, stage spraying plugging measures are carried out on the gob-side entry retaining walls of the 17202 working face. Figure 17 depicts the on-site spraying construction process. The specific steps of the construction process are outlined below:

(1) Substrate treatment. Use the pressurized air duct to flush the coal dust and dust on the surface of the roadway walls to ensure that the spray foam material is well-bonded to the roadway surface.
(2) Worker protection. Use equipment such as gas masks and protective clothing to protect construction workers from injury.
(3) Equipment connection. After connecting the air source, connect the two suction pipes to the A and B barrels, respectively. Pay special attention to the A cylinder for A material and the B cylinder for B material. Do not mix them.
(4) Spraying construction. Spray with spray gun at a constant speed along the bottom of the roadway bottom plate from the bottom up to spray, the first spraying distance of about 5 mm or so, to be fully reactive material for the second surface spraying, for the unsprayed area or the need to strengthen the spraying position to make up the spray.
(5) Cleaning. Immediately after stopping construction, wash with water to thoroughly clean the residual slurry.

5.2. Effect Analysis

To evaluate the impact of air leakage on the gob-side entry retaining section, a smoke pipe is employed to examine the air leakage pathway before and after sealing. This process is illustrated in Figure 18.

As can be seen from Figure 16, if the gob-side entry retaining is not sprayed, smoke can be affected by air leaks from the gob area and will continue to spread outward vertically, eventually flowing out with the return airflow. After spraying, the smoke is barely affected by the air escaping from the gob area, and there is no obvious change in the vertical direction of the gob-side entry retaining wall. Therefore, it can be observed that, following the application of spraying and plugging techniques on the gob-side entry retaining wall, there is a clear reduction in the degree of air leakage from the gob area.

To conduct additional testing on the efficacy of spraying and plugging, SF_6 tracer gas is employed to measure the extent of air leakage within the gob-side entry retaining section, with the outcomes presented in Table 6.

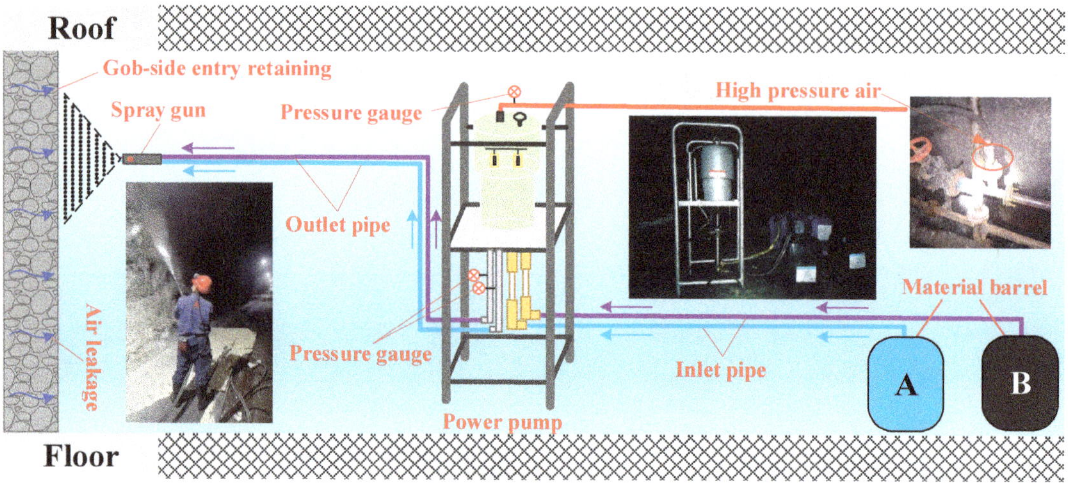

Figure 17. Spraying construction process of gob-side entry retaining wall.

Figure 18. Changes in smoke migration before and after plugging air leakage. (**a**) Unplugged test. (**b**) Test after plugging.

The test results show that the SF_6 tracer gas concentration does not change much during the sampling period as the distance between the SF_6 sampling point and the SF_6 release point increases.

Table 6. Measurement of air leakage and analysis of SF_6 concentration after plugging the remaining lane section.

Release Point	Sampling Points	Release Amount (mL/min)	Detection Concentration (ppm)
The intersection between the 17202 working face and the gob-side entry retaining wall	$3^{\#}$	1200	67.38
	$4^{\#}$		66.92
	$5^{\#}$		66.77
	$6^{\#}$		66.68
	$7^{\#}$		66.63
	$8^{\#}$		66.61

The volume of air leakage and the air leakage rate of the 17202 working plane following the application of gob-side entry retaining wall spraying and plugging are determined using Formulas (6) and (7). The specific calculations and results can be found in Table 7.

Table 7. Determination yields the air leakage results and the air leakage rate for the gob-side entry retaining after plugging.

Measurement Section	Air Leakage Volume (m^3/min)	Air Leakage Rate (%)	Total Air Leakage Volume (m^3/min)
$3^{\#}$~$4^{\#}$	12.24	0.68	20.59
$4^{\#}$~$5^{\#}$	4.03	0.22	
$5^{\#}$~$6^{\#}$	2.43	0.13	
$6^{\#}$~$7^{\#}$	1.35	0.07	
$7^{\#}$~$8^{\#}$	0.54	0.03	

After testing, the degree of air leakage is relatively high in sections $3^{\#}$~$4^{\#}$, with the air leakage amount reaching 12.24 m^3/min and the air leakage rate being 0.68%, while there is no obvious air leakage in the rest of the sections. Evidently, with the gradual spraying and plugging of the gob-side entry retaining wall, the airflow leakage from the gob-side entry retaining section is dramatically reduced. The gob-side entry retention section of the 17202 working plane has an air leakage of 20.59 m^3/min, which represents 0.87% of the total air volume in the gob-side entry retaining section.

6. Conclusions

1. SF_6 tracer gas is used to measure the approximate migration route of the airflow within the goaf of the 17202 working face, which flows into the working plane from the main air inlet lane and then leaks into the goaf, and, after a while, the airflow leaks out of the gap in the gob-side entry retaining wall. The total air leakage of the gob-side entry retaining section of the 17202 working face is 171.59 m^3/min, representing 7.3% of the overall airflow within the gob-side entry retaining section.
2. The simulation finds that the airflow from the working plane near the main intake airflow roadway leaks into the goaf and flows to its upper part, reaches a certain height, and then shifts horizontally to the gob-side entry retaining, and when the airflow is close to the gob-side entry retaining, it flows to the lower part of the zone and leaks out from the gob-side entry retaining wall. $z = 1$ plane, the air leakage route in the goaf shows an L-shape distribution, and the maximum width of the oxidation zone is 58.7 m~151.8 m. After the blocking measures are taken, the airflow migration route in the gob area becomes a U-shape distribution, and the maximum width of the oxidation zone reaches 42.8 m~80.7 m.
3. Following the application of sprayed material to seal the air leakage, the volume of air leakage within the gob-side entry retaining section decreases from 171.59 m^3/min to 20.59 m^3/min. This results in the overall air volume in the gob-side entry retaining section decreasing from 7.3% to 0.78%. It significantly minimizes air leakage in the

gob area, decreases the risk of CSC within the gob area, and guarantees the safe operation of the mine.

Author Contributions: Z.Z.: Conceptualization, Methodology, and Writing—Original Draft Preparation; X.C.: Supervision and Writing—Review and Editing; Z.Y.: Supervision and Writing—Review and Editing; H.S.: Resources and Project Administration; D.H.: Investigation and Data Curation; J.W.: Investigation; H.Z.: Formal Analysis. All authors have read and agreed to the published version of the manuscript.

Funding: This research was funded by the National Natural Science Foundation of China (No. 51804245) and the Provincial Natural Science Foundation of Shaanxi (2022JQ-408).

Institutional Review Board Statement: Not applicable.

Informed Consent Statement: Not applicable.

Data Availability Statement: Data are contained within the article.

Conflicts of Interest: Author Hao Sun was employed by the company Heilongjiang Longmei Shuangyashan Mining Co., Ltd. The remaining authors declare that the research was conducted in the absence of any commercial or financial relationships that could be construed as a potential conflict of interest.

References

1. Danish, E.; Onder, M. Application of Fuzzy Logic for Predicting of Mine Fire in Underground Coal Mine. *Saf. Health Work* **2020**, *11*, 322–334. [CrossRef]
2. Kursunoglu, N.; Gogebakan, M. Prediction of spontaneous coal combustion tendency using multinomial logistic regression. *Int. J. Occup. Saf. Ergon.* **2022**, *28*, 2000–2009. [CrossRef] [PubMed]
3. Ray, S.K.; Singh, R.P. Recent Developments and Practices to Control Fire in Underground Coal Mines. *Fire Technol.* **2007**, *43*, 285–300. [CrossRef]
4. Deng, J.; Lei, C.; Xiao, Y.; Cao, K.; Ma, L.; Wang, W.; Laiwang, B. Determination and prediction on "three zones" of coal spontaneous combustion in a gob of fully mechanized caving face. *Fuel* **2018**, *211*, 458–470. [CrossRef]
5. Ozcelik, M. Spontaneous combustion of coal seams in the Bengiler coal mine in Turkey. *Euro-Mediterr. J. Environ. Integr.* **2023**, *8*, 665–674. [CrossRef]
6. Onifade, M. Countermeasures against coal spontaneous combustion: A review. *Int. J. Coal Prep. Util.* **2022**, *42*, 2953–2975. [CrossRef]
7. Onifade, M.; Genc, B.; Gbadamosi, A.R.; Morgan, A.; Ngoepe, T. Influence of antioxidants on spontaneous combustion and coal properties. *Process Saf. Environ.* **2021**, *148*, 1019–1032. [CrossRef]
8. Fernánez-Alaiz, F.; Castañón, A.M.; Gómez-Fernández, F.; Bernardo-Sánchez, A.; Bascompta, M. Analysis of the Fire Propagation in a Sublevel Coal Mine. *Energies* **2020**, *13*, 3754. [CrossRef]
9. Bascetin, A.; Brune, J.F.; Adiguzel, D. The study of permeability changes of a gob structure in an underground coal mine to prevent spontaneous combustion. *Int. J. Min. Reclam. Environ.* **2021**, *35*, 693–708. [CrossRef]
10. Tutak, M. Numerical research of oxidation zone variation in goaf of longwalls U-type system from borders and U-type system to the borders ventilated. *IOP Conf. Ser. Earth Environ. Sci.* **2019**, *221*, 012090. [CrossRef]
11. Islavath, S.R.; Deb, D.; Kumar, H. Numerical analysis of a longwall mining cycle and development of a composite longwall index. *Int. J. Rock Mech. Min.* **2016**, *89*, 43–54. [CrossRef]
12. Konicek, P.; Schreiber, J. Heavy rock bursts due to long-wall mining near protective pillars: A case study. *Int. J. Min. Sci. Technol.* **2018**, *28*, 799–805. [CrossRef]
13. Sakhno, I.G.; Sakhno, S.V.; Kamenets, V.I. Stress environment around head entries with pillarless gob side entry retaining through numerical simulation incorporating the two type of filling wall. *IOP Conf. Ser. Earth Environ. Sci.* **2022**, *1049*, 012011. [CrossRef]
14. Golubev, D.D. Using pillar-free mining technologies in gently dipping and self-ignitable coal seams. *Min. Inf. Anal. Bull.* **2020**, *7*, 64–77. [CrossRef]
15. Tao, Z.; Song, Z.; He, M.; Meng, Z.; Pang, S. Principles of the roof cut short-arm beam mining method (110 method) and its mining-induced stress distribution. *Int. J. Min. Sci. Technol.* **2017**, *28*, 391–396. [CrossRef]
16. Kim, B.; Walton, G.; Larson, M.K.; Berry, S. Experimental study on the confinement-dependent characteristics of a Utah coal considering the anisotropy by cleats. *Int. J. Rock Mech. Min.* **2018**, *105*, 182–191. [CrossRef] [PubMed]
17. Wang, Y.; Wang, Q.; Tian, X.; Wang, H.; Yang, J.; He, M. Stress and deformation evolution characteristics of gob-side entry retained by roof cutting and pressure relief. *Tunn. Undergr. Space Technol.* **2022**, *123*, 104419. [CrossRef]
18. Islavath, S.R.; Kasturi, P.K. Development of pillar extraction strategy for a coal panel and rib pillar stability index under the influence of the goaved-out panels. *Arab. J. Geosci.* **2023**, *16*, 13. [CrossRef]
19. Singh, A.K.; Singh, R.; Maiti, J.; Kumar, R.; Mandal, P.K. Assessment of mining induced stress development over coal pillars during depillaring. *Int. J. Rock Mech. Min.* **2011**, *48*, 805–818. [CrossRef]

20. Yadav, A.; Behera, B.; Sahoo, S.K.; Singh, G.S.P.; Sharma, S.K. An Approach for Numerical Modeling of Gob Compaction Process in Longwall Mining. *Min. Metall. Explor.* **2020**, *37*, 631–649. [CrossRef]
21. Zhou, X.; Jing, Z.; Li, Y. Research on controlling gas overrun in a working face based on gob-side entry retaining by utilizing ventilation type "Y". *Sci. Rep.* **2023**, *13*, 9199. [CrossRef] [PubMed]
22. Arasteh, H.; Saeedi, G.; Farsangi, M.A.E.; Esmaeili, K. A New Model for Calculation of the Plastic Compression Index and Porosity and Permeability of Gob Materials in Longwall Mining. *Geotech. Geol. Eng.* **2020**, *38*, 6407–6420. [CrossRef]
23. Janus, J. Numerical Investigation of Air Flow in Goaf While Mapping Its Flow Parameters. *Processes* **2023**, *11*, 987. [CrossRef]
24. Tutak, M.; Brodny, J.; Szurgacz, D.; Sobik, L.; Zhironkin, S. The Impact of the Ventilation System on the Methane Release Hazard and Spontaneous Combustion of Coal in the Area of Exploitation—A Case Study. *Energies* **2020**, *13*, 4891. [CrossRef]
25. Marzły, M.; Trzaskalik, P. Comparative Analysis of Methane Concentration Near the Junction of the Longwall and Top Road. *Mang. Syst. Prod. Eng.* **2019**, *27*, 166–173.
26. Szlązak, N.; Obracaj, D.; Swolkień, J. Enhancing Safety in the Polish High-Methane Coal Mines: An Overview. *Min. Metall. Explor.* **2020**, *37*, 567–579. [CrossRef]
27. Brodny, J.; Tutak, M. The Impact of the Strength of Roof Rocks on the Extent of the Zone with a High Risk of Spontaneous Coal Combustion for Fully Powered Longwalls Ventilated with the Y-Type System—A Case Study. *Appl. Sci.* **2019**, *9*, 5315.
28. Roghanchi, P.; Kocsis, K.; Sunkpal, M. Sensitivity analysis of the effect of airflow velocity on the thermal comfort in underground mines. *J. Sustain. Min.* **2016**, *15*, 175–180. [CrossRef]
29. Rezaei, M. Long-term stability analysis of goaf area in longwall mining using minimum potential energy theory. *J. Min. Environ.* **2018**, *9*, 169–182.
30. Miao, D.; Chen, X.; Ji, J.; Lv, Y.; Zhang, Y.; Sui, X. New Technology for Preventing and Controlling Air Leakage in Goaf Based on the Theory of Airflow Boundary Layer. *Processes* **2022**, *10*, 954. [CrossRef]
31. Saki, S.A.; Brune, J.F.; Khan, M.U. Optimization of gob ventilation boreholes design in longwall mining. *Int. J. Min. Sci. Technol.* **2020**, *30*, 811–817. [CrossRef]
32. Khattri, S.K.; Log, T.; Kraaijeveld, A. Tunnel Fire Dynamics as a Function of Longitudinal Ventilation Air Oxygen Content. *Sustainability* **2019**, *11*, 203. [CrossRef]
33. Kurnia, J.; Sasmito, A.; Mujumdar, A. Simulation of a novel intermittent ventilation system for underground mines. *Tunn. Undergr. Space Technol.* **2014**, *42*, 206–215. [CrossRef]
34. Menéndez, J.; Merlé, N.; Fernández-Oro, J.M.; Galdo, M.; de Prado, L.Á.; Loredo, J.; Bernardo-Sánchez, A. Concentration, Propagation and Dilution of Toxic Gases in Underground Excavations under Different Ventilation Modes. *Int. J. Environ. Res. Public Health* **2022**, *19*, 7092. [CrossRef] [PubMed]
35. Vives, J.; Bascompta, M.; Felipe, J.J.; Sanmiquel, L. Computational Fluid Dynamics (CFD) study to optimize the auxiliary ventilation system in an underground mine. *Dyna* **2022**, *89*, 84–91. [CrossRef]
36. Tian, Y.; Yang, C.; Sun, Q.; Chang, K.; Guo, Z. Air leakage law in goaf of the working face using U-type ventilation. *Coal Eng.* **2020**, *52*, 132–136.
37. Zhai, X.; Wang, B.; Jiang, S.; Zhang, W. Oxygen Distribution and Air Leakage Law in Gob of Working Face of U+L Ventilation System. *Math. Probl. Eng.* **2019**, *2019*, 8356701. [CrossRef]
38. Li, T.; Wu, B.; Lei, B.; Huang, Q. Study on air leakage and gas distribution in goaf of y-type ventilation system. *Energy Source Part A* **2020**, *5*, 1–14. [CrossRef]
39. Guo, L.; Yao, J.; Su, X. Influence of W-type and U-type ventilation methods on working face. *IOP Conf. Ser. Earth Environ. Sci.* **2020**, *546*, 042014. [CrossRef]
40. Zhu, X.; Wen, H. Numerical simulation study on the influence of air leakage on oxygen concentration in goafs of fully mechanized caving mining with shallow buried and large mining height. *Front. Earth Sci.* **2023**, *11*, 1138925. [CrossRef]
41. Zhang, F.; Liu, W.; Qin, Y.; Chu, X.; Xu, H.; Wu, F.; Li, Y. Optimization of coalbed methane recovery from extraction borehole using novel plastic spraying material: A field application and evaluation. *Process Saf. Environ.* **2022**, *169*, 534–546. [CrossRef]
42. Biswal, P.K.; Parida, D.; Mishra, G.; Sahoo, A.K. Study of air flow pattern in mine model gallery and its validation using CFD modelling. *World Sci. News* **2019**, *130*, 1–24.
43. Krukovskyi, O.; Kurnosov, S.; Makeiev, S.; Ryzhov, H.; Pilipenko, Y. Tamponage of massif by modern polymeric materials for isolating mined-out areas in the coal seams prone to spontaneous ignition. *IOP Conf. Ser. Earth Environ. Sci.* **2022**, *970*, 012046. [CrossRef]
44. Zhang, Y.; Zou, Q.; Guo, L. Air-leakage Model and Sealing Technique with Sealing–Isolation Integration for Gas-drainage Boreholes in Coal Mines. *Process Saf. Environ.* **2020**, *140*, 258–272. [CrossRef]
45. Xi, X.; Jiang, S.; Yin, C.; Wu, Z. Experimental investigation on cement-based foam developed to prevent spontaneous combustion of coal by plugging air leakage. *Fuel* **2021**, *301*, 121091. [CrossRef]
46. *AQ 1116-2020*; General Safety Specification of Polymer Material for Consolidation, Water Stopping, Void Filling and Sprayed Sealing at Coal Mine. Ministry of Emergency Management of the People's Republic of China: Beijing, China, 2020.

fire

MDPI

Article

Optimization of the Monitoring of Coal Spontaneous Combustion Degree Using a Distributed Fiber Optic Temperature Measurement System: Field Application and Evaluation

Fengjie Zhang [1,*], Dongyang Han [1], Yueping Qin [1,2,*], Shiyang Peng [1], Dawei Zhong [1], Fei Tang [1], Zhencai Xiang [3] and Hao Xu [4]

[1] School of Emergency Management and Safety Engineering, China University of Mining & Technology-Beijing, Beijing 100083, China; handy0324@163.com (D.H.); psyang0722@163.com (S.P.); zhong990212@163.com (D.Z.); 18325533132@163.com (F.T.)
[2] State Key Laboratory of Coal Resources and Safety Mining, China University of Mining & Technology-Beijing, Beijing 100083, China
[3] Chongqing Baian Technology Co., Chongqing 400026, China; xzc236@163.com
[4] College of Safety and Environmental Engineering, Shandong University of Science and Technology, Qingdao 266590, China; skdxh0904@sdust.edu.cn
* Correspondence: yczhangfengjie@163.com (F.Z.); qyp@cumtb.edu.cn (Y.Q.)

Abstract: Coal spontaneous combustion (CSC) in gob not only leads to wasted resources and casualties, but also produces a lot of waste gas that pollutes the underground environment. Mastering the degree of CSC helps ensure that timely and effective control measures are taken. The real-time and accurate monitoring of temperature, as the primary indicator of the extent of CSC, is difficult due to the harsh and hidden environment of gob, resulting in a reduced ability to anticipate and prevent CSC. In this work, a complete distributed optical fiber temperature sensing system (DTSS) set with strong anti-damage ability was developed. The optical cable is protected by internal parallel steel cables and external protective pipes, which greatly improve the system's reliability and integrity when used in gob. During its application in the Wangyun Coal Mine, an abnormal temperature rise in gob was discovered in time, and the effect of inhibitor spraying was monitored and evaluated. The degree of CSC in the gob was accurately identified, which shows that the work of coal mining can be targeted. This work is expected to improve early warning capability to prevent the risk of CSC in gob, and has promising applications.

Keywords: coal spontaneous combustion (CSC); distributed temperature sensing system; protective pipe; online monitoring; gob

1. Introduction

As a major fossil energy source, a small portion of coal is inevitably left in gob due to the limitations of mining technology [1,2]. The residual coal undergoes oxidation and gradually releases heat under the coupling effect of multiple physical fields [3,4]. The continuous accumulation of heat can easily lead to the occurrence of coal spontaneous combustion (CSC), which is difficult to extinguish in a short period of time [5,6]. CSC gravely threatens the safety of coal production and the safety of workers [7,8]. It not only leads to a large amount of wasted coal resources [9], but also destroys the underground geological and ecological environment [10]. When certain conditions are met, it may even cause secondary disasters such as gas explosions, thus causing large-scale casualties [11–13]. Therefore, corresponding measures must be taken to prevent the occurrence of CSC disasters in gob [14].

In order to implement targeted governance measures, it is particularly vital to grasp the degree of CSC in gob [15,16]. The main characteristics of CSC include rising temperature and the generation of gases such as alkane [17]. And CO is usually used as the main iconic gas for the early warning and prediction of CSC in gob [18,19]. The mapping relationship between iconic gases and the degree of CSC is generally determined through mathematical models established in the laboratory [20,21]. However, applying this mapping relationship in large-sized gob may raise doubts about the accuracy of the model due to numerous uncertain factors [22], and iconic gases are the products of a coal oxidation reaction, so it exhibits hysteresis when predicting CSC in gob [23]. Given that the direct effect of the oxidized exotherm of the coal body is to cause an increase in temperature, it is feasible and reasonable to use temperature indicators to anticipate CSC.

Due to the concealment and complexity of gob [24], the selection of temperature monitoring technology should be comprehensively considered from various aspects, such as the accuracy, continuity, stability, and intuitiveness of data results. The current commonly used probe-based temperature measurement method is point-of-contact, which has the disadvantages of long measurement time, a large workload, and the inability to continuously monitor. Most notably, the probe and cable are easily damaged, and it is not suitable to use it on a large scale [25]. A characteristic of methods such as isotope radon measurement [26] and magnetic and infrared induction [27,28] is that they cannot achieve directly contact in high-temperature locations, which leads the positioning and value of temperature anomaly points to be inaccurate. The radio wave detection method has high accuracy and low cost, but the transmission technology used in gob still needs further improvement [29]. In recent years, fiber optic sensing technology has been widely applied in temperature monitoring and warning in various industrial fields [30,31]. However, the complex underground environment of coal mines places higher demands on the safety performance of equipment. And in the application of coal mine gob with coal/rock caving and mining stress, determining how to prevent the negative effect of perception and the transmission of temperature caused by damage to optical cables is a challenge that hinders the application of optical fiber temperature measurement technology [32].

To address the above issues, herein, a new set of products and technologies are developed that can improve the safety of the temperature measurement method through the design of safety barriers. And a method is designed to improve its damage resistance from the internal and external structures of optical cables, which can ensure the accuracy and continuity of online temperature monitoring in complex environments with gob. This product and technology were successfully applied in the Wangyun Coal Mine in China and greatly improved the monitoring and warning capabilities for CSC hazards in gob.

2. Detection System

2.1. Technical Principle

The technology of distributed optical fiber temperature detection is achieved through Raman scattering [33,34]. Figure 1 shows the technical principle of the distributed optical fiber temperature sensing system (DTSS), which mainly includes a device host, a terminal computer, optical cables for temperature measurement, and a meta-plane of temperature perception.

The device host sends a laser pulse into the optical cable, and the interaction between the laser pulse and the fiber molecules generates Raman-scattered light [35,36]. A small fraction of it is called back-scattered light, which is oriented in the opposite direction to the incident light. The intensity of this back-scattered light has a certain correlation with the temperature of the scattering point in the optical cable. The higher the temperature, the higher the intensity of the back-scattered light. When the back-scattered light signal returns to the device host, it can be converted into a digital signal through demodulation technology and output on the terminal computer to display its temperature and position information. This technique can detect temperature anomalies in a timely manner, which facilitates the implementation of risk prevention measures.

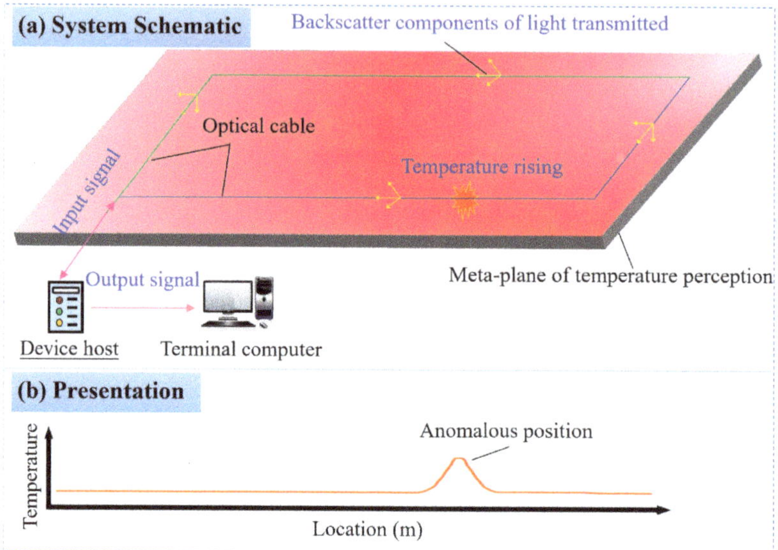

Figure 1. Technical principle of DTSS. (**a**) System schematic (**b**). Presentation.

2.2. Demodulation Process

Figure 2 shows the demodulation process of DTSS. The laser pulse is emitted by a laser with a high-power semiconductor. The laser pulse entering the external optical cable generates back-scattered optical signals and transmits them back to the signal receiver. Next, a wavelength division multiplexer (WDM) separates the temperature-insensitive Stokes light and temperature–sensitive anti–Stokes light from the back–scattered light, and amplifies their gain through a signal amplifier. A photoelectric converter demodulates the temperature information carried by the light into electrical signals, and finally, outputs them to an external computer [37,38].

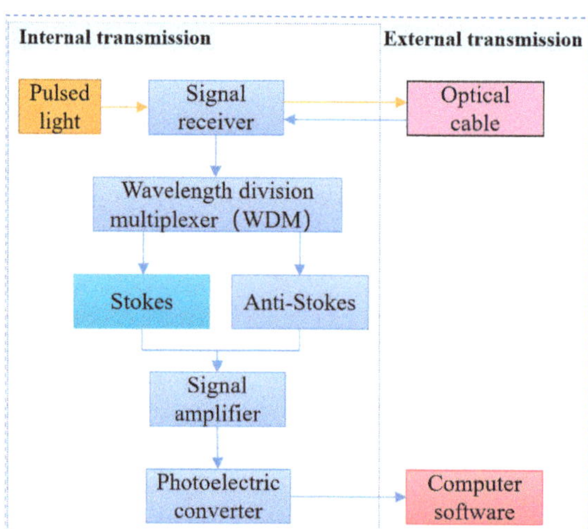

Figure 2. Demodulation process.

The specific principle of signal demodulation involves the following mathematical formula. The scattering photon frequencies of Stokes and anti−Stokes light are [39–41]:

$$\nu_s = \nu_0 - \Delta\nu \tag{1}$$

$$\nu_{as} = \nu_0 + \Delta\nu \tag{2}$$

where v_0 is the vibrational frequency of the optical fiber molecule, and Δv is the frequency shift value of Raman scattering, 1.32×10^{13} Hz.

The intensity of Stokes and anti−Stokes at L (measuring point) shall be expressed as Equations (3) and (4), respectively [42,43]:

$$I_s = E_0 k_s B \nu_s^4 \cdot \exp[-(a_0 - a_s) \cdot L] \times R_s \tag{3}$$

$$I_{as} = E_0 k_{as} B \nu_{as}^4 \cdot \exp[-(a_0 - a_{as}) \cdot L] \times R_{as} \tag{4}$$

$$R_s = (1 - \exp(-h\Delta\nu/kT))^{-1} \tag{5}$$

$$R_{as} = (\exp(h\Delta\nu/kT) - 1)^{-1} \tag{6}$$

where R_s and R_{as} are the coefficients related to the population of optical fiber molecules at low and high energy levels; E_0 is the intensity of the initial incident light in the optical fiber; k_s and k_{as} are the scattering coefficients related to the aperture (scattering cross−section) of the optical fiber; B is the back−scatter coefficient; ν_s and ν_{as} are the frequency of Stokes and anti−Stokes scattering light; α_0, α_s, and α_{as} are the average transmission loss of incident light, Stokes light, and anti−Stokes light in the optical fiber; L is the location of the measuring point; H is the Planck constant, $6.62607015 \times 10^{-34}$ J·s; k is the Boltzmann constant, 1.380649×10^{-23} J/K; and T is the temperature of the measuring point.

Comparing Equation (3) with Equation (4), the temperature information of L can be obtained:

$$F(T) = \frac{I_{as}}{I_s} = \frac{k_{as}}{k_s}\left(\frac{\nu_{as}}{\nu_s}\right)^4 \cdot \exp(-h\Delta\nu/kT) \cdot \exp[-(\alpha_{as} - \alpha_s)L] \tag{7}$$

The temperature information when the reference temperature is T_0 is:

$$F(T_0) = \frac{I_{as}}{I_s} = \frac{k_{as}}{k_s}\left(\frac{\nu_{as}}{\nu_s}\right)^4 \cdot \exp(-h\Delta\nu/kT_0) \cdot \exp[-(\alpha_{as} - \alpha_s)L] \tag{8}$$

Combining Equations (7) and (8), the temperature T of the measuring point can be obtained based on the reference temperature T_0:

$$\frac{1}{T} = \frac{1}{T_0} - \frac{k}{h\Delta\nu}\ln\frac{F(T)}{F(T_0)} \tag{9}$$

In the optical time domain, the position of the measuring point is [42]:

$$L = \frac{1}{2}vt = \frac{1}{2}\frac{c}{n}t \tag{10}$$

where v is the propagation speed of light in the optical fiber; c is the speed of light in a vacuum; n is the refractive index of the optical fiber; and t is the measuring time.

The temperature value and the position of a certain point in the optical cable can be obtained using the above demodulation method.

2.3. Description of DTSS

In this work, a new DTSS technology that can be used in coal mine gob is proposed [44,45]. Figure 3 shows the complete DTSS set. Figure 3a illustrates the fiber optic device host. The left side of the device host is the power supply, the right side is the signal transmission interface, and the screen in the middle of DTSS is a visualization window that can directly display the detected temperature data. Figure 3b illustrates the optical cable used for temperature measurement. The design of a flat appearance increases the temperature detection area; the design of parallel steel cable reinforcement enables the optical cable to have very strong compressive and impact resistance. Figure 3c illustrates the software interface displayed in the visualization window, which can display the position and temperature values of a certain point on the optical cable. The horizontal axis of the coordinate map represents the length of the optical cable, and the vertical axis represents the temperature value detected by the optical cable. Figure 3d illustrates the internal structure of the detection host, where the device has added safety barriers and backup batteries to improve safety performance and prevent the impact of power outages. In addition, the bias plate component used can conduct small signals and improve the signal−to−noise ratio.

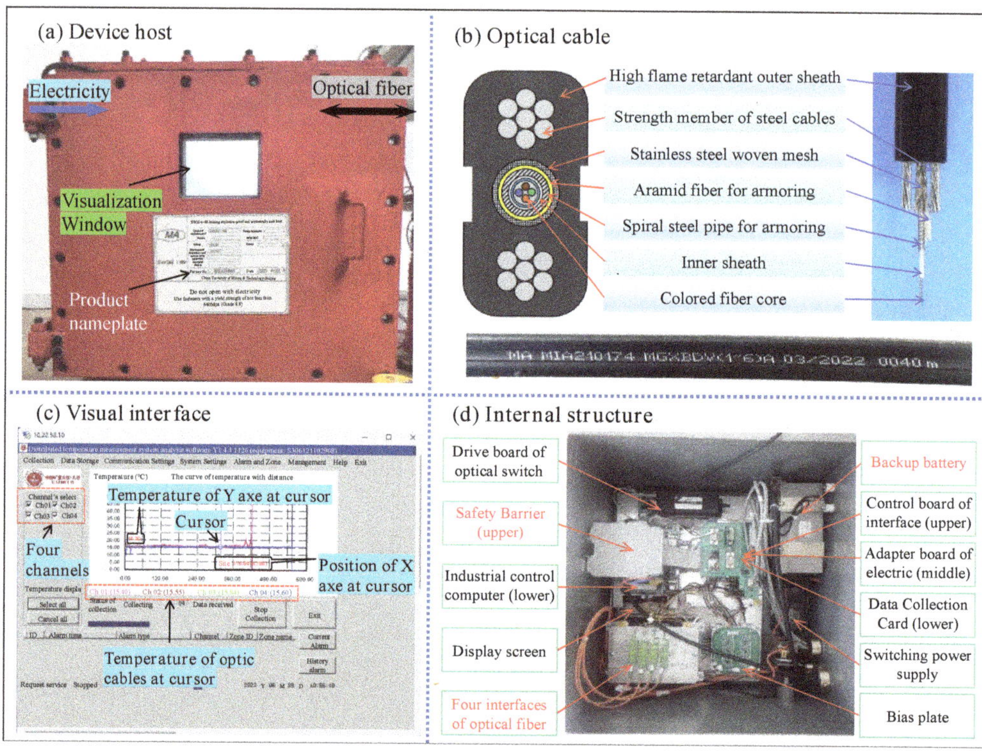

Figure 3. Complete DTSS. set (**a**) Device host. (**b**) Optical cable. (**c**) Visual interface. (**d**) Internal structure.

2.4. Performance Characteristics

This product has obtained a safety label certificate (No. MFE230006) for mining products and is classified as explosion−proof and intrinsically safe equipment.

Table 1 shows the performance parameters of the device host, with a high positioning accuracy of 0.4 m, a temperature measurement accuracy of 1.54 °C, and a reaction time of 2 s, which can be applied to multi−mode optical fiber with a diameter of 62.5/125 µm. It has four channels and can be connected to four optical cables. The working current is 800 mA.

Table 1. Performance parameters.

Performance Index	Parameters
Supply voltage	AC127V
Working current	800 mA
Number of channels	Four
Distance for temperature measurement	4 km for one channel
Scope of temperature	(0~100) °C
Spatial resolution	2 m
Accuracy of positioning	≤±0.4 m
Accuracy of temperature	1.54 °C
Response time	2 s
Wavelength of the detection light	1550 nm
Emission power of the detection light	≤0 dbm
Communication interface	RJ45, RS485
Type of fiber that can be connected	Multi−mode optical fiber: 62.5/125 μm
Inside connectors of optical fiber	E2000

3. Comparative Experiment of Casing on Optical Cable

3.1. Testing Method

The environment of underground gob is relatively harsh, and optical cables may break and be damaged under the high stress of falling rocks. We propose a protection method that uses high−pressure rubber hoses as casing pipes on the optical cables. To determine the effect of the protective pipe on the detection results, a comparison experiment was designed and performed without and with the protective pipe, and the changes in the temperature value and temperature sensing speed were analyzed.

The test design is shown in Figure 4. A silicone flexible electric heating film connected to a PID temperature controller is used as the heating strap for better contact and transfer of heat to the optical cable. The heating strap that wraps around the optical cable is fixed with several sealing strips. A DN19 high−pressure rubber hose with four layers of steel wire winding reinforcement was selected as the protective pipe. The protective pipe can fully meet the requirements for use on gob, with a maximum working pressure of 42.0 MPa and a minimum blasting pressure of 168.0 MPa.

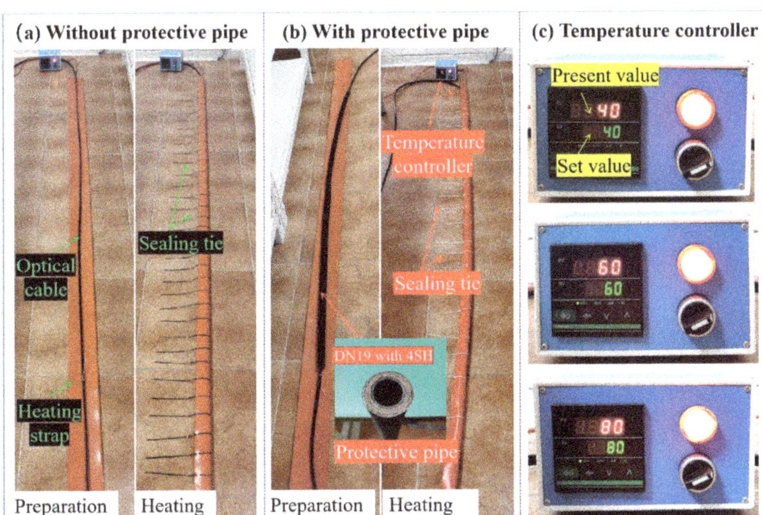

Figure 4. Test without and with protective pipe. (**a**) Without protective pipe. (**b**) With protective pipe, (**c**) Temperature controller.

The length of the optical cable for heating is 2 m, the length of the high−pressure rubber hose is 2 m, and the length of the heating strip wrapped externally is 3 m. The detection temperatures are set at 40 °C, 60 °C, and 80 °C, respectively.

During the experiment, PID was used to control the temperature to the present value, and then, the temperature detected by the optical cable within 70 min was observed. A set of data was exported every 1 min by the device host. And the heating strap temperature could increase to 60 °C and 80 °C sequentially after the first temperature detection was finished.

3.2. Analysis of Temperature Value Accuracy

Figure 5 indicates the changes in temperature values detected without and with the protective pipe. The horizontal axis is the length of the optical cable, and the vertical axis is the average temperature value of five tests. From Figure 5, the maximum detection temperature values for the three preset temperatures (40 °C, 60 °C, and 80 °C) without the protective pipe are 38.48 °C, 59.42 °C, and 79.27 °C, respectively. After using the protective pipe, the maximum detection temperature values for the three preset temperatures are 37.21 °C, 58.75 °C, and 79.41 °C, respectively. The differences between the two experiments are 0.87 °C, 0.67 °C, and 0.24 °C, and neither of them exceed 1 °C, as shown in Figure 5d. And the trend of their temperature changes is consistent, which indicates that the influence of the protective pipe on the temperature value is very small and can be ignored.

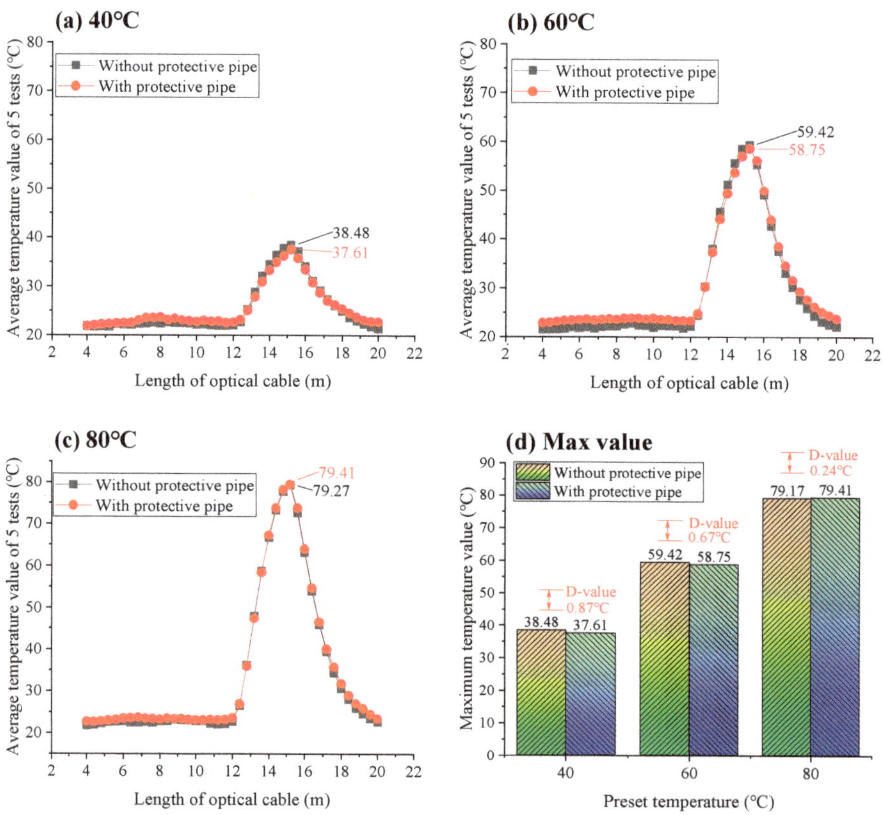

Figure 5. Comparison of temperature values without and with protective pipe. (**a**) 40 °C. (**b**) 60 °C. (**c**) 80 °C. (**d**) Max value.

In addition, it is worth noting that the lengths with temperature variations at the three different set temperatures were all 12–20 m, which is more than the heating length of 3 m. This length with increased temperature distribution is the response distance, which conforms to the temperature measurement principle of the temperature measurement system [46]. The test temperature only affects the value of the response temperature and does not change the response distance (length with increased temperature distribution).

3.3. Analysis of the Speed of Temperature Sensing

In order to understand the influence of the protective pipe on the temperature sensing speed, the temperature detected at the highest temperature measuring point within 70 min was analyzed over time without and with the protective pipe, as shown as Figure 6. The maximum temperature values without the protective pipe appear at 24 min, 38 min, and 56 min, respectively. After using the protective pipe, they are delayed to 47 min, 54 min, and 65 min, respectively, and delayed for 23 min, 16 min, and 9 min, respectively. In addition, the temperature rises rapidly in the first 10 min without the protective pipe. And after using the protective pipe, the overall temperature rise rate is very slow and there is no sharp increase.

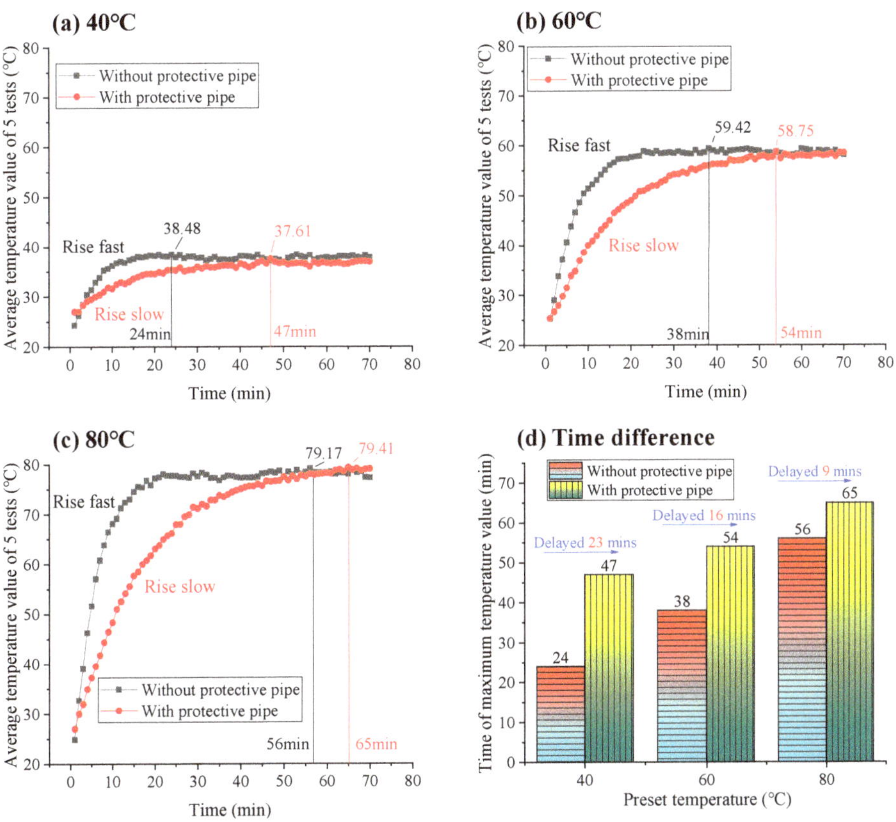

Figure 6. Comparison of temperature sensing speed without and with protective pipe. (**a**) 40 °C. (**b**) 60 °C. (**c**) 80 °C. (**d**) Time difference.

Overall, although the time required for the temperature detection value to stabilize after using the protective pipe increases, the final temperature value is basically unchanged, and this lag time can be ignored for large−scale gob. Therefore, using protective pipe to protect optical cables in gob will not affect the accuracy of the temperature values.

4. Application

4.1. Field Temperature Measuring Design

The Wangyun Coal Mine, located in Shanxi Province, China, has an annual output of 1.8 million tons. Currently, the 15th coal seam is being mined, with a thickness of 4.7 m and an oxygen absorption capacity of 0.84 cm^3/g, classified as type II spontaneous combustion coal seam. The cut length of the 15106 working face currently being mined is 180 m. In order to grasp the real-time spontaneous combustion status of residual coal in gob, the DTSS proposed in this work was used to monitor the temperature changes in the gob. Figure 7 illustrates the application design of the DTSS, with the 1# and 2# optical cables distributed on both sides of the roadway with the protective pipes. The DN19 high-pressure rubber hose with four layers of steel wire winding reinforcement was selected as the protective pipe, and its safety sign number is MEE150868. The protective pipes are each 10 m long, with a screw joint machined at each end. An optical cable is laid in the roadway, and then, the first protective pipe is placed on the optical cable at the working face and pulled to the entrance of the roadway. Then, this is repeated with the remaining protective pipes inserted. After all the protective pipes have been installed, the joint between the two protective pipes is tightened to complete the installation.

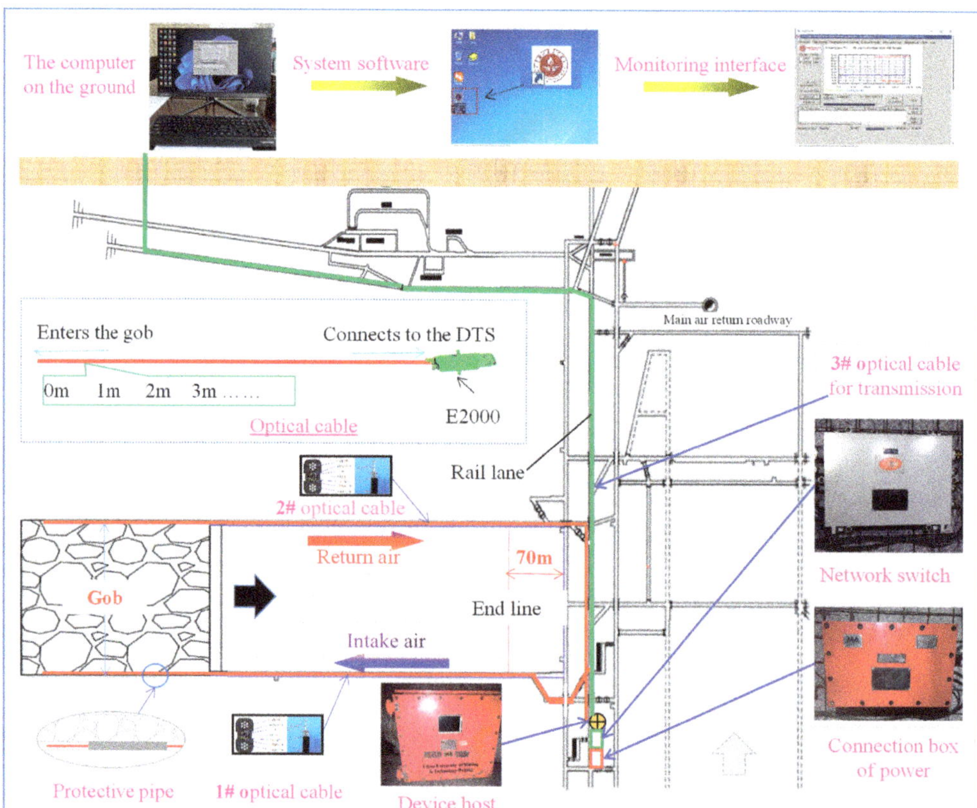

Figure 7. Application design of DTSS in gob.

One end of the optical cable with an E2000 fiber optic connector enters the device host. And as the working face advances, the other end, beginning a 0 m, is buried in the gob. The connection box of power supplies power to the device host. The temperature signal demodulated by the device host is transmitted to the ground computer through a network switch and the 3# transmission optical cable, which achieves the ground monitoring of temperature data in the underground gob.

4.2. Application Effect

Figure 8 shows the construction of each part of the DTSS on-site, and Figure 8a,b show the laying of optical cables with identification labels on them. Subsequently, the wiring work is completed using a fusion splicer. The connection of optical cables, electric cables, and network cables to the device host is completed in the wiring chamber shown in Figure 8d. The on-site layout of the DTSS is shown in Figure 8g.

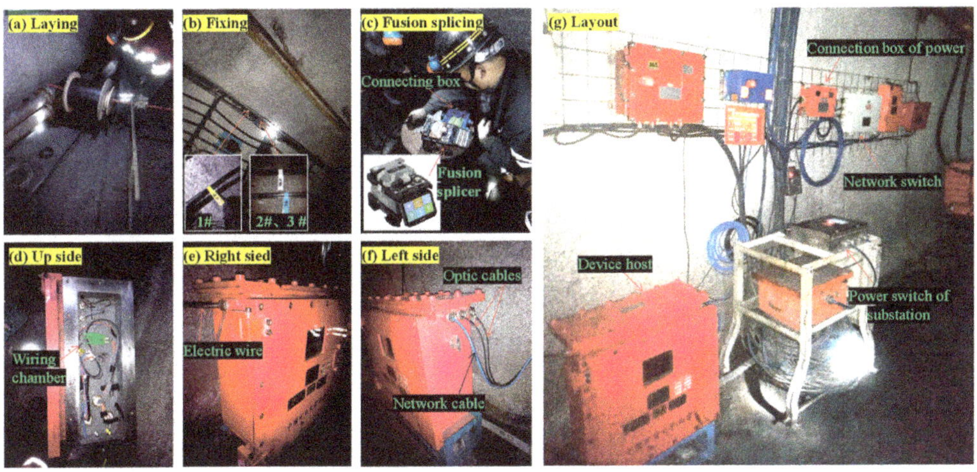

Figure 8. Construction effect of DTSS. (**a**) Laying. (**b**) Fixing. (**c**) Fusion splicing. (**d**) Top side. (**e**) Right side. (**f**) Left side. (**g**) Layout.

After the completion of the DTSS, the system software of the ground computer obtains temperature data from four channels, as shown in Figure 9. The optical cables used in this work are dual-mode with two cores (double insurance), so channel 1 and channel 2 display the temperature data of the 1# optical cable, while channel 3 and channel 4 display the temperature data of the 2# optical cable. The temperature data detected by the four channels has a certain degree of fluctuation, but the difference is not significant, and the temperature on the entire optical cable remains stable. Real-time data show that the temperature of the gob is between 15.45~16.97 °C, with an average value of 16.09 °C. It should be noted that there is an irregular fluctuation in temperature at the end of each channel, which is a normal disturbance phenomenon at the end of the optical cable and can be ignored.

4.3. Evaluation of the Effectiveness of Inhibitors

The above research indicate that the DTSS established in this work can accurately monitor temperature changes in gob areas. In addition, the DTSS can effectively support the prevention and control of natural fires. During the use of the DTSS, a geological formation appeared while the 15,106 working face advanced to the position of 347 m. This resulted in a decrease in forward speed and a decrease in daily forward length from 2.4 m to 0.5 m. A certain range of gob behind the working face has been in the oxidation zone for a long time. The temperature monitored by the optical cable of the DTSS on 26 June 2023 showed that the temperature in the gob at 18 m behind the working face (with an optical cable length of 399.6 m) was about 20 °C, which was 4 °C higher than normal.

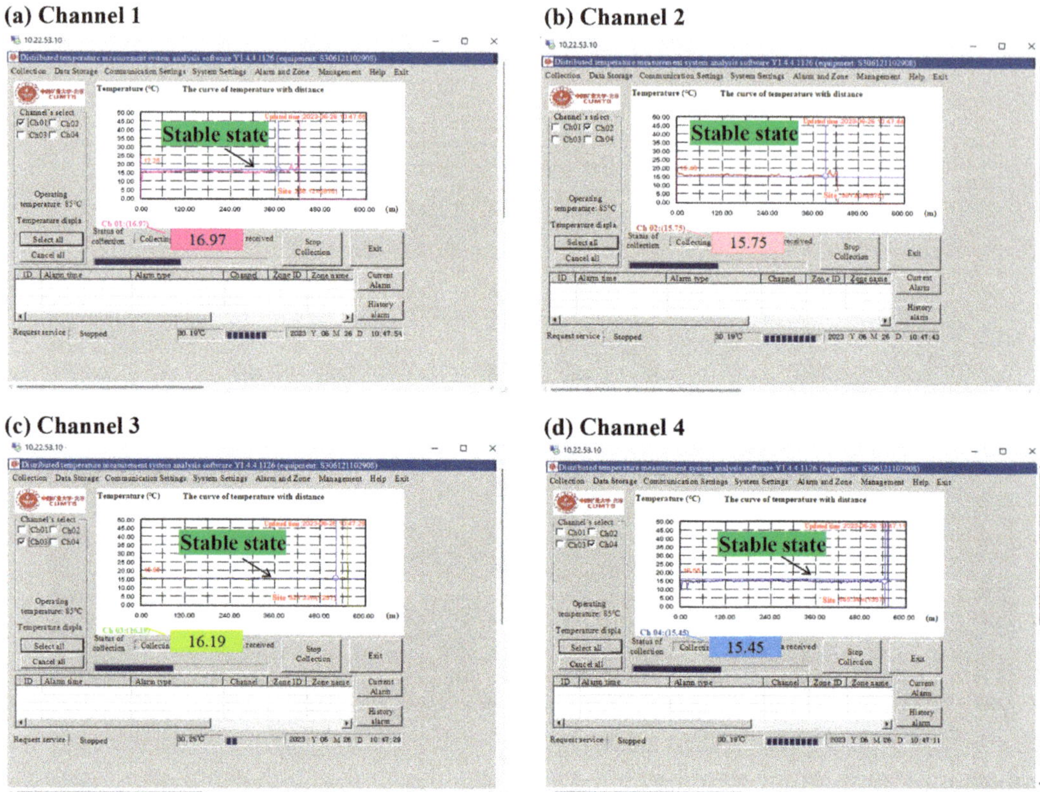

Figure 9. Results of temperature detection. (**a**) Channel 1. (**b**) Channel 2. (**c**) Channel 3. (**d**) Channel 4.

To control the increase in temperature after receiving the warning information issued by the DTSS, the method of using spray inhibitors in the gap between the hydraulic supports was employed to suppress the oxidation reaction of residual coal. Figure 10 shows the spraying process of the inhibitor solution. The inhibitor is industrial calcium chloride ($CaCl_2 \cdot 5H_2O$), with a concentration of 15%. The mixed inhibitor solution was delivered to the working surface through a pressure pump of model BZ-40/2.5 (manufactured by Shandong Changye Machinery Equipment Co., Ltd., China), and then, sprayed though a spraying gun from the gap between two hydraulic supports towards the gob, for at least 6 min each time, with a flow rate of no less than 35 L/min. Due to the temperature anomaly point detected by the 1# optical cable being close to the air intake side, inhibitor spraying was designed to be carried out at 60 m (34th hydraulic support) on the air intake side.

The working face advanced by 30 m within 65 days from 26 June 2023 to 31 August 2023. The effectiveness of the inhibitor solution was evaluated through temperature changes monitored by the 1# optical cable of the DTSS. Figure 11 shows the temperature values of the gob on the air intake side. In order to improve the reliability of data comparison, temperature data from 10:00 am, 14:00 pm, and 20:00 pm one day before and after spraying were selected. The data before spraying were selected from 26 June 2023, and the data after spraying were selected from 31 August 2023.

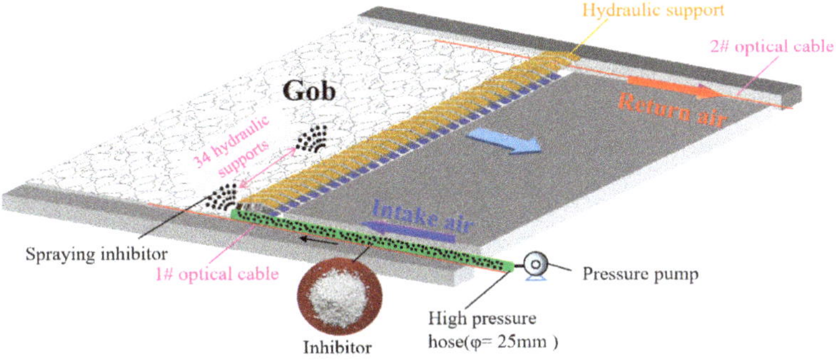

Figure 10. Spraying process of inhibitor.

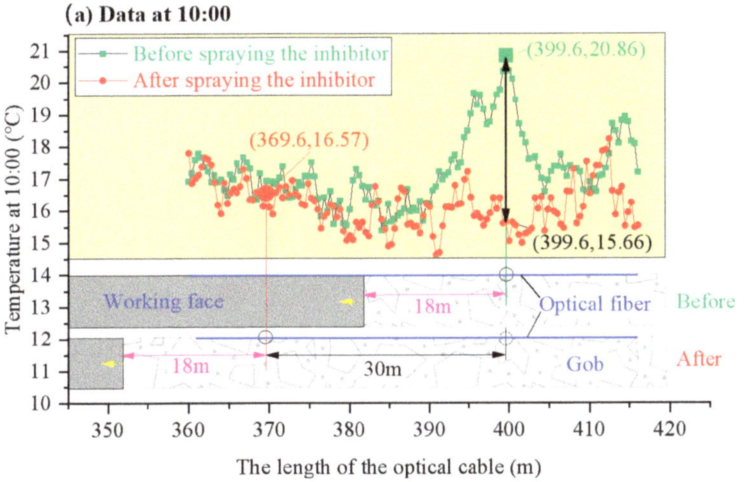

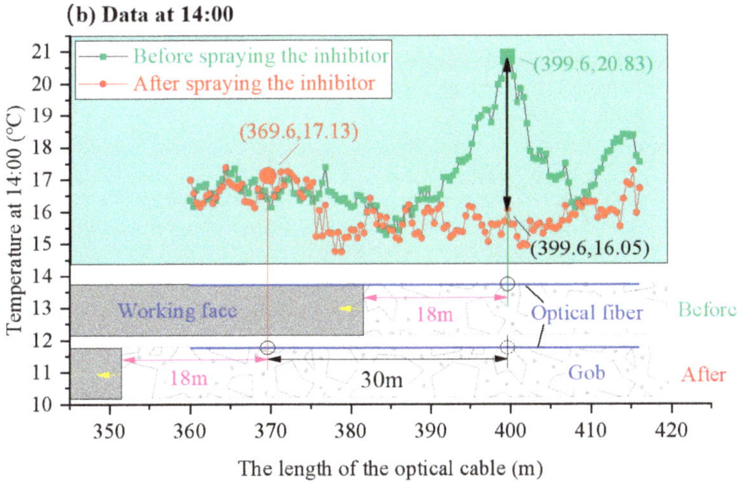

Figure 11. *Cont.*

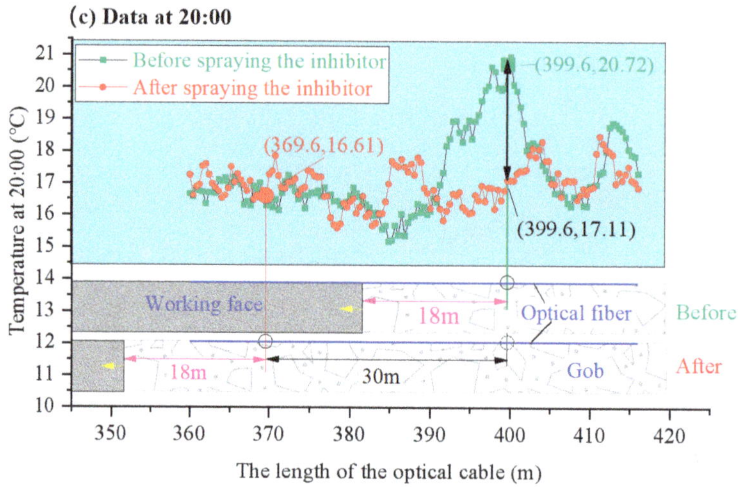

Figure 11. Comparison of temperature before and after praying the inhibitors after 65 days. (**a**) Data at 10:00. (**b**) Data at 14:00. (**c**) Data at 20:00.

What can be seen from Figure 11 is the following: ① Before spraying the inhibitor solution, the temperature of the gob at the position 18 m behind the working face (optical cable length 399.6 m) increased to 20.86 °C, 20.83 °C, and 20.72 °C, respectively. After using the inhibitor solution, the temperature of the gob with a relative delay of 18 m (cable length 369.6 m) was 16.57 °C, 17.13 °C, and 16.61 °C, respectively, decreasing by 4.29 °C, 3.7 °C, and 4.11 °C, respectively. The average temperature decreased from 20.8 °C to 16.77 °C, with a decrease of 19.38%. This means that the temperature of the gob behind the working face returned to a normal temperature. ② After 65 days of advancing the working face, the point at 399.6 m of the optical cable changed from 18 m to 48 m behind the working face, and the temperature at this position also decreased to 16.05 °C, 15.66 °C, and 17.11 °C, respectively. The average temperature at this point decreased from 20.8 °C to 16.27 °C, with a decrease of 21.78%. This indicates that the risk of coal spontaneous combustion in this area was eliminated. ③ These temperature changes indicate that the adopted inhibitor spray measures effectively suppressed the oxidation and exothermic effect of residual coal, and the DTSS used strongly supports the safety of production work in the coal mine.

This product and technology have greatly improved the monitoring and warning capabilities of CSC hazards in gob.

5. Conclusions

In this work, a monitoring and early warning system suitable for harsh environments in gob, called the distributed optical fiber temperature sensing system, was developed and successfully applied to monitoring and evaluating the degree of CSC in a coal mine in China. The main conclusions are summarized below:

1. The internal parallel steel cables and the external protective pipe improved the anti-damage ability of the optical cables. The technology of the protective pipe only increases the time required for detecting the actual temperature and does not affect the accuracy of temperature measurement. The temperature difference before and after use of the protective pipe is only 0.87 °C, 0.67 °C, and 0.24 °C, respectively. So, the protective pipe can be used with confidence.
2. The technology of the DTSS could improve early warning ability for preventing the risk of CSC in gob. This technology can detect abnormal temperature conditions in gob in a timely manner. Through the monitoring and evaluation of the DTSS, the average temperature at the same location in the gob after spraying the inhibitor

returned from 20.8 °C to 16.27 °C, with a decrease of 21.78%, which means the risk of CSC in the gob was eliminated.

3. This study only analyzes the application effect of DTSS technology in coal mine gob, which is not comprehensive enough. So, in the future, to enrich these research findings, comparative research on other measurement methods, such as temperature measurement methods of thermal resistance, will be carried out. In addition, we also hope to use the DTSS for in-depth research in different scenarios.

Author Contributions: Conceptualization, F.Z. and Y.Q.; methodology, F.Z.; formal analysis, F.Z., D.H. and H.X.; investigation, F.Z., D.Z. and S.P.; resources, Z.X.; data curation, D.H.; writing—original draft preparation, F.Z.; writing—review and editing, F.Z. and F.T.; project administration, Y.Q.; visualization, F.Z.; funding acquisition, F.Z. and Y.Q. All authors have read and agreed to the published version of the manuscript.

Funding: This research was funded by the research fund of the National Natural Science Foundation of China (52074303, 51874315); the Fundamental Research Funds for the Central Universities (2022JC-CXAQ06); and the project of the Basic Research Business Fee of China University of Mining and Technology (Beijing)—Top Innovative Talents Cultivation Fund for Doctoral Students (BBJ2023011). Additionally, we are grateful to the editor and anonymous reviewers for their valuable comments, which improved this paper significantly.

Institutional Review Board Statement: Not applicable.

Informed Consent Statement: Not applicable.

Data Availability Statement: Not applicable.

Acknowledgments: We are grateful to the editor and anonymous reviewers for their valuable comments, which improved this paper significantly.

Conflicts of Interest: The authors declare no conflict of interest.

References

1. Danish, E.; Onder, M. Application of Fuzzy Logic for Predicting of Mine Fire in Underground Coal Mine. *Saf. Health Work.* **2020**, *11*, 322–334. [CrossRef] [PubMed]
2. Hao, M.; Li, Y.; Song, X.; Kang, J.; Su, H.; Zhou, F. Hazardous areas determination of coal spontaneous combustion in shallow-buried gobs of coal seam group: A physical simulation experimental study. *Environ. Earth Sci.* **2019**, *78*, 39. [CrossRef]
3. Liu, W.; Qin, Y. Multi-physics coupling model of coal spontaneous combustion in longwall gob area based on moving coordinates. *Fuel* **2017**, *188*, 553–566. [CrossRef]
4. Jiang, X.; Yang, S.; Zhou, B.; Lan, L. The auto-oxidation characteristic of coal at different stages of the low-temperature oxidation process. *Fuel* **2023**, *352*, 129130. [CrossRef]
5. Liu, W.; Zhang, F.; Gao, T.; Chu, X.; Qin, Y. Efficient prevention of coal spontaneous combustion using cooling nitrogen injection in a longwall gob: An application case. *Energy* **2023**, *281*, 128214. [CrossRef]
6. Qiao, M.; Ren, T.; Roberts, J.; Yang, X.; Li, Z.; Wu, J. Insight into proactive inertisation strategies for spontaneous combustion management during longwall mining of coal seams with various orientations. *Energy Sources Part A Recover. Util. Env. Eff.* **2023**, *45*, 2788–2810. [CrossRef]
7. Saffari, A.; Sereshki, F.; Ataei, M.; Ghanbari, K. Presenting an engineering classification system for coal spontaneous combustion potential. *Int. J. Coal Sci. Technol.* **2017**, *4*, 110–128. [CrossRef]
8. Onifade, M.; Genc, B.; Gbadamosi, A.R.; Morgan, A.; Ngoepe, T. Influence of antioxidants on spontaneous combustion and coal properties. *Process Saf. Environ. Prot.* **2021**, *148*, 1019–1032. [CrossRef]
9. Si, R.; Li, R. Dynamic characteristics of low-concentration oxygen-containing gas explosion and prevention and control key technologies. *Coal Sci. Technol.* **2020**, *48*, 17–36. [CrossRef]
10. Ma, D.; Qin, B.; Gao, Y.; Jiang, J.; Feng, B. An experimental study on the methane migration induced by spontaneous combustion of coal in longwall gobs. *Process Saf. Environ. Prot.* **2021**, *147*, 292–299. [CrossRef]
11. Ranjith, P.G.; Zhao, J.; Ju, M.; De Silva, R.V.S.; Rathnaweera, T.D.; Bandara, A.K.M.S. Opportunities and Challenges in Deep Mining: A Brief Review. *Engineering* **2017**, *3*, 546–551. [CrossRef]
12. Zhang, L.; Wu, W.; Bian, Y.; Yang, M.; Luo, H. Influences of Gas Drainage Pipe Positions on Spontaneous Coal Combustion in the Gob: A Case Study of Baode Coal Mine in China. *Combust. Sci. Technol.* **2022**, *194*, 3357–3373. [CrossRef]
13. Zou, J.; Zhang, R.; Zhou, F.; Zhang, X. Hazardous Area Reconstruction and Law Analysis of Coal Spontaneous Combustion and Gas Coupling Disasters in Goaf Based on DEM-CFD. *ACS Omega* **2023**, *8*, 2685–2697. [CrossRef]

14. Shadab, A.; Ansari, M.T.I.; Raghuwanshi, S.K.; Kumar, S. Smoke Detection Using rGO-Coated eFBG Sensor for Early Warning of Coal Fire in Mines. *IEEE Sens. J.* **2023**, *23*, 2153–2160. [CrossRef]
15. Liu, C.; Zhang, R.; Wang, Z.; Zhang, X. Research on the fire extinguishing performance of new gel foam for preventing and controlling the spontaneous combustion of coal gangue. *Environ. Sci. Pollut. Res.* **2023**, *30*, 88548–88562. [CrossRef]
16. Deng, J.; Zhou, J.; Bai, Z.; Liu, L.; Wang, C. Effect of gas on microstructure and thermal reactivity of coal during low temperature oxidation. *Coal Sci. Technol.* **2023**, *51*, 304–312. [CrossRef]
17. Zhao, X.; Sun, P.; Yang, J.; Zhang, Z.; Zhang, L.; Tang, Y.; Sun, M.; Pang, T.; Liu, X.; Xu, Q. Online monitoring system of index gases concentration applied to coal sponta-neous combustion. *Meitan Xuebao/J. China Coal Soc.* **2021**, *46*, 319–327. [CrossRef]
18. Song, W.X.; Yang, S.Q.; Jiang, C.L.; Niu, J. Experimental research on the formation of CO during coal spontaneous combustion under the condition of methane-contained airflow. *J. China Coal Soc.* **2012**, *37*, 1320–1325. [CrossRef]
19. Cheng, J.; Ma, Y.; Lu, W.; Liu, G.; Cai, F. Using inverting CO critical value to predict coal spontaneous combustion severity in mine gobs with considering air leakages—A case study. *Process Saf. Environ. Prot.* **2022**, *167*, 45–55. [CrossRef]
20. Wang, K.; Li, Y.; Zhang, Y.; Deng, J. An approach for evaluation of grading forecasting index of coal spontaneous combustion by temperature-programmed analysis. *Environ. Sci. Pollut. Res.* **2023**, *30*, 3970–3979. [CrossRef]
21. Tan, B.; Shao, Z.; Guo, Y.; Zhao, T.; Zhu, H.; Li, C. Research on grading and early warning of coal spontaneous combustion based on correlation analysis of index gas. *China Saf. Sci. J.* **2021**, *31*, 33–39. [CrossRef]
22. Deng, J.; Xiao, Y.; Li, Q.; Lu, J.; Wen, H. Experimental studies of spontaneous combustion and anaerobic cooling of coal. *Fuel* **2015**, *157*, 261–269. [CrossRef]
23. Zhang, J. Current Situation and Development Trend of Coal Mine Beam Tube Monitoring System. *Saf. Coal Mines* **2019**, *50*, 89–92. [CrossRef]
24. Kong, D.; Pu, S.; Cheng, Z.; Wu, G.; Liu, Y. Coordinated Deformation Mechanism of the Top Coal and Filling Body of Gob-Side Entry Retaining in a Fully Mechanized Caving Face. *Int. J. Geomech.* **2021**, *21*, 04021030. [CrossRef]
25. Qin, Y.; Yan, L.; Liu, W.; Xu, H.; Song, Y.; Guo, W. Continuous monitoring system of gob temperature and its application. *Environ. Sci. Pollut. Res.* **2022**, *29*, 53063–53075. [CrossRef]
26. Wu, S.; Zhou, B.; Wang, J.; Yang, Q.; Dong, W.; Dong, Z. Isotope radon measurement method to identify spontaneous combustion regions in coal gangue hills: Case study for a coal mine in China. *Int. J. Coal Prep. Util.* **2023**, 1–13. [CrossRef]
27. Gao, Y.; Hao, M.; Wang, Y.; Dang, L.; Guo, Y. Multi-scale coal fire detection based on an improved active contour model from landsat-8 satellite and UAV images. *ISPRS Int. J. Geo-Inf.* **2021**, *10*, 449. [CrossRef]
28. Li, F.; Li, J.; Liu, X.; Meng, X. Coal fire detection and evolution of trend analysis based on CBERS-04 thermal infrared imagery. *Environ. Earth Sci.* **2020**, *79*, 1–15. [CrossRef]
29. Sun, Y.; Luo, X.; Gao, Y.; Ren, Q.; Wang, P.; He, Y. Application of the radioactive waveperspective technology to goaf detection in the coal mine resource integration. *J. Saf. Environ.* **2020**, *20*, 533–547. [CrossRef]
30. Sun, M.; Tang, Y.; Yang, S.; Li, J.; Sigrist, M.W.; Dong, F. Fire source localization based on distributed temperature sensing by a dual-line optical fiber system. *Sensors* **2016**, *16*, 829. [CrossRef]
31. Su, H.; Ou, B.; Yang, L.; Wen, Z. Distributed optical fiber-based monitoring approach of spatial seepage behavior in dike engineering. *Opt. Laser Technol.* **2018**, *103*, 346–353. [CrossRef]
32. Zhong, X.; Wang, J.; Zhou, K. Monitoring and early warning technology of coal spontaneous combustion in coal mines: Research status and intelligent development trends. *Ind. Mine Autom.* **2021**, *47*, 7–17. [CrossRef]
33. Shan, Y.; Ma, Y.; Fu, H.; Li, W.; Wang, C. Application of distributed optical fiber temperature measurement system in coal mine fire monitoring system. *Chin. J. Sens. Actuators* **2014**, *27*, 704–708. [CrossRef]
34. Hou, G.; Hu, Z.; Li, Z.; Zhao, Q.; Feng, D.; Cheng, C.; Zhou, H. Present situation and prospect of coal mine safety monitoring based on fiber bragg grating and distributed optical fiber sensing technology. *J. China Coal Soc.* **2023**, *48*, 96–110. [CrossRef]
35. Tereshchenko, E.R.; Shishlakov, V.F.; Gubanova, N.A.; Leonteva, A.B.; Manzuk, M.V. Application of Fiber Optic Temperature Sensor in the Development of a Distributed Data Acquisition System. In Proceedings of the 2021 IEEE Conference of Russian Young Researchers in Electrical and Electronic Engineering (ElConRus), St. Petersburg, Russia, 26–29 January 2021; Volume 2021, pp. 1111–1112. [CrossRef]
36. Du, C.; Hu, T.; Li, Y.; Wang, J. Feasibility study on distributed optical fiber monitoring of coal roadway roof. *Min. Saf. Environ. Prot.* **2022**, *49*, 81–88. [CrossRef]
37. Viveiros, D.; Ribeiro, J.; Ferreira, J.; Pinto, A.M.R.; Perez-Herrera, R.A.; Diaz, S.; Lopez-Gil, A.; Dominguez-Lopez, A.; Esteban, O.; Martins, H.F.; et al. Monitoring of coal waste piles with fiber optic sensing technology. In Proceedings of the 2015 3rd Experiment International Conference (exp. at'15), Ponta Delgada, Portugal, 2–4 June 2015; pp. 147–148. [CrossRef]
38. Ashry, I.; Mao, Y.; Wang, B.; Hveding, F.; Bukhamsin, A.Y.; Ng, T.K.; Ooi, B.S. A Review of Distributed Fiber-Optic Sensing in the Oil and Gas Industry. *J. Light. Technol.* **2022**, *40*, 1407–1431. [CrossRef]
39. Kus, J. Impact of underground coal fire on coal petrographic properties of high volatile bituminous coals: A case study from coal fire zone No. 3.2 in the Wuda Coalfield, Inner Mongolia Autonomous Region, North China. *Int. J. Coal Geol.* **2017**, *171*, 185–211. [CrossRef]
40. Kus, J. Oxidatively and thermally altered high-volatile bituminous coals in high-temperature coal fire zone No. 8 of the Wuda Coalfield (North China). *Int. J. Coal Geol.* **2017**, *176–177*, 8–35. [CrossRef]

41. Zhu, L.; Gu, W.; Chai, J.; Ma, Z.; Qiu, F. Evolution of mining-induced overburden deformation using distributed optical fiber. *J. Min. Strat. Control Eng.* **2022**, *4*, 013014. [CrossRef]
42. Zhang, Z.; Wang, J.; Yu, X.; Guo, N.; Wu, X.; Feng, H.; Insoo, S.; Sangki, O.; Yohee, K. The Research of Raman Type Distributed Optical Fiber Temperature Measuring Method. *J. Optoelectron.* **2001**, *12*, 596–600.
43. Wen, H.; Wu, K.; Ma, L.; Wang, W.; Wang, T. Application of Distributed Optical Fiber Temperature Measurement System in Monitoring Goaf Coal Spontaneous Combustion. *Saf. Coal Mines* **2014**, *5*, 100–105. [CrossRef]
44. Li, Z.; Li, F.; Wang, K.; Gao, X. Application of Distributed Optical Fiber Temperature Measurement System in Wanglou Coal Mine. *Coal Technol.* **2020**, *39*, 156–158. [CrossRef]
45. Su, H.; Zhou, F.; Song, X.; Qiang, Z. Risk analysis of spontaneous coal combustion in steeply inclined longwall gobs using a scaled-down experimental set-up. *Process Saf. Environ. Prot.* **2017**, *111*, 1–12. [CrossRef]
46. Ning, F.; Zhu, Y.; Cui, H.; Li, X.; Jin, Z. A Linear Correction Algorithm for Improving the Spatial Resolution of Distributed Fiber Optic Temperature Measurement System. *Acta Photonica Sin.* **2012**, *41*, 408–413. [CrossRef]

fire

MDPI

Article

Study on Spontaneous Combustion Characteristics and Early Warning of Coal in a Deep Mine

Caiping Wang [1,2], Yuxin Du [1,2], Yin Deng [1,2,*], Yu Zhang [1], Jun Deng [1,2], Xiaoyong Zhao [1] and Xiadan Duan [1,2]

[1] School of Safety Science and Engineering, Xi'an University of Science & Technology, Xi'an 710054, China; wcp9688@126.com (C.W.); hope1015d@163.com (Y.D.); zy40209@163.com (Y.Z.); dengj518@xust.edu.cn (J.D.); 15047430581@139.com (X.Z.); dxiadan@126.com (X.D.)

[2] Shaanxi Key Laboratory of Prevention and Control of Coal Fire, Xi'an 710054, China

* Correspondence: dengyym@126.com

Abstract: Due to high stress, high ground temperature, high moisture, and other factors in deep mines, the risk of coal spontaneous combustion (CSC) is enhanced, seriously affecting the safety of coal mining. To achieve early prediction of spontaneous combustion in the No. 3 coal seam at the Juye coalfield in the deep mine, this paper employs a temperature-programmed device to analyze the changing pattern of single-index gases and composite gas indices with temperature derived from the gas produced during csc. It also optimizes the index gas of coal sample spontaneous combustion. Simultaneously, the characteristics of coal temperature and a four-level warning indicator system for CSC are determined based on the analysis of indicator gas growth rate method, carbon-to-oxygen ratio, and the characteristics of the indicator gas. The composite index gases of the No. 3 coal seam in Juye coalfield are selected in the initial oxidation stage (Rco), accelerated oxidation stage (R_1, G_1), intense oxidation stage (R_2, G_1, G_3), and oxidative decomposition stage (G_3). This leads to the construction of a six-level warning system consisting of initial warning value, blue, yellow, orange, red, and black levels. Meanwhile, warning thresholds are also established.

Keywords: deep mining; coal spontaneous combustion; early warning indicator system; indicator gas

1. Introduction

The spontaneous combustion of coal has consistently been the most significant factor influencing coal mining, with over 85% of fire accidents in China being attributed to the spontaneous combustion of coal seams [1]. Simultaneously, coal resources are being progressively extracted from shallow to deep layers [2]. Mining depths have reached 1000–1500 m and are increasing at an annual rate of 10–30 m [3]. When coal enters deep mining, the combined influence of factors such as the increased stress on coal and rock, rising ground temperatures, and elevated water content in the goaf results in a complex occurrence environment for deep coal seams. This heightened complexity raises the challenges of coal mining and amplifies the risk of coal spontaneous ignition [4,5].

Currently, a considerable number of scholars have conducted simulations of deep CSC mining conditions, considering aspects such as surrounding rock stress, elevated ground temperatures, and moisture levels [6]. With the increase in ground stress, the frequency of coal rock fractures rises, ground temperatures elevate, and the risk of CSC becomes more pronounced. Pan et al. [7,8] found that ground stress increases with mining depth, leading to an elevated risk of coal oxidation. Additionally, coal was observed to oxidize more readily under conditions of air leakage. Chao et al. [9] found that the increase in axial stress initially promotes and later inhibits the spontaneous combustion of crushed coal. Meanwhile, as active mining depth increases, deep mines encounter the significant challenge of elevated ground temperatures. Niu et al. [10] found that high temperatures facilitate the accumulation of heat within coal, while the active structures on the coal surface become more reactive. This intensifies the propensity for CSC in deep mines. Jia et al. [11]

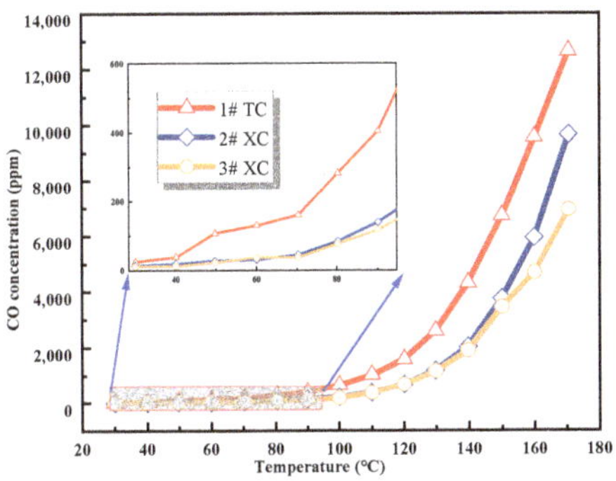

(**a**) CO concentration

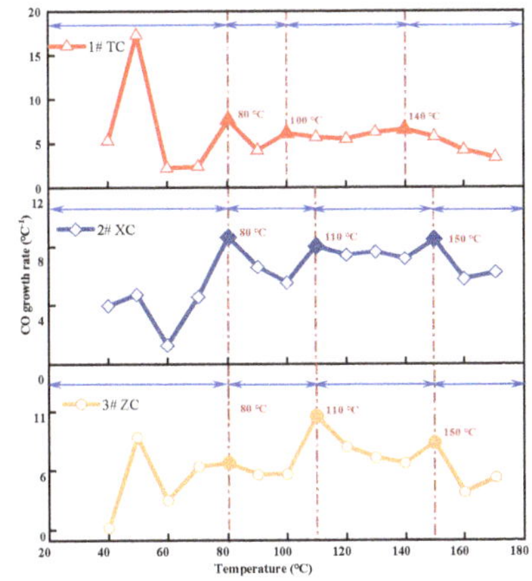

(**b**) CO growth rate

Figure 2. CO concentration and growth rate.

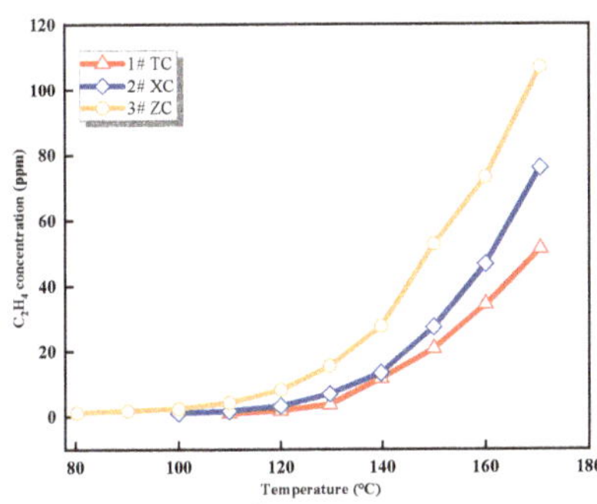

(**a**) C_2H_4 concentration

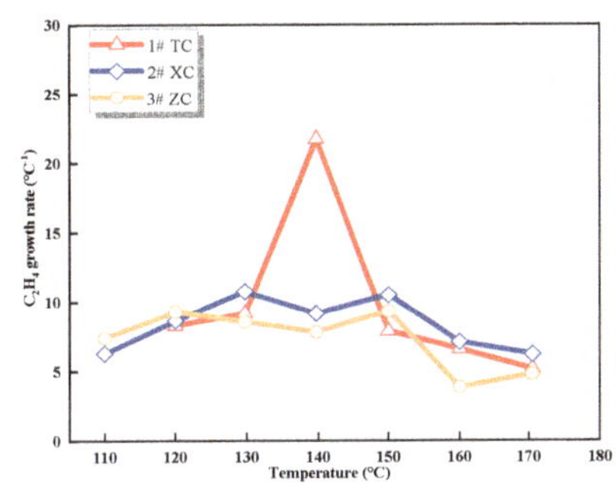

(**b**) C_2H_4 growth rate

Figure 3. C_2H_4 concentration and growth rate.

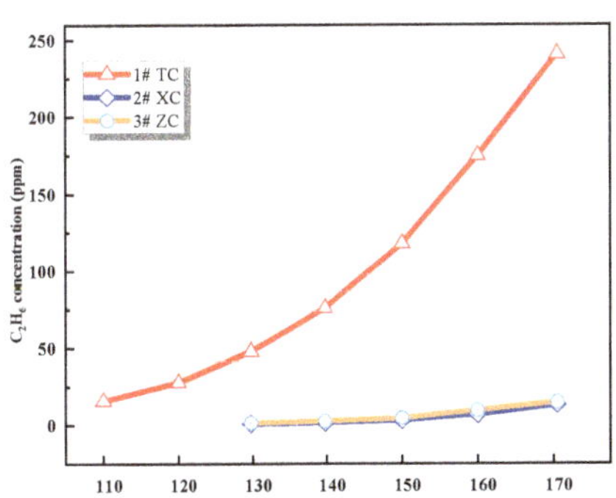

(**a**) C_2H_6 concentration

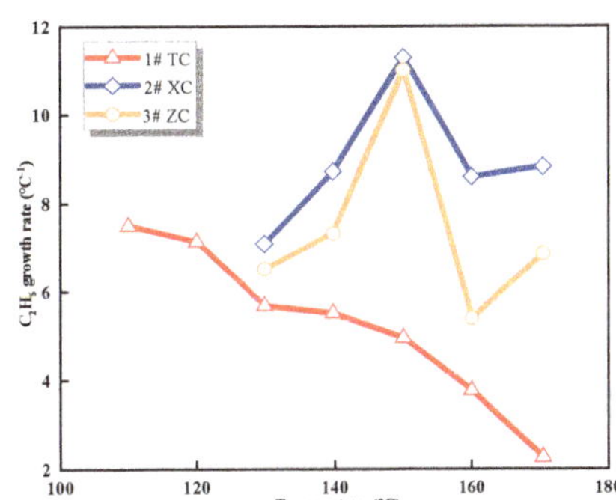

(**b**) C_2H_6 growth rate

Figure 4. C_2H_6 concentration and growth rate.

It is evident from these data that C_2H_4 is not initially formed during the experiment and only appears in small quantities around 80–110 °C. The concentration of C_2H_4 is highest in Zhao Lou coal, followed by New Julong coal, and lowest in Tang Kou coal. Therefore, C_2H_4 does not exist initially in the coal samples, and the generation of C_2H_4 gas at high temperatures can be attributed to the oxidative pyrolysis of coal. Similarly, a small amount of C_2H_6 appears around 110–130 °C, with the highest concentration observed in Tang Kou coal. Comparing the relationship curves between the growth rates of CO, C_2H_4, and C_2H_6 with temperature, when Tang Kou coal reaches 140 °C and New Julong coal and Zhao Lou coal reach 150 °C, the CO growth rate curve exhibits a third minor peak, and the growth rates of C_2H_4 and C_2H_6 also show the first peak. The characteristic temperatures of the three curves correspond well to each other.

3.2. Analysis of Changes in Composite Indicators

Because of the multitude of gas products generated during the coal oxidation process depending on a single gas as a predictive indicator for CSC, it is susceptible to the impact of airflow and environmental factors. Consequently, this approach leads to limited accuracy [30]. This inconvenience hampers the prediction of CSC. Therefore, this study selects composite indicators such as φ(CO)/φ(CO_2), hydrocarbon ratio, Graham coefficient (R_1, R_2, R_3), G_1, G_2, G_3, etc., to optimize the characteristic gases generated during the oxidation of the No. 3 coal seam in Juye coalfield. Gas indicators for different oxidation stages are determined, leading to the establishment of an early prediction index system suitable for the early-stage prediction of spontaneous combustion in the No. 3 coal seam in Juye coalfield.

3.2.1. φ(CO)/φ(CO_2) and Alkane Ratio

Due to the ability of φ(CO)/φ(CO_2) and alkane ratio to reduce the influence of underground airflow on gas concentration [31], in order to further validate the reliability of characteristic temperature points obtained through growth rate analysis, three coal samples with mixed particle sizes were selected for analysis. The growth rate of φ(CO)/φ(CO_2) was calculated using Equation (3) and is shown in Figure 5, while the growth rate of alkane ratio is shown in Figure 6.

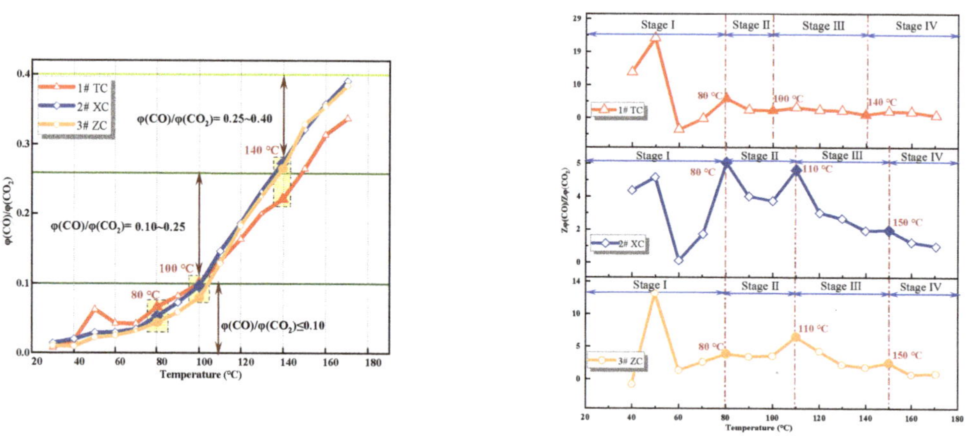

(**a**) φ(CO)/φ(CO_2) (**b**) Zφ(CO)/Zφ(CO2)

Figure 5. φ(CO)/φ(CO_2) and growth rate.

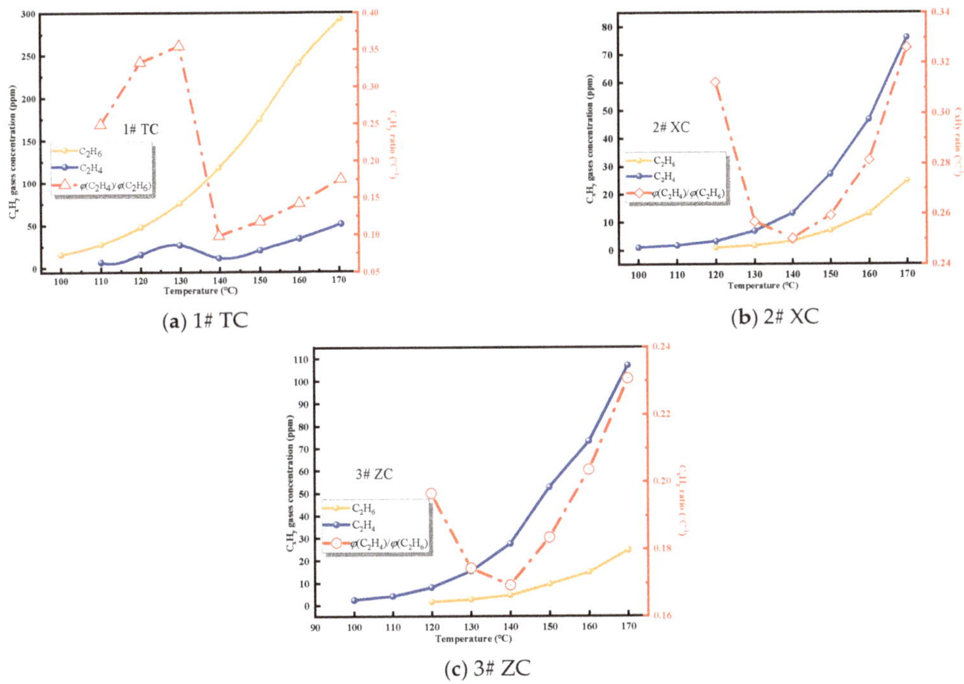

Figure 6. C_xH_y concentration and growth rate.

From Figure 5, it can be observed that the ratio of $\varphi(CO)/\varphi(CO_2)$ increases with the rise in temperature. As the temperature increases from ambient temperature to 60 °C, the ratio gradually increases but remains below 0.1, indicating a weak coal–oxygen composite reaction during this stage. After reaching 80 °C, the curve shows an upward trend, and the ratio reaches 0.1 at 100 °C. Beyond 100 °C, the ratio increases rapidly and fluctuates between 0.1 and 0.25, indicating intensifying oxidation. However, the growth rate decreases slightly due to the different rates of increase in CO and CO_2 volume fractions. Upon reaching 140–150 °C, the ratio increases rapidly again, rising from 0.25 to 0.4. Hence, during this stage, the formation of coal–oxygen complexes increases, with a significantly higher relative yield of CO compared to CO_2. This implies the occurrence of CSC [15]. The three peaks in the growth rate curve of $\varphi(CO)/\varphi(CO_2)$ correspond to points where the oxidation intensity intensifies, namely the critical temperature, desiccation temperature, and cleavage temperature points [32]. These points align well with the characteristic temperature points obtained in Table 2.

Table 2. Characteristic temperature and oxidation stage division for different coal samples.

Coal Samples	Oxidation Stage (°C)	T_1 (°C)	Accelerated Oxidation Stage (°C)	T_2 (°C)	Intense Oxidation Stage (°C)	T_3 (°C)	Oxidation Decomposition Stage (°C)
1# TC	40–80	80	80–100	100	100–140	140	140–170
2# XC	40–80	80	80–110	110	110–150	150	150–170
3# ZC	40–80	80	80–110	110	110–150	150	150–170

From Figure 6, it can be observed that the initial appearance temperatures of ethane and ethylene are 100 °C and 110 °C, respectively. This indicates that at this stage, chemical adsorption and reactions of coal start to dominate. With increasing temperature, the production of ethylene and ethane increases. The ratio of alkene to alkane fluctuates, indicating an alternating change in the production of ethylene and ethane. This further signifies the intensification of coal oxidation reactions and a more apparent possibility of CSC [29,32].

After a comprehensive comparison of the growth rates and the relationship between the indicator gases and the temperature, it was determined that the critical temperature (T_1) for the three coal samples was near 80 °C, the desiccation temperature (T_2) range was near 100–110 °C, and the cleavage temperature (T_3) range was 140–150 °C. Based on the characteristic temperatures and the three-step coal–oxygen composite reaction, Wang et al. [33,34] classified the low-temperature oxidation process of coal into the stages of slow oxidation, accelerated oxidation, and intense oxidation. According to the aforementioned criteria, the low-temperature oxidation process of Tang Kou coal can be divided into the initial oxidation stage (30–80 °C), accelerated oxidation stage (80–110 °C), intense oxidation stage (110–140 °C), and oxidation decomposition stage (140–170 °C). For the classification of the other two coal samples, please refer to Table 2.

3.2.2. Analysis of Fire Hazard Index Variation

The Graham fire hazard index, proposed by British scholar Graham, consists of three components: the first fire hazard index (R_1), the second fire hazard index (R_2), and the third fire hazard index (R_3). These indices are calculated based on the changes in CO_2 concentration ($+\Delta CO_2$), CO concentration ($+\Delta CO$), and O_2 concentration ($--\Delta O_2$) during the coal oxidation process. The fire hazard index fire coefficient cannot be affected by certain objective factors, and it is easy to select the best indicators, excluding the influence of external factors on the determination of the risk of spontaneous combustion of coal due to the dilution of air leakage from the working face and the mining area. The specific calculation formulas are as follows:

$$R_1 = +\Delta CO_2/(-\Delta O_2) \times 100\% \tag{4}$$

$$R_2 = +\Delta CO/(-\Delta O_2) \times 100\% \tag{5}$$

$$R_3 = +\Delta CO/(+\Delta CO_2) \times 100\% \tag{6}$$

Based on the characteristic temperatures, coal oxidation is divided into four stages: the initial oxidation stage (Stage I), the accelerated oxidation stage (Stage II), the intense oxidation stage (Stage III), and the oxidation decomposition stage (Stage IV). By applying the aforementioned formulas, the experimental data were processed and analyzed to calculate the three fire hazard indices for each coal sample. The relationships between the indices and temperature were then plotted, as shown in Figure 7a–c.

During the initial oxidation stage, the fire hazard index exhibits significant fluctuations (as shown in Figure 7). These fluctuations could be attributed to the desorption of native gases from the coal and the release of CO_2 that was adsorbed in the fractures and pores of the fragmented coal. This observation aligns with the results in Figure 1a, which depict a phenomenon of decreasing and then increasing CO_2 concentration as the coal temperature rises before reaching the critical temperature. This further validates the credibility of this hypothesis. As a result, the fire hazard index is not a suitable parameter during the initial oxidation stage. Moving beyond this stage, the R_1 value of Tang Kou coal shows a clear upward trend, marked by two peak values and notable fluctuations. In contrast, the R_1 values of the other two coal samples exhibit fluctuations within the range of 9.97% to 26.08% with relatively minor variations. Following this stage, the R_3 values of all three coal samples stabilize and fluctuate within the range of 14.19% to 58.52%. However, their correlation with temperature is not evident. Therefore, both R_1 and R_3 are not suitable indicators for predicting CSC.

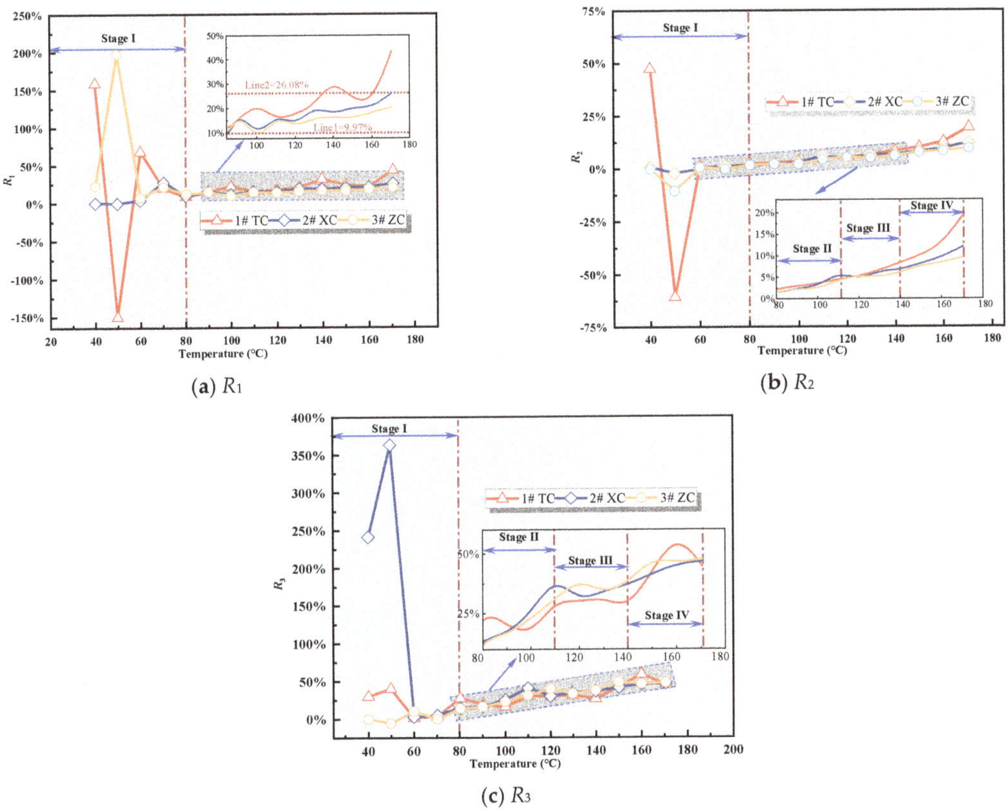

Figure 7. Fire hazard index.

However, after the initial oxidation stage, the R_2 value of Tang Kou coal shows an increasing trend with temperature. It remains stable at around 2% between 60 and 80 °C. Starting from 110 °C, the R_2 value increases rapidly and exceeds 5% at around 140 °C. Subsequently, it exhibits exponential growth with temperature, indicating a significant increase. Therefore, based on the corresponding temperature from the experimental data, when the R_2 value exceeds 2%, the temperature of Tang Kou coal has reached or exceeded the spontaneous combustion critical temperature, entering the stage of CSC. When the R_2 value exceeds 5%, the coal–oxygen composite reaction becomes intense, entering the stage of intense oxidation. When the R_2 value surpasses 8%, the coal sample undergoes vigorous oxidation, entering the stage of oxidation decomposition. Under the experimental conditions, the corresponding coal temperature exceeds 150 °C, indicating the need for timely measures. Similarly, the R_2 values of New Julong coal and Zhao Lou coal also remain stable between 60 and 80 °C and exhibit monotonic growth from 80 °C to 170 °C. Therefore, R_2 can serve as an indicator for CSC during the latter three stages for the three coal samples [35].

3.2.3. Analysis of Changes in Composite Gas Indicators

This study utilizes composite processing of different individual gas indicators to characterize the oxidation intensity of coal and its relationship with temperature, further enhancing the indicators for CSC early warning. The specific calculation formula for the composite indicators is as follows:

$$G_1 = \varphi(CO) + \Delta CO_2, \tag{7}$$

$$G_2 = \frac{G_1}{-\Delta O_2}, \tag{8}$$

$$G_3 = \frac{G_1}{\varphi(O_2)}. \tag{9}$$

Whether it can serve as an indicator for determining the degree of CSC, the calculation formula for the growth rate of composite gas indicators is introduced to analyze their variations:

$$k = \frac{G^{i+1} - G^{i}}{t_{i+1} - t_i}. \tag{10}$$

In the equation, t_i and t_{i+1} represent the temperature at adjacent experimental time points in degrees Celsius; G^i and G^{i+1} represent the composite indicator parameters corresponding to t_i and t_{i+1}. k denotes the growth rate of composite gas indicators. The growth rate is analyzed based on the gas concentration at adjacent temperature points. By determining the trend and monotonicity of the growth rate of composite indicator parameters, the indicators that correspond to the degree of CSC development are identified. Figures 8 and 9 depict the relationship curves and growth rates with temperature.

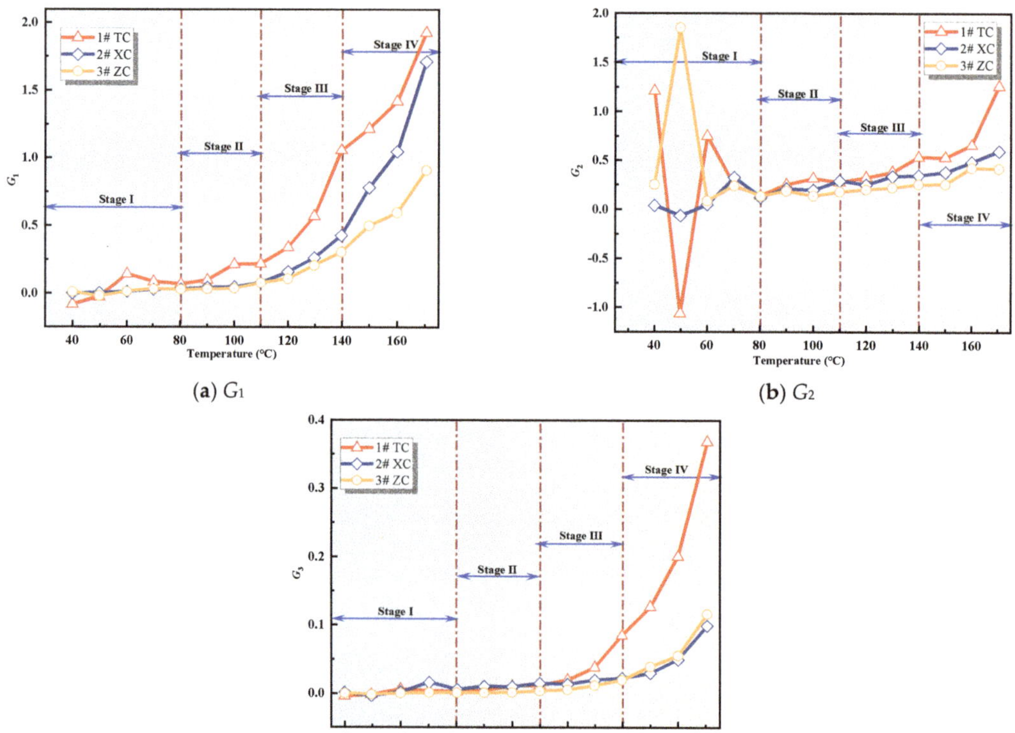

(**a**) G_1 (**b**) G_2 (**c**) G_3

Figure 8. Compound indicator gas concentration.

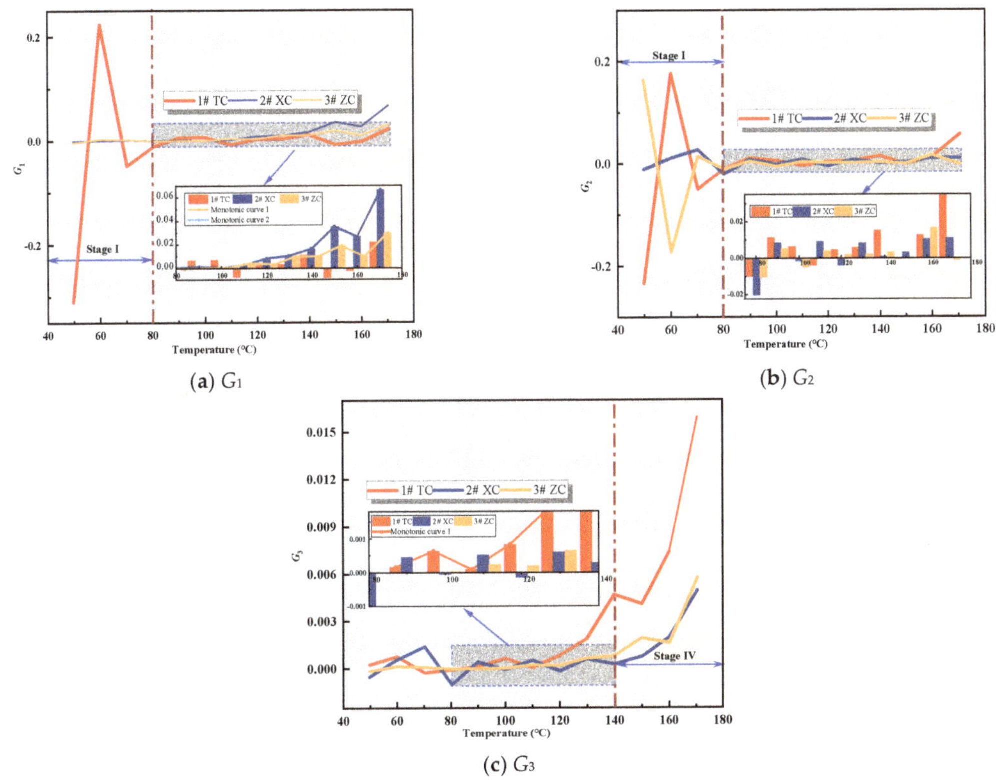

Figure 9. Compound Indicator gas growth rate.

From Figure 8, it can be observed that the values of G_1 and G_3 for all three coal samples show an increasing trend with the rise in temperature. Once the coal temperature reaches the devolatilization temperature, its values exhibit significant growth, indicating an intensification of coal oxidation reactions. However, the value of G_2 shows significant fluctuations during the initial oxidation stage, and after this stage, G_2 for all three coal samples exhibits fluctuating increases. Additionally, Figure 9 illustrates the variation trend of the growth rates of the composite indicators for the three coal samples. In the initial oxidation stage (Stage I), the growth rates of G_2 for all three coal samples exhibit large fluctuations, and in the subsequent three stages after the critical temperature, they are non-monotonic. Therefore, G_2 is not suitable as an indicator for CSC.

The growth rate of G_1 for the three coal samples exhibits significant fluctuations and is non-monotonic during the initial oxidation stage. However, after the critical temperature, the growth rates of G_1 for the New Julong coal and Zhao Lou coal are consistently positive and show a monotonically increasing trend with temperature. Therefore, G_1 can serve as an auxiliary indicator coefficient during the acceleration oxidation stage to the oxidative decomposition stage, in addition to the fire index coefficient R_2. The growth rate of G_3 for Tang Kou coal initially decreases slightly and then exhibits a monotonically increasing trend after the critical temperature, remaining consistently positive within the temperature range of 80 °C to 170 °C. Therefore, in this temperature range, in addition to the fire index coefficient R_2, G_3 can also be used as an auxiliary indicator to assess the degree of CSC. Moreover, for the New Julong coal and Zhao Lou coal samples, the growth rates of G_3 during the oxidative decomposition stage are all positive and show a clear correlation with temperature, with higher temperatures corresponding to larger G_3 values. Therefore, G_3

can also be considered as a CSC indicator coefficient during the oxidative decomposition stage. In summary, different indicators can be selected for CSC prediction during different oxidation stages for the three coal samples. Refer to Table 3 for specific details.

Table 3. CSC indicator gas preference.

Coal Samples	Characteristic Temperature (°C)	Coal Seam Temperature Range (°C)	Preferred Metrics
1# TC	40	(40–80]	R_{co}
2# XC	40	(40–80]	R_{co}
3# ZC	40	(40–80]	R_{co}
1# TC	80	(80–100]	R_2, G_1, G_3
2# XC	80	(80–110]	R_2, G_1
3# ZC	80	(80–110]	R_2, G_1
1# TC	100	(100–140]	R_2, G_1, G_3
2# XC	110	(110–150]	R_2, G_1
3# ZC	110	(110–150]	R_2, G_2
1# TC	140	(140–170]	R_2, G_1, G_3
2# XC	150	(150–170]	R_2, G_1, G_3
3# ZC	150	(150–170]	R_2, G_1, G_3

3.3. Classification and Warning Indicator System for CSC Hazard

The classification and warning system for the CSC hazard is formulated by analyzing the fluctuation trends and characteristic temperatures of CSC indicators. This system meticulously delineates the stages of coal spontaneous ignition, identifies the warning indicators along with their corresponding thresholds for each stage, and establishes a comprehensive classification and warning framework for spontaneous combustion within the No. 3 coal seam of Juye coalfield. This framework offers theoretical guidance for accurately identifying spontaneous combustion hazards within the No. 3 coal seam of Juye coalfield.

The carbon oxide ratio values ($\varphi(CO)/\varphi(CO_2)$) at the characteristic temperature moments were used as the warning thresholds, denoted as R_I, R_{II}, R_{III}, R_{IV} and the warning levels were classified as blue, yellow, orange, and red [36]. Based on the data from the programmed temperature rise test, the characteristic temperatures of each coal sample were identified using the growth rate analysis method mentioned earlier. By combining them with the carbon oxide ratio value, the classification criteria for warning levels were reconstructed. Please refer to Table 4 for details.

Table 4. Juye No. 3 coalfield spontaneous combustion graded warning indicators and grade classification criteria.

Early Warning Level	Coal Samples	Early Warning Temperature Range (°C)	R_{CO} (($\varphi(CO)/\varphi(CO_2)$))
Blue alert	1# TC	(40–80]	$R_I \leq 0.066$
	2# XC	(40–80]	$R_I \leq 0.053$
	3# ZC	(40–80]	$R_I \leq 0.044$
Yellow alert	1# TC	(80–100]	$0.066 < R_{II} \leq 0.101$
	2# XC	(80–110]	$0.053 < R_{II} \leq 0.147$
	3# ZC	(80–110]	$0.044 < R_{II} \leq 0.130$
Orange alert	1# TC	(110–140]	$0.101 < R_{III} \leq 0.221$
	2# XC	(110–150]	$0.147 < R_{III} \leq 0.322$
	3# ZC	(110–150]	$0.130 < R_{III} \leq 0.329$
Red alert	1# TC	(140–170]	$R_{IV} > 0.221$
	2# XC	(150–170]	$R_{IV} > 0.322$
	3# ZC	(150–170]	$R_{IV} > 0.329$

Using single–indicator gases and the selected composite–indicator gases, the threshold curve for classifying and issuing early warnings regarding spontaneous combustion hazards in the No. 3 coal seam of Juye coalfield was constructed, as illustrated in Figure 10.

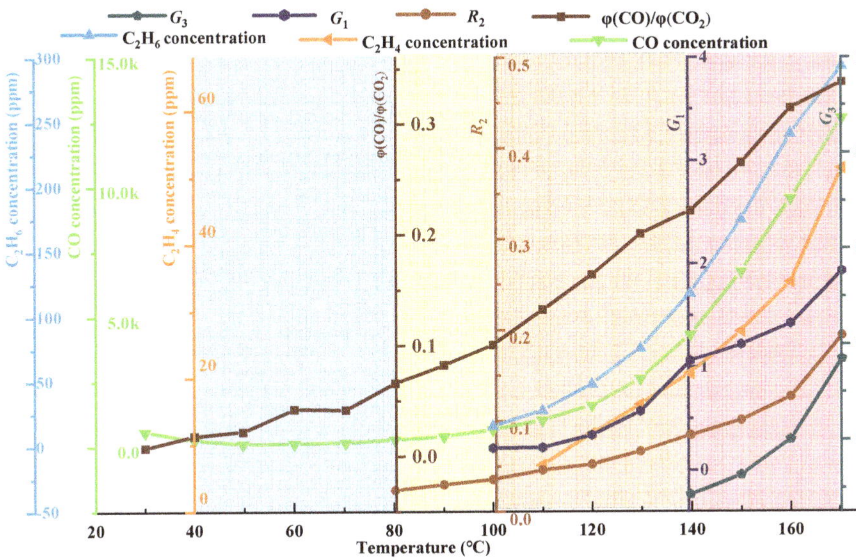

(a) 1# TC

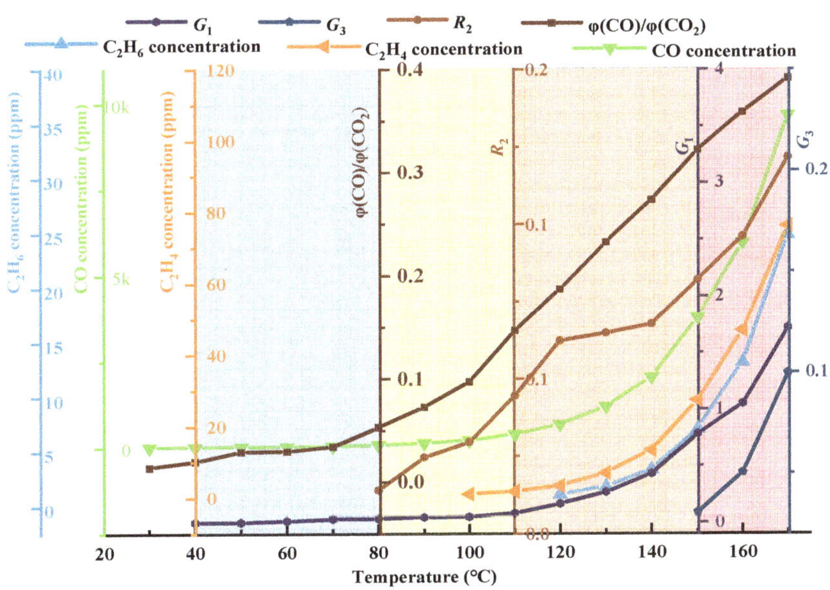

(b) 2# XC

Figure 10. *Cont.*

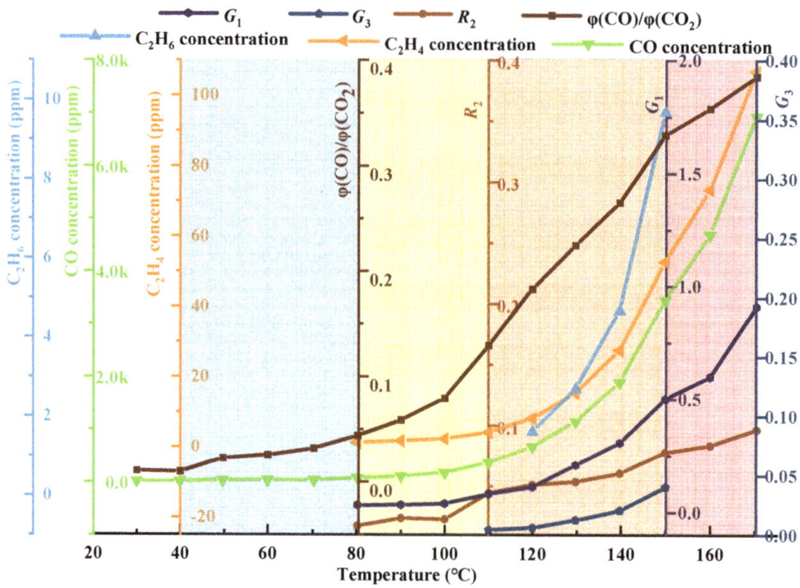

(c) 3# ZC

Figure 10. Coal seam spontaneous combustion risk classification warning threshold curve.

From Figure 10a–c, it can be observed that the low−temperature oxidation process of coal is divided into four warning levels, namely blue, yellow, orange, and red, based on characteristic temperatures and the carbon oxide ratio. These levels correspond to the initial oxidation stage, accelerated oxidation stage, intense oxidation stage, and oxidation decomposition stage of coal, respectively. The selected composite indicator gases G_1 and G_3, as well as the second fire coefficient R_2, are marked in each warning level region. Combined with the temperature and concentration curves of CO, C_2H_4, and C_2H_6, they collectively form the threshold curve for the classification and early warning of spontaneous combustion hazard in coal. Overall, the gas generation of coal shows a nonlinear increase, indicating that the oxidation of coal is a heterogeneous reaction process, and the reaction models differ in each stage. In the blue warning level, which represents the gas distribution characteristics of coal before reaching the critical temperature, the CO concentration changes are not significant, indicating a relatively weak tendency for spontaneous combustion in the three coal samples. In the yellow warning level, with coal temperatures between the critical temperature and the drying−cracking temperature, the CO concentration starts to increase, and C_2H_4 and C_2H_6 begin to appear, indicating an increased tendency for spontaneous combustion. In the orange warning level, with coal temperatures between the drying-cracking temperature and the decomposition temperature, there is a significant increase in CO and C_2H_6 concentrations, indicating an irreversible tendency for spontaneous combustion. The red warning level represents the region where the coal temperature reaches the decomposition temperature. The generation of CO, C_2H_6, and other gases increases exponentially. When the No. 3 coal seam of Juye coalfield is in this warning level, the occurrence and development of spontaneous combustion in coal intensify.

Based on the refined theory of CSC stages and regulations regarding CSC monitoring, indicative gases, critical values, ignition precursors, and fire management [37], combined with the optimized CSC indicator gases in Table 3, the classification criteria for CSC warning levels in Table 4, and the reconstructed threshold curve for the classification and early warning of spontaneous combustion hazard in the No. 3 coal seam of Juye coalfield in

Figure 10, a hierarchical warning system and indicator thresholds for CSC hazard were established. Please refer to Table 5 for details.

Table 5. Spontaneous combustion risk classification warning system and index thresholds for No. 3 coal seam in Juye coalfield.

Coal Samples	Characteristic Temperature (°C)	Coal Seam Temperature Range (°C)	Preferred Metrics	Judgment Threshold
	40	(40–80]	R_{CO}	$I_1 = \{R_I \leq 0.066\}$
	80	(80–100]	R_2, G_1, G_3	$I_2 = I_1 \cap \{0.024 < R_2 < 0.036\}$
	100	(100–140]	R_2, G_1, G_3, C_2H_4	$I_3 = I_2 \cap \{0.036 < R_2 < 0.084\} \cap \{0.217 < G_1 < 1.058\} \cap \{0.011 < G_3 < 0.081\} \cap \{\varphi(C_2H_4) > 0\}$
	140	(140–170]	R_2, G_1, G_3, C_2H_4	$I_4 = I_3 \cap \{0.084 < R_2 < 0.194\} \cap \{0.085 < G_3 < 0.369\} \cap \{11.675 < \varphi(C_2H_4) < 51.34\}$
		>170	$\varphi(C_2H_4)/\varphi(C_2H_6)$ rapid growth	$I_5 = I_4 \cap \{\varphi(C_2H_2) > 0\}$
	40	(40–80]	R_{CO}	$I_1 = \{R_I \leq 0.053\}$
	80	(80–110]	R_2, G_1	$I_2 = I_1 \cap \{0.014 < R_2 < 0.062\}$
	110	(110–150]	R_2, G_1, C_2H_6	$I_3 = I_2 \cap \{0.0062 < R_2 < 0.082\} \cap \{0.080 < G_1 < 0.783\} \cap \{\varphi(C_2H_6) > 0\}$
	150	(150–170]	R_2, G_1, G_3, C_2H_4	$I_4 = I_3 \cap \{0.082 < R_2 < 0.122\} \cap \{0.783 < G_1 < 1.716\} \cap \{0.036 < G_3 < 0.100\} \cap \{\varphi(C_2H_6) > 0\} \cap \{1.777 < \varphi(C_2H_4) < 24.84\}$
		>170	$\varphi(C_2H_4)/\varphi(C_2H_6)$ rapid growth	$I_5 = I_4 \cap \{\varphi(C_2H_2) > 0\}$
	40	(40–80]	R_{CO}	$I_1 = \{R_I \leq 0.044\}$
	80	(80–110]	R_2, G_1	$I_2 = I_1 \cap \{0.017 < R_2 < 0.045\} \cap \{0.031 < G_1 < 0.081\}$
	110	(110–150]	R_2, G_1, G_3, C_2H_6	$I_3 = I_2 \cap \{0.045 < R_2 < 0.078\} \cap \{0.081 < G_2 < 0.502\} \cap \{\varphi(C_2H_4) > 0\}$
	150	(150–170]	R_2, G_1, C_2H_4	$I_4 = I_3 \cap \{0.078 < R_2 < 0.097\} \cap \{0.502 < G_1 < 0.912\} \cap \{\varphi(C_2H_4) > 0\} \cap \{0.184 < \varphi(C_2H_4)/\varphi(C_2H_6) < 0.231\}$
		>170	$\varphi(C_2H_4)/\varphi(C_2H_6)$ rapid growth	$I_5 = I_4 \cap \{\varphi(C_2H_2) > 0\}$

4. Conclusions

(1) The growth rate of CO and C_2H_4 gas concentration was calculated according to the growth rate analysis method and verified by analyzing the growth rate of $\varphi(CO)/\varphi(CO_2)$; it was determined that the critical temperature of the No. 3 coal seam in the Juye coalfield was near 80 °C, the dry cracking temperature was in the range of 100–110 °C, and the fissure temperature was about 130–150 °C.

(2) The variation of single and composite indicator gases during the coal oxidation process was analyzed, and predictive indicators were selected for different oxidation stages. R_{CO}, R_2, G_1, and G_3 were identified as predictive indicators for Tang Kou coal, New Julong coal, and Zhao Lou coal in the initial oxidation and accelerated oxidation stages. R_2, G_1, and G_3 were selected as predictive indicators in the intense oxidation and oxidative decomposition stages.

(3) Based on the characteristic temperatures of coal and in combination with the carbon oxide ratio, a four-level warning system consisting of blue, yellow, orange, and red levels was established for the No. 3 coal seam in the Juye coalfield. The composite indicator gases G_1 and G_2, as well as the second fire hazard coefficient R_2, were annotated in their respective warning level regions. Along with the concentration curves of CO and C_2H_6, these indicators were used to construct the threshold curves

for the classification and warning of the self-ignition hazard in the No. 3 coal seam of the Juye coalfield.

(4) Based on the refined theory of self-ignition stages in coal, as well as regulations regarding coal self-ignition monitoring, indicative gases, critical values, fire initiation precursors, and fire management, the self-ignition hazard classification and warning system, along with the threshold values, were reconstructed for the No. 3 coal seam in the Juye coalfield. The system was based on the classification criteria and threshold curves for self-ignition risk levels. It established a six-level warning system, including the initial warning level, blue, yellow, orange, red, and black levels, and their respective indicator thresholds, to predict the occurrence of CSC.

Author Contributions: C.W.: Conceptualization, Methodology, Writing—Original draft preparation; Y.D. (Yuxin Du): Supervision, Writing—Reviewing and Editing, Validation; Y.D. (Yin Deng): Supervision, Data curation; Y.Z.: Supervision, Data curation; J.D.: Supervision. X.Z.: Investigation; X.D.: Investigation. All authors have read and agreed to the published version of the manuscript.

Funding: This work was supported by the National Natural Science Foundation of China (No. 5197-4234, No. 52174201), Shaanxi Science Fund for Distinguished Young Scholars (2022JC-30), and Xinjiang Autonomous Region Key Research and Development Project (2022B03031-1).

Institutional Review Board Statement: Not applicable.

Informed Consent Statement: Not applicable.

Data Availability Statement: Not applicable.

Conflicts of Interest: The authors declare that they have no known competing financial interest of personal relationships that could have appeared to influence the work reported in this paper.

References

1. Ni, G.H.; Xie, H.C.; Li, Z.; Zhuansun, L.X.; Niu, Y.Y. Improving the permeabilityof coal seam with pulsating hydraulic fracturing technique: A case study in changping coal mine. *Process Saf. Environ.* **2018**, *117*, 565–572.
2. Chen, X.J.; Li, L.Y.; Wang, L.; Qi, L.L. The current situation and prevention and control countermeasures for typical dynamic disasters in kilometer-deep mines in China. *Saf. Sci.* **2019**, *114*, 229–236. [CrossRef]
3. Li, X.B.; Gong, F.Q.; Tao, M.; Dong, L.J.; Du, K.; Ma, C.D.; Zhou, Z.L.; Yin, T.B. Failure mechanism and coupled static-dynamic loading theory in deep hard rock mining: A review. *J. Rock Mech. Geotech. Eng.* **2017**, *9*, 767–782. [CrossRef]
4. Belle, B.; Biffi, M. Cooling pathways for deep Australian longwall coal mines of the future. *Int. J. Min. Sci. Technol.* **2018**, *28*, 865–875. [CrossRef]
5. Bukowski, P. Water hazard assessment in active shafts in upper silesian coal basin mines. *Mine Water Environ.* **2011**, *30*, 302–311. [CrossRef]
6. Ranjith, P.G.; Zhao, J.; Ju, M.H.; De Silva, R.V.; Rathnaweera, T.D.; Bandara, A.K. Opportunities and challenges in deep mining: A brief review. *Engineering* **2017**, *3*, 546–551. [CrossRef]
7. Pan, R.K.; Ma, Z.H.; Yu, M.G.; Chao, J.K.; Li, C.; Wang, J. Study on the mechanism of coal oxidation under stress disturbance. *Fuel* **2020**, *275*, 117901. [CrossRef]
8. Wang, K.; Li, K.N.; Du, F. Study on Prediction of Coal-Gas Compound Dynamic Disaster Based on GRA-PCA-BP Model. *Geofluids* **2021**, *2021*, 3508806. [CrossRef]
9. Chao, J.K.; Chu, T.X.; Yu, M.G.; Han, X.F.; Hu, D.M.; Liu, W.; Yang, X.L. An experimental study on the oxidation kinetics characterization of broken coal under stress loading. *Fuel* **2021**, *287*, 119515. [CrossRef]
10. Niu, H.Y.; Sun, Q.Q.; Bu, Y.C.; Chen, H.Y.; Yang, Y.X.; Li, S.P.; Sun, S.W.; Mao, Z.H.; Tao, M. Study of the microstructure and oxidation characteristics of residual coal in deep mines. *J. Clean. Prod.* **2022**, *373*, 133923. [CrossRef]
11. Jia, H.; Yang, Y.; Ren, W.; Kang, Z.; Shi, J. Experimental study on the characteristics of the spontaneous combustion of coal at high ground temperatures. *Combust. Sci. Technol.* **2022**, *194*, 2880–2893. [CrossRef]
12. Odintsev, V.N.; Miletenko, N.A. Water inrush in mines as a consequence of spontaneous hydrofracture. *J. Min. Sci.* **2015**, *51*, 423–434. [CrossRef]
13. Wang, K.; Hu, L.H.; Deng, J.; Zhang, Y.N. Multiscale thermal behavioral characterization of spontaneous combustion of pre-oxidized coal with different air exposure time. *Energy* **2023**, *262*, 125397. [CrossRef]
14. Adamus, A.; Šancer, J.; Guřanová, P.; Zubíček, V. An investigation of the factors associated with interpretation of mine atmosphere for spontaneous combustion in coal mines. *Fuel Process. Technol.* **2011**, *92*, 663–670. [CrossRef]
15. Wojtacha-Rychter, K.; Smoliński, A. A study of dynamic adsorption of propylene and ethylene emitted from the process of coal self-heating. *Sci. Rep.* **2019**, *9*, 18277. [CrossRef]

16. Baris, K.; Kizgut, S.; Didari, V. Low-temperature oxidation of some Turkish coals. *Fuel* **2012**, *93*, 423–432. [CrossRef]
17. Onifade, M.; Genc, B. Spontaneous combustion of coals and coal-shales. *Int. J. Min. Sci. Technol.* **2018**, *28*, 933–940. [CrossRef]
18. Singh, A.K.; Singh, R.V.K.; Singh, M.P.; Chandra, H.; Shukla, N.K. Mine fire gas indices and their application to Indian underground coal mine fires. *Int. J. Coal Geol.* **2007**, *69*, 192–204. [CrossRef]
19. Baris, K.; Aydin, H.; Didari, V. Statistical modeling of the effect of rank, temperature, and particle size on low-temperature oxidation of Turkish coals. *Combust. Sci. Technol.* **2010**, *183*, 105–121. [CrossRef]
20. Kong, B.; Niu, S.Y.; Cao, H.M.; Lu, W.; Wen, J.M.; Yin, J.L.; Zhang, W.R.; Zhang, X.L. Study on the application of coal spontaneous combustion positive pressure beam tube classification monitoring and early warning. *Environ. Sci. Pollut. Res.* **2023**, *30*, 75735–75751. [CrossRef]
21. Wang, C.P.; Deng, Y.; Xiao, Y.; Deng, J.; Shu, C.M.; Jiang, Z.G. Gas-heat characteristics and oxidation kinetics of coal spontaneous combustion in heating and decaying processes. *Energy* **2022**, *250*, 123810. [CrossRef]
22. Gbadamosi, A.R.; Onifade, M.; Genc, B.; Rupprecht, S. Analysis of spontaneous combustion liability indices and coal recording standards/basis. *Int. J. Min. Sci. Technol.* **2020**, *30*, 723–736. [CrossRef]
23. Wang, K.; Li, Y.; Zhang, Y.N.; Deng, J. An approach for evaluation of grading forecasting index of coal spontaneous combustion by temperature-programmed analysis. *Environ. Sci. Pollut. Res.* **2023**, *30*, 3970–3979. [CrossRef] [PubMed]
24. Mukherjee, M.; Misra, S. A review of experimental research on Enhanced Coal Bed Methane (ECBM) recovery via CO_2 sequestration. *Earth-Sci. Rev.* **2018**, *179*, 392–410. [CrossRef]
25. Lu, H.; Li, J.L.; Lu, W.; Xu, Z.; Li, J.H.; He, Q.L. Variation laws of CO_2/CO and influence of key active groups on it during low-temperature oxidation of coal. *Fuel* **2023**, *339*, 127415. [CrossRef]
26. Wieckowski, M.; Howaniec, N.; Smolinski, A. Natural desorption of carbon monoxide during the crushing of coal simulating natural rock mass pressure. *Sci. Total Environ.* **2020**, *736*, 139639. [CrossRef]
27. Zhou, B.Z.; Yang, S.Q.; Jiang, X.Y.; Cai, J.W.; Xu, Q.; Song, W.X.; Zhou, Q.C. The reaction of free radicals and functional groups during coal oxidation at low temperature under different oxygen concentrations. *Process Saf. Environ. Prot.* **2021**, *150*, 148–156. [CrossRef]
28. Liu, Q.Q.; Sun, L.L.; Liu, Z.Y.; Wang, G.; Ma, J.Y. Effects of air volume and pre-oxidation on re-ignition characteristics of bituminous coal. *Energy* **2023**, *265*, 126124. [CrossRef]
29. Wojtacha-Rychter, K.; Smoliński, A. Coal oxidation with air stream of varying oxygen content and flow rate-Fire gas emission profile. *Fire Saf. J.* **2020**, *116*, 103182. [CrossRef]
30. Liang, Y.T.; Zhang, J.; Wang, L.C.; Luo, H.Z.; Ren, T. Forecasting spontaneous combustion of coal in underground coal mines by index gases: A review. *J. Loss Prev. Process Ind.* **2019**, *57*, 208–222. [CrossRef]
31. Guo, Q.; Ren, W.X.; Lu, W. A Method for Predicting Coal Temperature Using CO with GA-SVR Model for Early Warning of the Spontaneous Combustion of Coal. *Combust. Sci. Technol.* **2020**, *194*, 523–538. [CrossRef]
32. Wang, C.P.; Zhao, X.Y.; Bai, Z.J.; Deng, J.; Shu, C.M.; Zhang, M. Comprehensive index evaluation of the spontaneous combustion capability of different ranks of coal. *Fuel* **2021**, *291*, 120087. [CrossRef]
33. Rua, M.O.B.; Aragon, A.J.D.; Baena, P.B. A study of fire propagation in coal seam with numerical simulation of heat transfer and chemical reaction rate in mining field. *Int. J. Min. Sci. Technol.* **2019**, *29*, 873–879.
34. Zhang, D.; Cen, X.; Wang, W.F.; Deng, J.; Wen, H.; Xiao, Y.; Shu, C.M. The graded warning method of coal spontaneous combustion in Tangjiahui Mine. *Fuel* **2021**, *288*, 119635. [CrossRef]
35. Zhao, J.Y.; Zhang, Y.L.; Song, J.J.; Guo, T.; Deng, J.; Shu, C.M. Oxygen distribution and gaseous products change of coal fire based upon the semi-enclosed experimental system. *Energy* **2023**, *263*, 125721. [CrossRef]
36. Zhao, J.Y.; Ming, H.Q.; Song, J.J.; Lu, S.P.; Xiao, Y.Y.; Zhang, Y.L.; Shu, C.M. Preoptimal analysis of phase characteristic indicators in the entire process of coal spontaneous combustion. *J. Loss Prev. Process Ind.* **2023**, *84*, 105131. [CrossRef]
37. Sun, Y.; Wang, S.G.; Wei, L.; Ying, J.Z.; Cao, Y.J.Z.; Li, J.L. Coal spontaneous combustion characteristics based on constant temperature difference guidance method. *Process Saf. Environ.* **2019**, *131*, 223–234.

fire

MDPI

Article

Evaluation and Optimization of Multi-Parameter Prediction Index for Coal Spontaneous Combustion Combined with Temperature Programmed Experiment

Xuefeng Xu [1,2,*] and Fengjie Zhang [3,*]

[1] School of Energy Science and Engineering, Henan Polytechnic University, Jiaozuo 454000, China
[2] Henan Key Laboratory for Green and Efficient Mining & Comprehensive Utilization of Mineral Resources, Henan Polytechnic University, Jiaozuo 454000, China
[3] School of Emergency Management and Safety Engineering, China University of Mining & Technology, Beijing 100083, China
* Correspondence: 10460120025@hpu.edu.cn (X.X.); yczhangfengjie@163.com (F.Z.)

Abstract: Coal spontaneous combustion (CSC) is a serious threat to the safe mining of coal resources, and the selection of suitable gas indicators to predict the CSC state is crucial for the prevention and control of coal mine fires. In this paper, the temperature-programmed experiment of CSC was first carried out to analyze the gas components and compositions in the oxidative pyrolysis process of three coal samples (lignite, long-flame coal, and lean coal) with different coalification degrees. Subsequently, the spontaneous combustion tendency of these three coal samples was evaluated. Finally, through the variation of gas concentration, gas concentration ratio, and fire coefficient with coal temperature, the indicators suitable for predicting the spontaneous combustion of coal were preferred, and a multi-parameter indicator system was established to make a comprehensive judgment on the spontaneous combustion status of coal. The results show that coal rank is negatively correlated with oxygen consumption rate. The higher the coalification degree of coal, the slower the oxidation reaction and the later the characteristic temperature point appears. The lignite selected in this experiment is a type of coal that is more prone to spontaneous combustion than long-flame coal and poor coal, and the CO concentration, C_2H_6/CH_4, and second fire coefficient R_2 can be used as the main indicators for predicting CSC, while the other gases, olefin-alkane ratio and fire coefficient can be used as auxiliary indicators. To some extent, the research content can effectively and accurately determine the stage and degree of coal spontaneous combustion, which has a certain guiding role in predicting CSC.

Keywords: coal spontaneous combustion; temperature programmed system; indicator gases; forecast indicators; fire coefficient

1. Introduction

During coal mining, coal spontaneous combustion (CSC) disasters is often accompanied, which not only causes waste of coal resources, but also seriously threatens the safety of workers and coal production [1–5]. Coal spontaneous combustion (CSC) is a multivariate self-accelerating exothermic process, which is mainly the result of the compounding of coal and oxygen molecules [6,7]. Different gases are produced at different stages of the coal auto-ignition process, and there are certain laws between the gas types and concentrations and the coal temperature. These laws can be utilized to predict the state of different stages of coal auto-ignition [8–10]. Therefore, an in-depth understanding of the relationship between gas products and temperature in the process of coal self-heating is of great significance for fire prevention as well as for ensuring the safe development of the coal industry.

Scientists have carried out a special study on the gas produced during theCSC process to estimate the degree of coal oxygen reaction and the stage of coal spontaneous combustion in goaf. At the low-temperature stage, the generation of CO and CO_2 is generally considered to be caused by the direct decomposition of stable oxygen-containing complexes containing intrinsic oxygen-containing functional groups and a small number of combustion reactions. At the high-temperature stage, coal has a violent oxidation reaction, and more olefin and alkane gases are generally produced by pyrolysis [11]. In addition, accurate measurement of the reduction of oxygen concentration can also be applied to determine the spontaneous combustion state of coal [12]. Some scholars believe that some single gases, such as CO, CO_2, C_2H_4, and C_2H_6, can be used as indicator gases for predicting spontaneous combustion [4,13–15]. Subsequently, the CO/CO_2 ratio, C_2H_4/C_2H_6 ratio, $\Delta CO/\Delta CO_2$ ratio, and C_2H_6/C_2H_2 ratio are also considered to be able to predict spontaneous combustion of coal [16–18]. Based on these indicator gases, other scholars have proposed the Graham fire coefficient [Wangcaiping2021], Willet's ratio [19], C/H ratio [20], and Graham's ratio [21] to predict the self-heating state of coal. Zhao et al. pointed out that the concentration of CO and CO_2 fluctuated with the change of ventilation, and the reliability of the gas as an indicator of spontaneous combustion of coal was poor [22]. To sum up, there are many indicators to distinguish coal's spontaneous combustion status at present, but the prediction indicators are different in different environments. It is easy to use wrong indicators to predict coal spontaneous combustion, resulting in misjudgment. Coal spontaneous combustion is a complex physical and chemical reaction process, and gases will affect each other. A single index gas cannot effectively predict coal spontaneous combustion [23,24]. Therefore, it is necessary to establish a multi-parameter index system to fully and accurately understand the state and development of CSC.

This paper analyzes the types and concentrations of gases produced during the oxidation of lignite, long-flame coal, and lean coal based on temperature-programmed experiments and shows the variation of gas concentration ratios with temperature. Then, the spontaneous combustion tendency of these three coal samples was characterized. Combined with the variation law of three parameters, namely, index gas concentration, ratio of gas concentration and fire coefficient, with coal temperature, the indexes suitable for predicting spontaneous combustion of coal are optimized. A multi-parameter index system was established to evaluate the spontaneous combustion status of coal more accurately and reasonably.

2. Experiment

2.1. Materials

Three coal samples of different coalification degrees were selected from coal mines in China, with coal ranks ranging from low to high being, lignite, long-flame coal, and lean coal.

Fresh coal samples were collected from the coal mining face and sealed and packed for transportation back to the laboratory. After that, the oxidized layer on the surface of the coal samples was peeled off, and then each type of coal was crushed by a jaw crusher and sieved to find enough coal samples to be used in the experiments. The coal particles were then dried in a drying oven. Proximate analysis and ultimate analysis experiments of each coal sample were carried out to understand the coal quality characteristics of the three coal samples. The results are shown in Table 1.

Table 1. Proximate analysis and ultimate analysis of coal.

No.	Coal Rank	Proximate Analysis				Ultimate Analysis				
		M_{ad} (%)	A_d (%)	V_{daf} (%)	FC_{ad}	C	O	N	H	S
1#	Lignite	26.35	10.86	40.21	67.19	65.12	23.28	1.67	4.06	1.34
2#	Long-flame coal	10.23	9.14	31.73	75.93	72.48	17.43	1.52	5.12	0.72
3#	Lean coal	8.19	7.05	18.62	78.69	75.95	10.06	1.89	5.61	1.03

Note: M_{ad} is the moisture (air drying basis); A_d is the ash (dry basis); and V_{daf} is the volatile (dry ash-free); FC_{ad} is fixed carbon.

Among them, the moisture, volatile, and ash content of lignite are higher than the other two coal samples. Lignite has the highest oxygen content and the lowest carbon and hydrogen content. The types and contents of minerals in coal samples with different coalification degrees show obvious differences, and the differences in minerals in coal also have a certain effect on the low-temperature oxidation of coal. The particle size of coal sample required for the temperature-programmed test is 0.18~0.38 mm. Before the experiment, coal particles were weighed with a balance and placed in a drying environment of 40 °C for 48 h. After that, they were taken out and put into sealed bags for later use. The mass of coal samples in each group of experiments was 50 g. Prepare nitrogen cylinders of 99.99% purity and standard oxygen cylinders of 20.96% standard oxygen concentration for the experiment.

2.2. Methods

In this work, the oxidation characteristics of coal auto-ignition were tested using a temperature-programmed test. The temperature-programmed experiment system consists of three parts: an air supply system, a programmed heating system, and a gas analysis system. The air supply system contains compressed air bottles, pressure-reducing valves, flowmeters, pressure gauges, etc., which are connected to each other in turn by latex tubes. The temperature-programmed system consists of a thermostat and a temperature-programmed control device, which is equipped with a preheating tube and a coal sample tank. The temperature accuracy is controlled at about 0.1 °C. The gas analysis system consists of a gas chromatograph and a data analysis device. The specific temperature-programmed experimental system is shown in Figure 1.

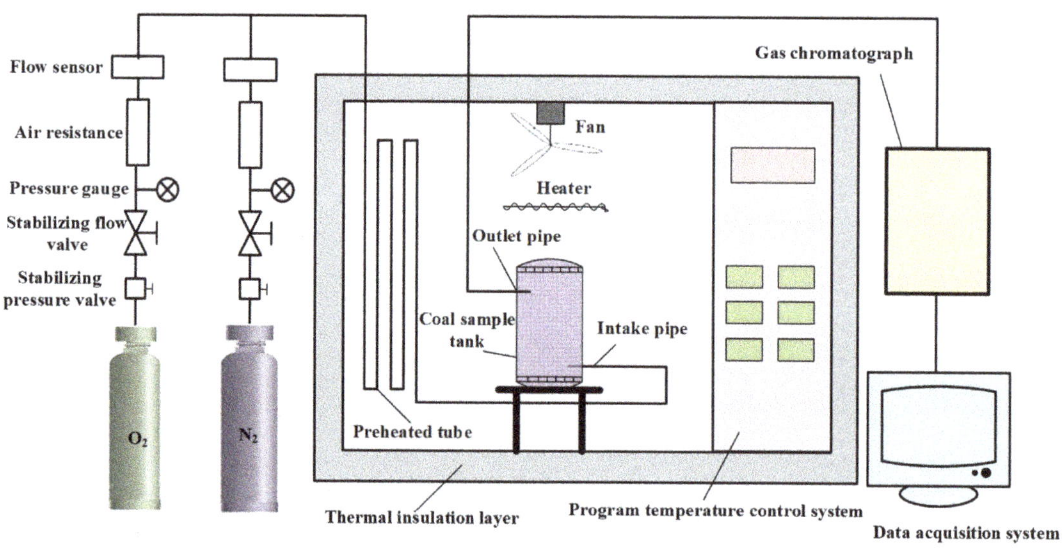

Figure 1. Temperature-programmed experimental system.

In a temperature-programmed oxidation experiment environment, the oxidation processing of coal itself will release heat, and the heat accumulation will lead to an increase in coal temperature [25,26]. Then, the experimental data are recorded by instrument, and the variation laws of coal temperature, the output of marker gas, and oxygen consumption with temperature are studied. Finally, the stage of spontaneous combustion to which the coal belongs is inferred, and early prediction of coal natural ignition is carried out. The main operation process of temperature programmed experiment can be summarized as follows:

- After checking and ensuring good gas tightness and proper functioning of all instruments, coal samples are placed into the coal sample tank, and a layer of asbestos with a thickness of 2~3 mm above the coal samples is evenly applied to prevent blocking of the gas path.
- The gas flow required for the experiment was adjusted; that is, dry air (oxygen concentration: 20.96%) was injected into the coal sample tank at a steady flow rate of 100 mL/min.
- The temperature is set through the temperature control system. The initial temperature is set to 30 °C, and the heating rate is 1 °C/min until it is heated to 200 °C. Here, ensure that the heating rate of each experiment is the same so as to improve the comparability of experimental results. The temperature sensor is used to monitor tank temperature and coal temperature, respectively, and the two are compared to obtain the crossing point temperature.
- The gas composition and concentration at different temperatures are monitored by gas chromatograph. We monitor the outlet gas concentration and the gas concentration produced by pyrolysis every 10 °C increment in temperature. Finally, the experimental results are recorded and saved in the computer terminal of the data collection system.

3. Results and Analysis

3.1. Outlet Oxygen Concentration and Oxygen Consumption Rate

Three coal samples with different degrees of deterioration were tested for oxidation at elevated temperatures, and their outlet oxygen volume fractions are shown in Figure 2a. Then, the oxygen consumption rate of each coal sample was calculated according to the calculation formula of oxygen consumption rate, as shown in Figure 2b.

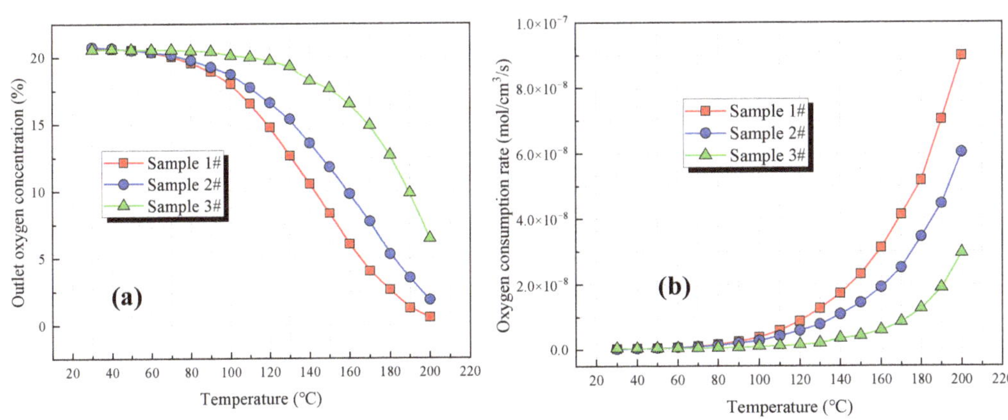

Figure 2. Variation of outlet oxygen concentration and oxygen consumption rate with temperature (**a**) outlet oxygen concentration (**b**) oxygen consumption rate.

It can be seen from Figure 2a that the oxygen concentration at the outlet of coal samples with different coalification degrees decreases with the increase in temperature. This indicates that the higher the temperature is, the more oxygen is involved in the reaction and the more violent the reaction is. When the temperature is fixed, the oxygen concentration at the outlet of the coal sample tank decreases with the increase in the coal rank. This indicates that the lower the rank of coal, the more intense the oxidation reaction of coal. In addition, Figure 2b more clearly shows that the oxygen consumption rate is positively correlated with temperature and negatively correlated with coal rank [22,27,28]. From the analysis of moisture, the moisture of each coal sample decreases with the increase in coalification degree. The water adsorbed in the pores inside the coal particles can evaporate only after the temperature reaches more than 100 °C and after a period of time. When the water evaporates, more pore structures in the coal are exposed to the air, accelerating the consumption of oxygen. Therefore, the higher the moisture content of the coal sample, the faster the oxygen consumption rate to some extent.

3.2. Gas Product Concentration

3.2.1. CO and CO_2

In the coal mining face, CO gas is released in large amounts and has good sensitivity, so CO is often selected as the index gas to divide different stages of the low-temperature oxidation process of coal [3,29]. Here, the variation of CO and CO_2 concentrations generated during the oxidation process is statistically shown in Figure 3. With the increase in temperature, CO and CO_2 concentrations of coal samples with different coalification degrees show an increasing trend with the increase in temperature. At the initial temperature of 30 °C, CO was detected in all kinds of coal, indicating that the coal had been slowly oxidized with O_2 at this time, and the CO gas was present throughout the whole process of temperature increase and oxidation. There is a sudden change in the process of CO production with the increase in temperature. This abrupt temperature point is the critical temperature of each coal sample, which is 60 °C, 70 °C, and 80 °C from lignite, long-flame coal, and lean coal. It can be observed that the higher the degree of coal deterioration, the greater the critical temperature point. The coal samples with high rank will make the critical temperature point that appears in the process of heating and oxidizing lag behind.

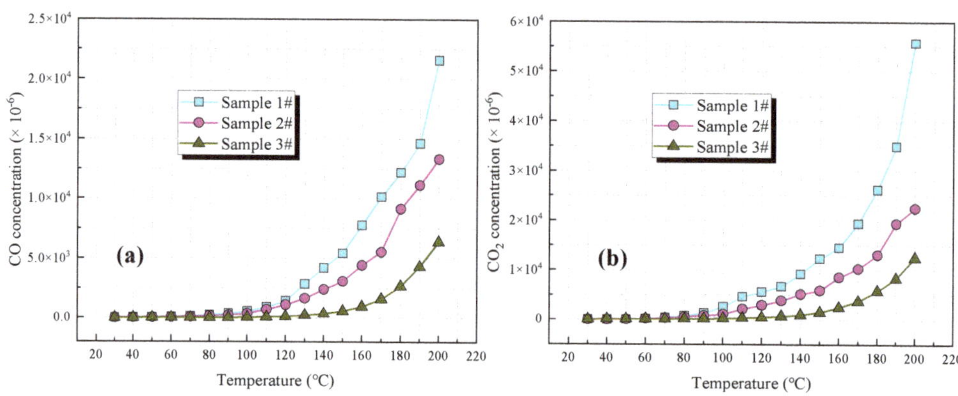

Figure 3. Variation of CO and CO_2 concentration with temperature (**a**) CO (**b**) CO_2.

3.2.2. CH_4 and C_2H_6

The variation trend of CH_4 and C_2H_6 gas concentration generated by temperature-programmed oxidation of each coal sample with temperature is shown in Figure 4. Overall, the CH_4 and C_2H_6 concentrations of each coal sample increase with the increase in temperature. Only the temperature points at which the two gases are generated are different. CH_4 is always present at all stages of temperature rise, which is due to the fact that the coal itself contains a small amount of CH_4 gas. The temperature points at which C_2H_6 gas is generated from lignite, long-flame coal, and lean coal are 90 °C, 100 °C, and 120 °C, respectively. The appearance of these characteristic temperature points can be considered that the spontaneous combustion and oxidation process of coal has entered an accelerated stage. At the same time, it also confirms the law that the higher the deterioration degree of coal, the more lagging the characteristic temperature. In addition, it can be seen from Figure 4a that the CH_4 concentration of Sample 2# of long-flame coal is higher than that of the other two coal samples because the CH_4 content adsorbed in the coal sample is higher. This presents that the CH_4 concentration under different coalification degrees and different environments is also quite different, and it is not appropriate to use a single CH_4 gas to predict the CSC process [9].

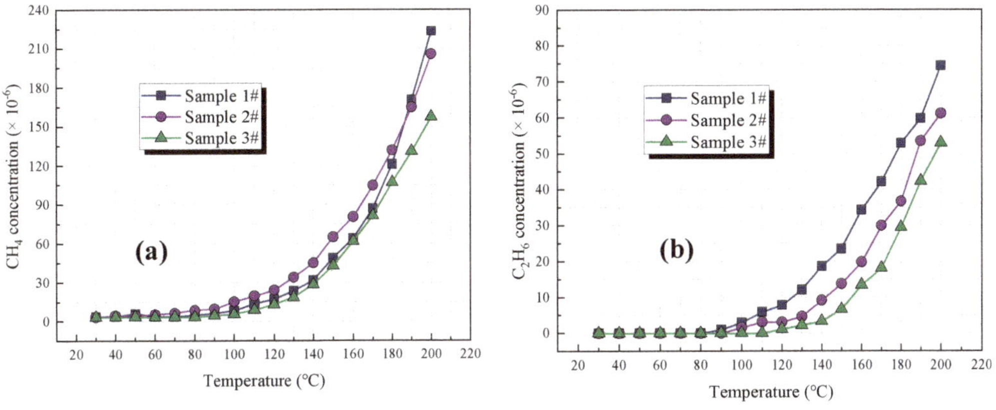

Figure 4. Variation of CH_4 and C_2H_6 concentrations with temperature (**a**) CH_4 (**b**) C_2H_6.

3.2.3. C_2H_4

Figure 5 illustrates the variation of the olefin gas C2H4 produced during spontaneous combustion of coal with temperature. It can be observed that C_2H_4 cannot be monitored at normal temperature, and the presence of C_2H_4 can be monitored only when a certain temperature value is reached. The temperature of C_2H_4 in coal samples with different coalification degrees is different. The temperature points of C_2H_4 gas generated by lignite, long-flame coal, and lean coal are 110 °C, 120 °C, and 150 °C, respectively. These temperatures can be referred to as dry cracking temperatures and symbolize the entry of the coal samples into the accelerated oxidation stage. After these temperature points, the C_2H_4 gas of each coal sample shows an exponential growth trend with the increase in temperature. For lignite and long-flame coals, a turning point in C_2H_4 concentration with increasing temperature occurs when the temperature is at 160 °C when the rate of C_2H_4 production becomes greater, and the oxidation reaction is more intense. For lean coal, the temperature at this transition point is about 180 °C. If this phenomenon is found, timely measures need to be taken to prevent the further expansion of spontaneous combustion.

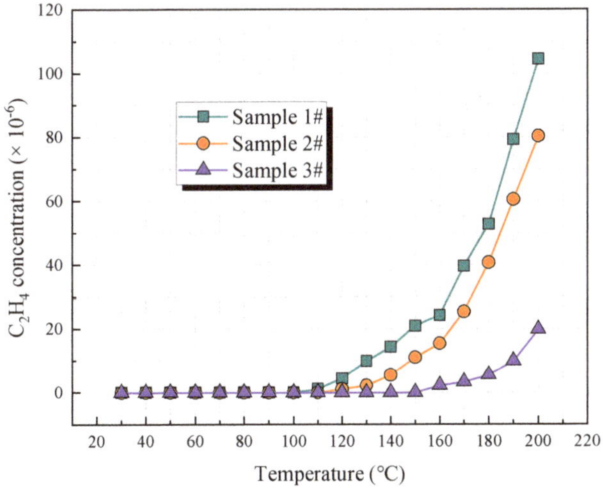

Figure 5. Variation of C_2H_4 concentration with temperature.

3.3. Change of Product Gas Concentration Ratio

The above only analyzes the change law of single index gas, and it is unreliable to predict the degree of coal spontaneous combustion only according to this [12,30]. Therefore, it is necessary to investigate the variation law of some composite index gases. The composite index gas here mainly refers to the ratio between the concentrations of main gases released during the natural oxidation and heating process of coal, including CO_2/CO (α), C_2H_6/CH_4 (β), C_2H_4/C_2H_6 (γ), and C_2H_4/CH_4 (δ).

3.3.1. CO_2/CO

Figure 6 shows the variation law of CO_2/CO (α) with temperature. The CO_2/CO of coal samples with different coalification degrees are different. For long-flame coal, the CO_2/CO of the whole temperature range is less than 5. Before about 50 °C, α decreases with the increase in temperature, indicating that the rate of CO production in the low-temperature oxidation stage is greater than that of CO_2. At this stage, the coal sample is in the slow oxidation stage, and the generated CO and CO_2 gases are less. Between 50 °C and 110 °C, α increases with the increase in temperature, indicating that the production rate of CO_2 is greater than that of CO. At this stage, the oxidation reaction rate of coal is increasing. After 110 °C, α decreases with the increase in temperature, and the production rate of CO is greater than that of CO_2. Coal spontaneous combustion enters an accelerated stage, and the oxidation reaction is intense. Lignite also shows a similar law with long-flame coal, but the characteristic temperature points of lignite should be compared with long-flame coal in advance. Additionally, for lean coal, α increases first and then decreases with the increase in temperature. However, this increasing trend is only in the short temperature stage of 30–40 °C. This is due to the higher rank of lean coal and better pore development, in which some CO_2 gas is adsorbed. At the beginning of heating up, CO_2 is desorbed out in a short time, resulting in the production rate of CO_2 being greater than that of CO. After 40 °C, the production rate of CO_2 is lower than that of CO.

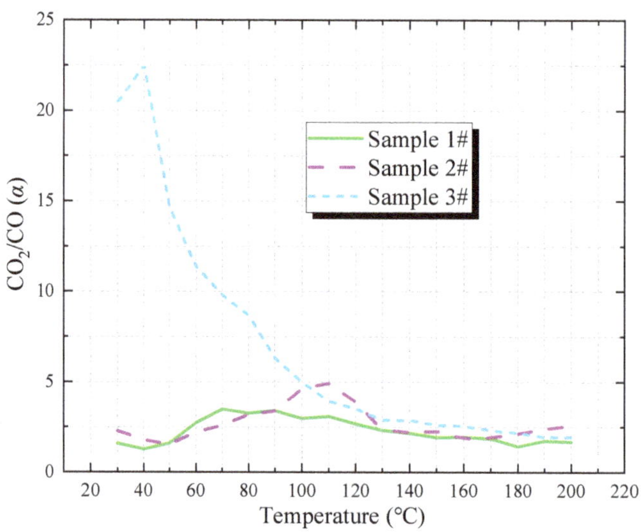

Figure 6. Variation of CO_2/CO with temperature.

3.3.2. C_2H_6/CH_4

Figure 7 demonstrates the variation of C_2H_6/CH_4 (β) with temperature. The β of lignite shows a tendency to increase and then decrease with increasing temperature, while the β of long-flame coal and poor coal generally increases with increasing temperature. Before 180 °C, the magnitude of β value is basically in the order of lignite > long-flame coal > lean coal. For lignite, after 80 °C, the production rate of C_2H_6 is greater than that of CH_4, and the β value increases rapidly, exceeding 0.3 at 100 °C. At this time, the coal sample entered the accelerated oxidation stage. Between 100 °C and 140 °C, the increase rate of β slows down, which is due to the release of large amounts of CH4 gas at high temperatures. At 140 °C, the value of β is about 0.6. After 140 °C, the production rate of C_2H_6 was lower than that of CH_4, and β showed a decreasing trend. In general, the β of lignite fluctuates greatly with temperature, and the regularity is not obvious. This could be attributed to lignite's unique coal structure and mineral composition. For long-flame coal and lean coal, the temperature point at which β begins to increase is different, 90 °C and 110 °C, respectively. They have abrupt temperature points at 130 °C and 140 °C, respectively. This means that a strong coal-oxygen complex reaction has taken place in the coal sample, and C_2H_6 gas is rapidly generated.

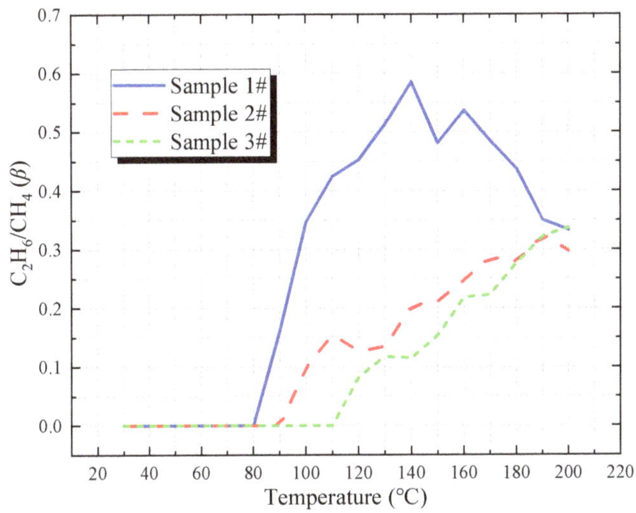

Figure 7. Variation of C_2H_6/CH_4 with temperature.

3.3.3. C_2H_4/C_2H_6 and C_2H_4/CH_4

Figure 8 demonstrates the variation of C_2H_4/C_2H_6 (γ) and C_2H_4/CH_4 (δ) with temperature. As can be seen from Figure 8a, γ shows an overall upward trend with the deepening of oxidation reaction. This indicates that the production rate of C_2H_4 is stably higher than that of C_2H_6. However, the change of γ of lignite with temperature is not obvious and fluctuates greatly. The γ of the other two coal samples basically increases with the increase in temperature. As for the long-flame coal, the temperature mutation points appeared at 120 °C, 170 °C, and 190 °C. However, the temperature mutation point of lean coal was found at 160 °C and 180 °C, respectively. In addition, it can be found from Figure 8b that δ of lignite basically shows a rapid increase and tends to be stable with temperature but still fluctuates in a small range after 130 °C. For lignite, however, a rapidly increasing mutation point occurs around 110 °C, which indicates that the coal enters a rapid oxidation stage at this stage. For both long-flame coal and lean coal, δ increases with temperature. This indicates that the production rate of C_2H_4 is stably higher than that of CH_4. Some temperature mutation points were also observed for long-flame coal and poor coal, which were 130 °C and 160 °C for long-flame coal and 160 °C and 180 °C for lean coal. In conclusion, the olefin and alkanes ratios show obvious segmentation characteristics with temperature, and they can be used as indicators to judge the spontaneous combustion process of coal to some extent.

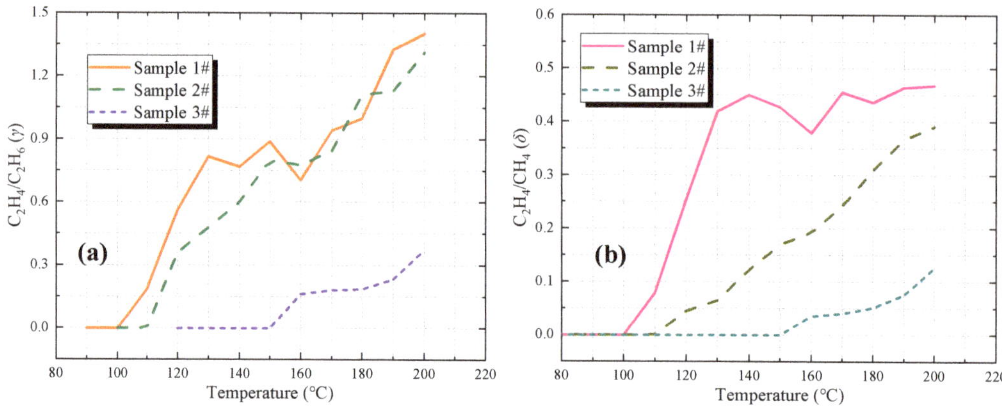

Figure 8. Changes in C_2H_4/C_2H_6 and C_2H_4/CH_4 with temperature (**a**) C_2H_4/C_2H_6 (**b**) C_2H_4/CH_4.

4. Establishment of Prediction Index System of Coal Spontaneous Combustion

4.1. Determination of Coal Spontaneous Combustion Tendency

CSC tendency is an inherent characteristic of coal spontaneous combustion, and it is the primary index for evaluating the risk of natural ignition of coal. It characterizes the possible degree of spontaneous combustion before coal mining and reflects the correlation between the physical and chemical properties of coal and its spontaneous combustion characteristics. According to the standard [31], the determination indexes of spontaneous combustion tendency of three coal samples with different coalification degrees in this paper are calculated, and their spontaneous combustion hazard grades are further evaluated. The calculation equation of the CSC tendency determination index is [31]:

$$I = \phi\left(\varphi_{C_{O_2}} \times I_{C_{O_2}} + \varphi_{T_{cpt}} \times I_{T_{cpt}}\right) - Z_1 \tag{1}$$

where, I is the determination index of coal spontaneous combustion tendency; ϕ is the amplification factor, $\phi = 40$; C_{O_2} is the oxygen concentration at the outlet of the coal sample tank when the temperature of the coal sample reaches 70 °C, %; $\varphi_{C_{O_2}}$ is the weight of low-temperature oxidation stage, $\varphi_{C_{O_2}} = 0.6$; $\varphi_{T_{cpt}}$ is the weight of the accelerated oxidation stage, $\varphi_{T_{cpt}} = 0.4$; $I_{C_{O_2}}$ is the oxygen concentration index at the outlet of the coal sample tank when the coal sample reaches 70 °C; $I_{T_{cpt}}$ is the temperature index at the intersection of coal samples under temperature-programmed condition; Z is the correction factor, $Z = 300$.

The calculation equation of $I_{C_{O_2}}$ is:

$$I_{C_{O_2}} = \frac{C_{O_2} - Z_{O_2}}{Z_{O_2}} \times 100 \tag{2}$$

where, Z_{O_2} is the calculation factor of oxygen concentration at the outlet of the coal sample tank, $Z_{O_2} = 15.5\%$.

The calculation equation of $I_{T_{cpt}}$ is:

$$I_{T_{cpt}} = \frac{T_{cpt} - Z_{cpt}}{Z_{cpt}} \times 100 \tag{3}$$

where, Z_{cpt} is the calculation factor of the temperature at the crossing point, $Z_{cpt} = 140$ °C.

The measured temperature at the crossing point of each coal sample and oxygen volume fraction at 70 °C were substituted into Equations (1)–(3) to calculate the determination index of each coal sample, as shown in Figure 9. After that, we obtained the spontaneous combustion propensity class of each coal sample according to the classification index of coal spontaneous combustion propensity, as shown in Table 2.

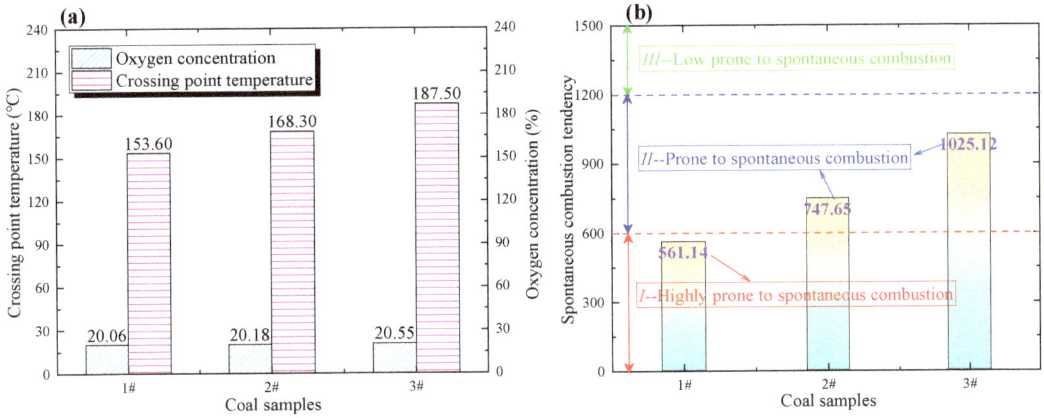

Figure 9. Spontaneous combustion tendency of each coal sample (**a**) Crossing point temperature (**b**) Determination index.

Table 2. Statistics of spontaneous combustion tendency grade and index of coal samples.

Coal Samples	Oxygen Volume Fraction at 70 °C (%)	Crossing Point Temperature (°C)	Determination Index of Spontaneous Combustion Tendency	Spontaneous Combustion Tendency Grade
Lignite	20.06	153.61	561.14	I
Long-flame coal	20.18	168.32	747.65	II
Lean coal	20.55	187.54	1025.12	II

Note: I stands for highly prone to spontaneous combustion; II indicates prone to spontaneous combustion.

From Figure 9 and Table 2, it can be concluded that the higher the coal rank is, the higher the cross point temperature and the higher the Determination index of spontaneous combustion tendency. The spontaneous combustion tendency grades of lignite, long-flame coal, and lean coal are I, II, and II, respectively, which demonstrates that these three kinds of coal samples have the danger of spontaneous combustion, and the lower the coal rank, the greater the danger of spontaneous combustion. Therefore, it is necessary to predict the process of these three kinds of coal samples to better prevent and control mine fires.

4.2. Prediction Index Optimization of Coal Spontaneous Combustion

4.2.1. Index Gases Concentration and Their Ratios

Through the analysis of the variation law of the generated gases in Section 3, it is clear that some indicator gases can be applied to predict the spontaneous combustion of coal, whereas some cannot. When selecting indicators to predict the spontaneous combustion of coal, attention should be paid to the selection of easy-to-monitor gases, and there is a good regularity between the concentration of gas indicators and coal temperature.

During the coal heating process, there is a good regularity between CO concentration and coal temperature. Generally, there will be abrupt changes at critical temperature and drying temperature. Therefore, CO concentration can be used as the main observation index to measure the stage and degree of coal spontaneous combustion. Since the emanation of C_2H_4 and C_2H_6 is unstable in the low-temperature oxidation stage, the concentration of these two gases shows an increasing trend when the temperature exceeds 120 °C, which demonstrates that the macromolecules in the coal body at this stage are not stable, the concentration of C_2H_4 and C_2H_6 is not stable in the low-temperature oxidation stage. This indicates that a large amount of gas is released from the cracking of macromolecular branch chains in coal at this stage, so it can be concluded that the coal temperature has exceeded the drying temperature. Therefore, C_2H_4 and C_2H_6 concentrations can be used as auxiliary gas indicators. Considering that some CH_4 and CO_2 occur in coal, the content of CH_4 and CO_2 varies greatly in coal under different environments and coalification degrees; it is sometimes impossible to determine whether the changing trend of CH_4 and CO_2 is caused by oxidation reaction or desorption effect [9,24,32]. Therefore, both gases are somehow not good enough to predict the spontaneous combustion of coal.

The overall CO_2/CO, C_2H_6/CH_4, C_2H_4/C_2H_6, and C_2H_4/CH_4 show obvious phasing characteristics with temperature, and some characteristic temperature points appear. It is possible to determine whether the coal has entered the stage of dry cracking and accelerated oxidation based on their trends. However, the oxidative decomposition of coal produces alkanes and olefin gases, and their generation mechanism is not clear at present. In the low temperature oxidation stage, alkanes and olefin gases are not stable and cannot react correctly with the coal temperature. In general, the ratio of alkane to olefin gas concentration can only predict the process of coal spontaneous combustion at high temperatures. Therefore, CO_2/CO, C_2H_6/CH_4, C_2H_4/C_2H_6, and C_2H_4/CH_4 can be used as indicators to assist in judging the state and degree of coal spontaneous combustion oxidation [17].

4.2.2. Fire Coefficient Index

Single gas prediction index is affected by airflow, gas extraction location, and so on, which results in misjudgment of the spontaneous combustion state of coal. For this reason, Graham put forward the concept of the fire coefficient in 1914. The percentage of fire coefficient can be calculated by the increment of CO_2 concentration ($+\Delta CO_2$), CO concentration increment ($+\Delta CO$), and reduction of O_2 concentration ($-\Delta O_2$), including the first fire coefficient (R_1), the second fire coefficient (R_2), and the third fire coefficient (R_3). The calculation formula of the specific fire coefficient is [14,33,34]:

$$R_1 = +\Delta CO_2/(-\Delta O_2) \times 100\% \quad (4)$$

$$R_2 = +\Delta CO/(-\Delta O_2) \times 100\% \quad (5)$$

$$R_3 = +\Delta CO/(+\Delta CO_2) \times 100\% \quad (6)$$

Combining with Equations (4)–(6) to analyze the data measured in the test, the three fire coefficients of the coal samples with three different degrees of coalification were calculated separately, as shown in Figure 10.

From Figure 10a, it can be clearly seen that at the initial stage of heating up, the R_1 value increases with the increase in temperature and then reaches a peak value. The peak values of different coal samples correspond to different temperature points. The corresponding temperature points of lignite, long-flame coal, and lean coal are about 60 °C, 70 °C, and 70 °C, respectively. After these temperature points, the rate of oxygen consumption is greater than the rate of CO_2 production under the faster coal-oxygen complexation, resulting in a decrease in the R_1 value. However, at around 120~130 °C, the value of R_1 begins to increase sharply. This is because the coal enters a rapid oxidation phase where the rate of increase in CO_2 concentration is greater than the decrease rate of O_2 concentration.

As can be observed from Figure 10b, the value of R_2 between 30 °C and 140 °C is relatively small, basically at about 10%. Taking lignite as an example, the R_2 value exceeds 10% at about 60 °C, which means that the coal temperature has exceeded the critical temperature and entered the stage of spontaneous combustion. Between 140 °C and 150 °C, the R_2 value exceeds 25%, and the coal-oxygen recombination is strong and enters the dry cracking stage. When the R_2 value exceeds 50% between 170 and 180 °C, it means that the coal sample has undergone a violent oxidation reaction, and corresponding countermeasures should be taken. However, long-flame coal and lean coal show a similar pattern with lignite, only the corresponding temperature turning point is different.

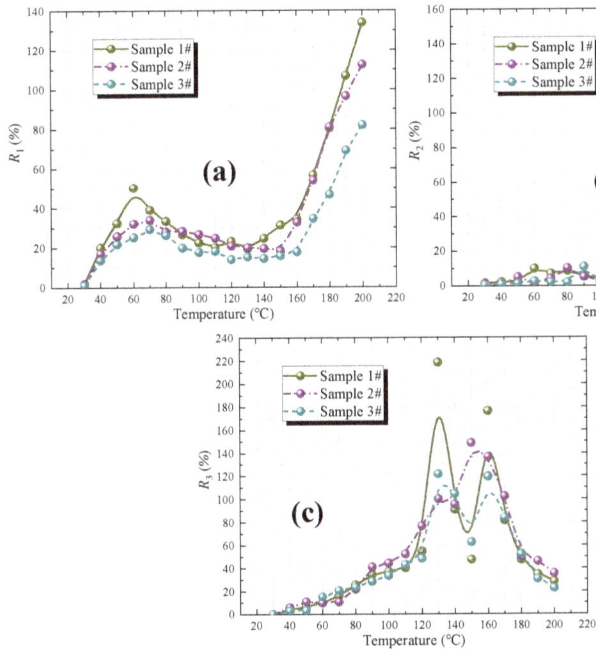

Figure 10. Variation curves of fire coefficient of different coal samples with temperature (**a**) R_1 (**b**) R_2 (**c**) R_3.

Furthermore, from Figure 10c, it can be found that the R_3 value of lignite gradually increases with the increase in coal temperature from 30 °C to about 110 °C. After 110 °C, the value of R_3 increases rapidly and reaches the first peak value. It indicates that the combination of coal and oxygen increases, and the coal enters the stage of dry cracking. After that, a peak value of R_3 appeared at 160 °C and exceeded 100%. The fluctuation of R_3 with temperature is large, and there is no obvious pattern, so it is appropriate to use the R_3 value as an auxiliary indicator. Overall, fire coefficient R_2 can comprehensively predict the entire oxidation stage of coal from low temperature to high temperature, and it is suitable to be used as the main indicator to predict coal spontaneous combustion, while R_1 and R_3 are used as auxiliary indicators.

4.3. Establishment of Prediction Index System of Coal Spontaneous Combustion

Based on the above analysis, a multi-index system can be developed to predict the spontaneous combustion state of coal, which is shown in Table 3. CO concentration, C_2H_6/CH_4, and the second fire coefficient R_2 are the main gas indexes for predicting the spontaneous combustion of coal. C_2H_4 concentration, C_2H_6 concentration, CO_2/CO, C_2H_6/CH_4, C_2H_4/C_2H_6, C_2H_4/CH_4, the first fire coefficient R_1, and the third fire coefficient R_3 were used as auxiliary gas indexes. When applying this index system, the spontaneous combustion of coal should be evaluated according to the parameters measured in the test and combined with the actual measurement in the field. During this period, relevant gas index parameters should be constantly modified to match the actual coal seam ignition situation and finally effectively guide the coal mine fire prevention and control work.

Table 3. Optimization of multi-parameter indexes for coal spontaneous combustion prediction.

Prediction Indicators	Gas Concentration			Gas Concentration Ratio					Fire Coefficient		
	CO	C_2H_4	C_2H_6	C_2H_6/CH_4	CO_2/CO	C_2H_6/CH_4	C_2H_4/C_2H_6	C_2H_4/CH_4	R_2	R_1	R_3
Main indicators	Yes	No	No	Yes	No	No	No	No	Yes	No	No
Auxiliary indicators	No	Yes	Yes	No	Yes	Yes	Yes	Yes	No	Yes	Yes

5. Conclusions

(i) During the heating process of the coal sample, with the increase in temperature, the oxygen concentration at the outlet decreases gradually, while the oxygen consumption rate and the concentration of each generated gas gradually increase. Oxygen consumption rate is negatively correlated with coal rank. The higher the coalification degree of coal, the more lagging some characteristic temperature points in the spontaneous combustion process.

(ii) Determination indexes of the spontaneous combustion tendency of three coal samples with different coalification degrees were determined, and their spontaneous combustion hazard levels were evaluated. Lignite is classified as Class I susceptible to spontaneous combustion, while long-flame coal and lean coal are classified as Class II capable of spontaneous combustion. Three kinds of fire coefficients were calculated to determine spontaneous combustion of coal. Taking lignite as an example, the range of critical temperature points in the spontaneous combustion process evaluated by the second fire coefficient R_2 is 60~70 °C, and the range of dry cracking temperature points is 140~150 °C.

(iii) A multi-parameter index system was established to predict the spontaneous combustion state of coal, in which the main indexes include CO concentration, C_2H_6/CH_4, and the second fire coefficient R_2. Auxiliary indicators include: C_2H_4 concentration, C_2H_6 concentration, CO_2/CO, C_2H_6/CH_4, C_2H_4/C_2H_6, C_2H_4/CH_4, the first fire coefficient R_1 and the third fire coefficient R_3. In the actual situation, continuous correction of the relevant gas index parameters is required to comprehensively determine the spontaneous combustion of coal and provide support for fire prevention and suppression work in coal mines.

Author Contributions: Conceptualization, X.X.; methodology, X.X. and F.Z.; investigation, X.X. and F.Z.; resources, X.X. and F.Z.; data curation, X.X.; writing—original draft preparation, X.X.; writing—review and editing, F.Z.; visualization, F.Z.; funding acquisition, X.X. All authors have read and agreed to the published version of the manuscript.

Funding: This research is funded by the research fund of Henan Key Laboratory for Green and Efficient Mining and Comprehensive Utilization of Mineral Resources (Henan Polytechnic University) (No. KCF2207), funded by the general project of the National Natural Science Foundation of China (No. 51274087 and 51674098), and funded by the key scientific and technological project of Henan Province (No. 152102210315).

Institutional Review Board Statement: Not applicable.

Informed Consent Statement: Not applicable.

Data Availability Statement: Not applicable.

Conflicts of Interest: The authors declare that they have no conflict of interest.

References

1. Xue, D.; Hu, X.; Cheng, W.; Wei, J.; Zhao, Y.; Shen, L. Fire Prevention and Control Using Gel-Stabilization Foam to Inhibit Spontaneous Combustion of Coal: Characteristics and Engineering Applications. *Fuel* **2020**, *264*, 116903. [CrossRef]
2. Wen, H.; Guo, J.; Jin, Y.; Wang, K.; Zhang, Y.; Zheng, X. Experimental Study on the Influence of Different Oxygen Concentrations on Coal Spontaneous Combustion Characteristic Parameters. *Int. J. Oil Gas Coal Technol.* **2017**, *16*, 187–202. [CrossRef]
3. Deng, J.; Xiao, Y.; Li, Q.; Lu, J.; Wen, H. Experimental Studies of Spontaneous Combustion and Anaerobic Cooling of Coal. *Fuel* **2015**, *157*, 261–269. [CrossRef]
4. Lu, P.; Liao, G.X.; Sun, J.H.; Li, P.D. Experimental Research on Index Gas of the Coal Spontaneous at Low-Temperature Stage. *J. Loss Prev. Process Ind.* **2004**, *17*, 243–247. [CrossRef]
5. Xu, H.; Qin, Y.; Yang, D.; Zhang, F.; Wu, F.; Chu, X. Modeling of Diffusion Kinetics during Gas Adsorption in a Coal Seam with a Dimensionless Inversion Method. *Fuel* **2022**, *326*, 125068. [CrossRef]
6. Bai, Z.; Wang, C.; Deng, J. Analysis of Thermodynamic Characteristics of Imidazolium-Based Ionic Liquid on Coal. *J. Therm. Anal. Calorim.* **2020**, *140*, 1957–1965. [CrossRef]
7. Qin, Y.; Song, Y.; Liu, W.; Wei, J.; Lv, Q. Assessment of Low-Temperature Oxidation Characteristics of Coal Based on Standard Oxygen Consumption Rate. *Process Saf. Environ. Prot.* **2020**, *135*, 342–349. [CrossRef]
8. Baris, K.; Kizgut, S.; Didari, V. Low-Temperature Oxidation of Some Turkish Coals. *Fuel* **2012**, *93*, 423–432. [CrossRef]
9. Xie, J.; Xue, S.; Cheng, W.; Wang, G. Early Detection of Spontaneous Combustion of Coal in Underground Coal Mines with Development of an Ethylene Enriching System. *Int. J. Coal Geol.* **2011**, *85*, 123–127. [CrossRef]
10. Li, J.L.; Lu, W.; Liang, Y.T.; Qi, G.S.; Kong, B.; Hu, X.M. Variation of CO_2/CO Ratio during Pure-Oxidation of Feed Coal. *Fuel* **2020**, *262*, 116588. [CrossRef]
11. Xiao, Y.; Ma, L.; Wang, Z.P.; Wen, H.; Deng, J. Law of Index Gases Adsorption and Condensability of Coal Spontaneous Combustion. *Meitan Xuebao/J. China Coal Soc.* **2007**, *32*, 1014–1018.
12. Wang, J.; Zhang, Y.; Xue, S.; Wu, J.; Tang, Y.; Chang, L. Assessment of Spontaneous Combustion Status of Coal Based on Relationships between Oxygen Consumption and Gaseous Product Emissions. *Fuel Process. Technol.* **2018**, *179*, 60–71. [CrossRef]
13. Wen, H.; Yu, Z.; Fan, S.; Zhai, X.; Liu, W. Prediction of Spontaneous Combustion Potential of Coal in the Gob Area Using Co Extreme Concentration: A Case Study. *Combust. Sci. Technol.* **2017**, *189*, 1713–1727. [CrossRef]
14. Guo, J.; Wen, H.; Zheng, X.; Liu, Y.; Cheng, X. A Method for Evaluating the Spontaneous Combustion of Coal by Monitoring Various Gases. *Process Saf. Environ. Prot.* **2019**, *126*, 223–231. [CrossRef]
15. Liu, H.; Wang, F.; Ren, T.; Qiao, M.; Yan, J. Influence of Methane on the Prediction Index Gases of Coal Spontaneous Combustion: A Case Study in Xishan Coalfield, China. *Fuel* **2021**, *289*, 119852. [CrossRef]
16. Itay, M.; Hill, C.R.; Glasser, D. A Study of the Low Temperature Oxidation of Coal. *Fuel Process. Technol.* **1989**, *21*, 81–97. [CrossRef]
17. Ma, L.; Zou, L.; Ren, L.F.; Chung, Y.H.; Zhang, P.Y.; Shu, C.M. Prediction Indices and Limiting Parameters of Coal Spontaneous Combustion in the Huainan Mining Area in China. *Fuel* **2020**, *264*, 116883. [CrossRef]
18. Niu, H.; Deng, X.; Li, S.; Cai, K.; Zhu, H.; Li, F.; Deng, J. Experiment Study of Optimization on Prediction Index Gases of Coal Spontaneous Combustion. *J. Cent. South Univ.* **2016**, *23*, 2321–2328. [CrossRef]
19. Willet, H.L. The Interpretation of Samples from behind Stoppings with a View to Reopening. *Trans. Inst. Min. Eng.* **1952**, *111*, 629–651.
20. Ghosh, A.K.; Banerjee, D.D. Use of Carbon–Hydrogen Ratio as an Index in the Investigation of Explosions and Underground Fires. *J. Mines Met. Fuels* **1967**, *15*, 334–340.
21. Graham, J.I. The Absorption of Oxygen by Coal: Part V. The Influence of Temperature on Rates of Absorption by Different Parts of the Barnsley Bed. *Trans. Inst. Min. Eng.* **1914**, *48*, 521–534.
22. Zhao, J.; Deng, J.; Wang, T.; Song, J.; Zhang, Y.; Shu, C.M.; Zeng, Q. Assessing the Effectiveness of a High-Temperature-Programmed Experimental System for Simulating the Spontaneous Combustion Properties of Bituminous Coal through Thermokinetic Analysis of Four Oxidation Stages. *Energy* **2019**, *169*, 587–596. [CrossRef]
23. Xue, S.; Dickson, B.; Wu, J. Application of 222Rn Technique to Locate Subsurface Coal Heatings in Australian Coal Mines. *Int. J. Coal Geol.* **2008**, *74*, 139–144. [CrossRef]
24. Zhang, Y.; Wu, J.; Chang, L.; Wang, J.; Li, Z. Changes in the Reaction Regime during Low-Temperature Oxidation of Coal in Confined Spaces. *J. Loss Prev. Process Ind.* **2013**, *26*, 1221–1229. [CrossRef]
25. Zhao, J.; Wang, W.; Fu, P.; Wang, J.; Gao, F. Evaluation of the Spontaneous Combustion of Soaked Coal Based on a Temperature-Programmed Test System and in-Situ FTIR. *Fuel* **2021**, *294*, 120583. [CrossRef]
26. Zhao, J.; Deng, J.; Song, J.; Shu, C.M. Effectiveness of a High-Temperature-Programmed Experimental System in Simulating Particle Size Effects on Hazardous Gas Emissions in Bituminous Coal. *Saf. Sci.* **2019**, *115*, 353–361. [CrossRef]
27. Yang, Y.; Li, Z.; Si, L.; Hou, S.; Li, Z.; Li, J. Study on Test Method of Heat Release Intensity and Thermophysical Parameters of Loose Coal. *Fuel* **2018**, *229*, 34–43. [CrossRef]
28. Nie, B.; Yan, H.; Liu, P.; Chen, Z.; Peng, C.; Wang, X.; Yin, F.; Gong, J.; Wei, Y.; Lin, S.; et al. Experimental Investigation on Spontaneous Combustion Oxidation Characteristics and Stages of Coal with Different Metamorphic Degrees. *Environ. Sci. Pollut. Res.* **2023**, *30*, 8269–8279. [CrossRef]
29. Zhang, Y.; Wang, J.; Wu, J.; Xue, S.; Li, Z.; Chang, L. Modes and Kinetics of CO_2 and CO Production from Low-Temperature Oxidation of Coal. *Int. J. Coal Geol.* **2015**, *140*, 1–8. [CrossRef]

30. Song, Z.; Kuenzer, C. Coal Fires in China over the Last Decade: A Comprehensive Review. *Int. J. Coal Geol.* **2014**, *133*, 72–99. [CrossRef]
31. Standardization Administration of China. *The Test Method of Oxidation Kinetics for the Propensity of Coal to Spontaneous Combustion*; Standardization Administration of China: Beijing, China, 2008.
32. Singh, A.K.; Singh, R.V.K.; Singh, M.P.; Chandra, H.; Shukla, N.K. Mine Fire Gas Indices and Their Application to Indian Underground Coal Mine Fires. *Int. J. Coal Geol.* **2007**, *69*, 192–204. [CrossRef]
33. Liang, Y.; Zhang, J.; Wang, L.; Luo, H.; Ren, T. Forecasting Spontaneous Combustion of Coal in Underground Coal Mines by Index Gases: A Review. *J. Loss Prev. Process Ind.* **2019**, *57*, 208–222. [CrossRef]
34. Wang, C.; Zhao, X.; Bai, Z.; Deng, J.; Shu, C.; Zhang, M. Comprehensive Index Evaluation of the Spontaneous Combustion Capability of Different Ranks of Coal. *Fuel* **2021**, *291*, 120087. [CrossRef]

fire

MDPI

Article

The Effects of Solvent Extraction on the Functional Group Structure of Long-Flame Coal

Jun Guo [1,2], Yanping Quan [1,*], Hu Wen [1,2], Xuezhao Zheng [1,2], Guobin Cai [1,2] and Yan Jin [1,3]

1 College of Safety Science and Engineering, Xi'an University of Science and Technology, Xi'an 710054, China; guojun@xust.edu.cn (J.G.); wenhu@xust.edu.cn (H.W.); zhengxuezhao@xust.edu.cn (X.Z.); 19220214086@stu.xust.edu.cn (G.C.); 15258821069@163.com (Y.J.)
2 Key Laboratory of Western Mine and Hazard Prevention, Ministry of Education of China, Xi'an 710054, China
3 Sinopec Sales Co., Ltd. Taizhou Petroleum Branch, Taizhou 318000, China
* Correspondence: 21220226136@stu.xust.edu.cn

Abstract: The functional group structures of coal molecules are one of the most important factors affecting spontaneous combustion. However, it is difficult to determine the exact effects of such structures. Extraction technology is able to modify the functional groups in coal as a means of inhibiting spontaneous combustion reactions. The present work extracted coal from the Caojiatan mine in northern Shaanxi, China, with various solvents. The extraction effectiveness of these solvents was found to decrease in the order of dioctyl sulfosuccinate (AOT) > water > n-hexane > cyclohexane + AOT + ethanol > cyclohexane > ethanol > methanol. With the exception of the AOT, the concentration of functional groups in the extracted coal was decreased compared with that in a control specimen extracted using only water. Ethanol, n-hexane, and methanol provided the optimal extraction efficiencies in terms of capturing coal molecules with aromatic structures, aliphatic structures, and oxygen-containing groups, respectively. The results of this work are expected to assist in future research concerning the extraction of coal molecules with specific functional groups. This work also suggests new approaches to the active prevention and control of spontaneous combustion during the mining, storage, and transportation of coal.

Keywords: coal spontaneous combustion; targeted extraction; functional groups

1. Introduction

Coal mining efficiency has been greatly improved by the adoption of retreating subsidence, high-rise grouting, and longwall mining together with other technologies. However, these techniques have also led to increasingly complex and harsh underground mining environments that are associated with spontaneous combustion and frequent underground fires [1–4]. Unfortunately, at present, the prediction and analysis of such incidents is challenging and new technologies together with in-depth theoretical research are required to identify risk factors [5,6]. Assessments of these factors would allow the effective prevention and control of coal spontaneous combustion. In particular, there is a need to study the characteristics of spontaneous combustion in coal seams.

In recent years, the rapid development of microscopic-scale analytic technologies has allowed the reaction characteristics of coal molecules to be evaluated. Such analyses are commonly based on the use of solid-liquid extraction to separate various mixtures. This separation technology is widely used in many fields, including medicine and biology, as well as in the food and chemical industries. As an example, Li et al. [7] extracted corn whiskers with both ethyl acetate and methanol to obtain DL-epiloliolide, N-p-hydroxy-cis-coumaroyltyramine, β-Sitosterul, and other compounds. Wang et al. [8] extracted extracellular proteins and alkaline phosphatase from thallus with n-butanol and used this extract to inhibit the activity of Staphylococcus aureus. Hu et al. [9] used various polar solvents (ethyl acetate and n-butanol) to extract nuclear biomass components from hawthorn and studied the antioxidant properties of these substances. Wang [10] obtained anti-aging compounds from dandelion, chicory root, and Perilla leaves using a supercritical extraction method.

The successful application of extraction in various industries suggests that this technique could also be applied to the prevention and control of coal fires. Because the molecular structure of coal is based on the condensation polymerization of benzene rings with alkyl side chains and coal also contains bridging bonds and other structures, it should be possible to use various solvents to separate coal molecules according to their physical and chemical properties [11–15]. Subsequent analyses of the microscopic coal structures could provide information concerning the spontaneous combustion mechanism. In prior work, Zhang et al. [16] used a variety of solvents, including cyclohexane and carbon disulfide, to carry out extraction trials and determined the capacities of these solvents. Zhang also employed Fourier transform infrared (FTIR) spectroscopy to determine the roles of various functional groups in the spontaneous combustion of coal. Ma et al. [17,18] extracted Hefeng (a location name) coal with petroleum ether, methanol, carbon disulfide, and acetone using a continuous five stage process involving ultrasonication together with either single or mixed solvents. The molecular structures of the extracted materials were ascertained using gas chromatography-mass spectrometry. The combination of acetone and carbon disulfide was found to promote the dissolution of coal molecules with hydroxy groups. Deng et al. [19,20] used an imidazole-based ionic liquid to extract coal specimens while degrading various functional groups. Thermogravimetric data indicated that [BMIM][BF_4] had the strongest effect, based on the reaction mechanism shown in Figure 1. Marek et al. [21] proposed that alcohols could be the most effective extraction solvents and performed trials using n-heptane, toluene, n-butanol, and three other isomeric alcohols with five Polish lignite specimens. The results showed that n-butanol provided the highest degree of extraction. The results of this work also indicated that the extraction efficiency was maximized only when the solvent was able to extract specific functional groups so as to weaken the spontaneous combustion of the extracted coal sample.

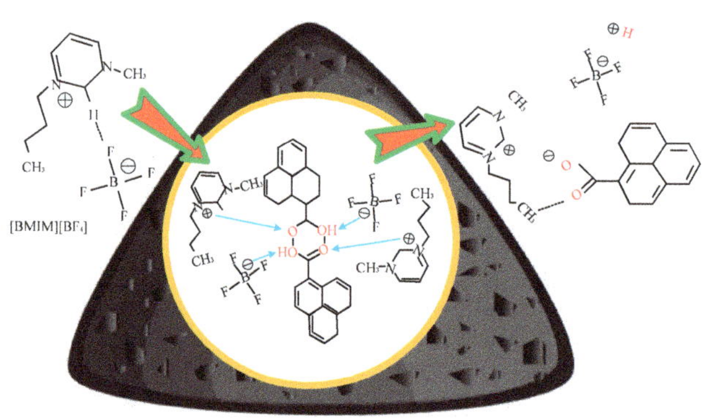

Figure 1. Diagram of the reaction between an ionic liquid and functional groups in coal molecules. (Figure source: Author's own work).

Based on this prior work, it is evident that the active functional groups such as oxygen-containing functional groups and alkyl side chain groups in coal are the main causes of spontaneous combustion of coal. The bridge bonds of coal molecules can be broken through extraction technology, resulting in functional groups being stripped. Finally, the spontaneous combustion of coal will be weakened. However, the effects of this extraction process under specific conditions require further study [22,23]. In the present work, the effects of different extraction solvents on the functional group structure of coal are compared based on trials involving pure solvents and solvent mixtures. This research is expected to provide theoretical guidance for subsequent research concerning the molecular structure of coal and the effects of extraction.

2. Experimental Section

2.1. Selection of Extractants

The spontaneous combustion of coal is an extremely complex process involving the reactions of various functional group structures, with certain structures being more reactive [24]. For this reason, suitable extraction agents must be identified based on the specific groups being targeted. The main categories of coal structures comprise the following. Firstly, macromolecular aromatic structures that form the main carbon chain skeletons in coal molecules will actively participate in combustion. Secondly, aliphatic side-chains will play a key role in terms of heat generation by participating in cracking reactions with oxygen. Lastly, oxygen-containing structures, such as hydroxy or phenolic groups, may also undergo reactions. During combustion, oxygen adsorption reactions generate energy and so intensify the initial combustion reaction. Coal is made of condensed ring polymers primarily comprising aromatic and nonaromatic moieties connected through bridging bonds [25]. The solvent extraction of coal is intended to remove active groups such as these by degrading the connecting structures, such that spontaneous combustion reactions are unlikely to occur. Table 1 summarizes a number of functional groups found in coal molecules along with suitable extractants for such groups.

Table 1. Extractants for coal and their operational principles [26–31].

The Name of the Reagent	For the Group Structure	Action Principle
carbon disulfide	associated structures (e.g., macromolecular aromatic hydrocarbons)	It affects charge transfer and breaks hydrogen bonds
cyclohexane	–C–O–C–, –OH and aliphatic hydrocarbon	Sabotaging the association between molecules
N, N-dimethyl acetamide	aliphatic hydrocarbon radical	High polarization; it has both electron donors and acceptors
acetone	aliphatic hydrocarbon–CH_3, –C–O–C–, –CH_3, –CH_2 and heteroatomic compounds	Like dissolves like
ethanol	structure of aliphatic hydrocarbons	Dissolve part of aliphatic structure or high polarity side chain group unit
ethylenediamine	–NH_2, –CH_2, –OH and other hydrogen bonds contain structures	It has a strong ability to supply electrons and break hydrogen bonds
methanol	aliphatic side chains and highly polar groups	It has strong solubility for polar compounds
petroleum ether	aliphatic hydrocarbons, aromatic hydrocarbons and heteroatomic compounds	Alkyl substitution reaction
n-hexane	hydrogen bonds with polar molecules	Absorption of organic matter
ionic liquid	different functional groups can be extracted by simple preparation and modification of ionic structures	Destruction of hydrogen bonds, branched carbonyl groups and other structures; Reduce or disperse oxygen-containing functional groups
reverse micelles	polar structures such as proteins	Reverse micellar nanoaggregates are formed by directional arrangement of hydrophilic or hydrophobic structures to control group flow

In theory, methanol will undergo alcoholysis reactions with coal molecules under relatively mild conditions [32]. The associated rupture of ester, hydrogen, or oxygen-containing bridging bonds by hydrogen supplied from the methanol will disrupt the linkages between aromatic rings in the coal macromolecules. The hydroxyl group and non-polar alkyl moiety in the molecular structure of ethanol also enable this solvent to dissolve both ionic compounds and non-polar substances, while n-hexane is commonly used in the food industry for the extraction of oils from various foods or spices [33]. Cyclohexane is able to dissolve a wide range of organic compounds. Reverse micelles extraction [34] is able to selectively solubilize polar compounds via the internal water cores of nano-scale aggregates and subsequently transfer the extracted material to the aqueous phase. The exchange of materials across the interface between the two phases is blocked by a layer of these aggregates to produce a separation effect, as shown in Figure 2. As an example, the surfactant dioctyl sulfosuccinate (AOT) has a solubilization effect (Increased solubility) based on the cleavage of hydrogen bonds formed by amide groups and is capable of extracting various analytes or improving extraction efficiency over a wide range of polarity [35].

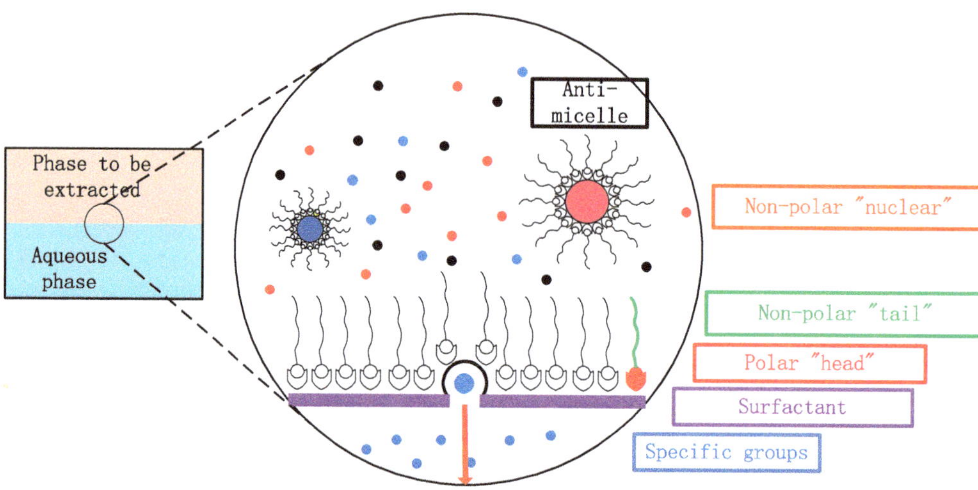

Figure 2. The reverse micelle extraction process. (Figure source: Author's own work).

The present work examined the extraction of coal by n-hexane, cyclohexane (CYH), ethanol (AE), methanol (MT), and a surfactant (AOT), either alone or combined.

2.2. Extraction Procedures

Different types of coal will exhibit variations in pore structure, chemical composition, and functional groups, and long flame coal is more readily extracted compared with lignite, which has a high degree of metamorphism. Long flame coal accounts for the majority of China's raw coal production. Therefore, in this study, raw long flame coal from the Caojiatan mine in northern Shaanxi, China, was employed as a model material (the location of Caojiatan Mine is shown in Figure 3). Fresh coal samples were collected from the working face of the mining area and then sealed and transferred to the laboratory. After unpacking, the outer oxide layer of the raw coal was removed and the coal was pulverized [36]. Both an industrial analysis and elemental analysis of three batches of coal were performed according to the GB/T 212-2008 standard, with the results presented in Table 2.

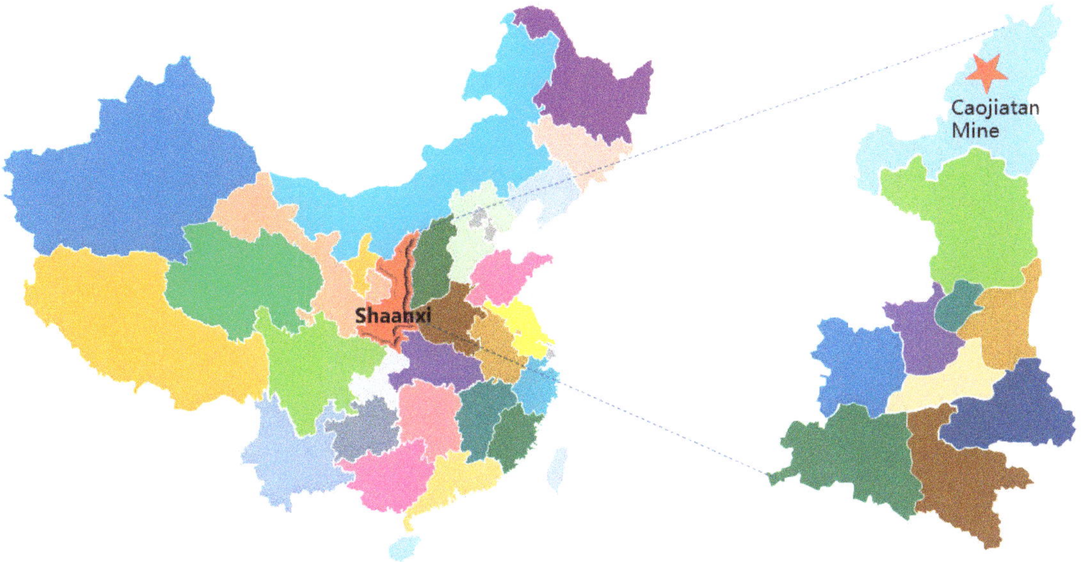

Figure 3. The location of Caojiatan Mine.

Table 2. The results of industrial analysis and elemental analysis of coal samples.

Caojiatan (CJT) Raw Coal Sample	Proximate Analysis/%				Elemental Analysis/%			
	Moisture Content M_{ad}	Ash Content A_d	Volatile Matter Content V_d	Fixed Carbon Content FC_{ad}	C	H	N	O
CJT1	4.72	3.65	35.63	57.85	75.93	4.987	1.135	17.948
CJT2	4.66	3.62	36.17	57.41	75.62	4.966	1.122	18.292
CJT3	4.70	3.57	36.40	57.21	75.91	4.900	1.268	17.922
average value	4.69	3.61	36.06	57.49	75.82	4.951	1.175	18.054

The pulverized coal samples were subsequently screened to obtain specimens with 0–0.12, 0.12–0.15, 0.15–0.18, 0.18–0.25, and 0.25–0.6 mm particle sizes, each with a mass of 200 g. The coal samples in each group were then sealed. A total of 10 groups of extraction solvents (each with a volume of 5 L) were prepared, each containing 20% solute (meaning that 1 kg of coal was added to the 5 L solvent). Each 1 kg coal sample was transferred to a wide-mouth flask along with the solvent, followed by thorough stirring. The flask was subsequently stoppered and allowed to sit under dark conditions for 4 h. Following this static extraction, the coal was collected and dried at 60 °C under a pressure of 0.9 MPa. After drying, the extracted coal was characterized. A small amount of the dry extracted coal sample was removed, weighed, and ground to a particle size of 100–180 mesh then transferred to a sealed tank to be tested. The specific process is shown in Figure 4.

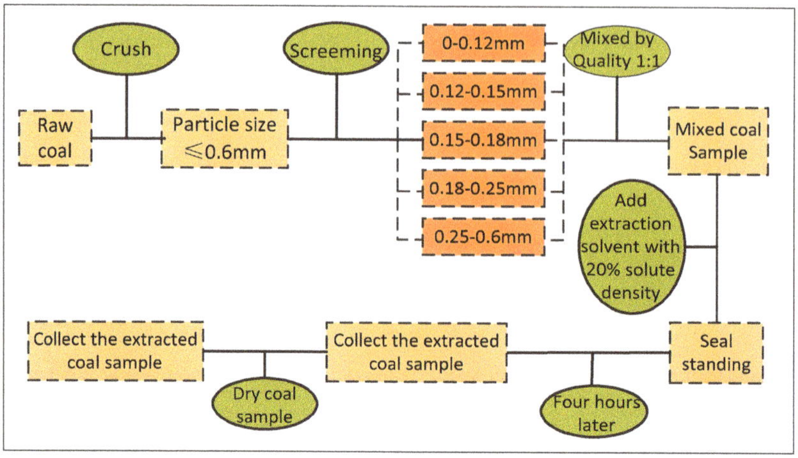

Figure 4. Flow chart summarizing the experimental procedure.

2.3. Extraction Validation

To ensure that the coal samples had been extracted effectively, the extent of extraction was calculated for each specimen based on the mass before and after extraction. The relevant equation is

$$W_i = \frac{(1 - M_{ad}) \times M_1 - M_2}{(1 - M_{ad} - A_d) \times M_1} \times 100\% \quad (1)$$

where W_i is the extent of extraction (%), M_{ad} is the moisture content, A_d is the ash content, M_1 is the mass of the raw coal (g), and M_2 is the mass of the extracted coal sample (g). Single solvent extractions and extractions with synergistic mixtures of solvents (compared to a single solvent) were performed along with a water extraction as a control (decrease error).

2.4. Quantitative Analysis of Functional Groups

FTIR spectra were acquired to quantitatively determine the extent to which various functional groups were extracted, using a Vertex 70V spectrometer with each coal specimen first processed into a KBr pellet. The peak fitting function of the Origin software package was used to process the spectra acquired from each extracted coal sample.

3. Extraction Effectiveness

3.1. Extraction Results

Table 3 summarizes the extraction efficiencies associated with the various solvents.

Table 3. Extraction efficiencies of each solvent.

Test Group	Quality of Raw Coal/g	Quality of the Extracted Coal Sample/g	Extraction Rate/%
MT	1045	953	4.457
CYH	1038	966	2.359
N-Hexane	1022	951	2.449
CYH + AOT + AE	1036	1077	−9.430
AE	1035	960	2.749
AOT	1018	1115	−15.505

It is evident that the AOT provided a negative extraction rate while the extraction efficiencies of the other solvents decreased in the order of MT > AE > n-hexane > CYH. The majority of these common solvents had a limited extraction effect considering the large quantity of coal that was extracted in each trial. Therefore, it is possible that not all functional group structures in the coal samples were completely extracted and lower efficiencies were obtained compared with those obtainable from smaller samples.

The dried specimen following extraction with the AOT contained a white crystalline material mixed among the extracted coal. It is thought that the AOT filled the internal pores of the coal and so was not completely removed, resulting in an apparent increase in mass and the negative extraction value. The present data also indicate that the extraction effects of solvent mixtures were superior to those of single solvents.

3.2. Analysis of Infrared Spectra

As noted, coal molecules are primarily made of macromolecular structures comprising crosslinked chains of aromatic rings. These aromatic backbones also contain side chains and functional groups and are connected via hydrogen bonds and van der Waals forces [37]. In the present work, FTIR spectroscopy was employed to ascertain which functional groups were present in the various specimens. Spectra were acquired from the raw coal samples and six groups of extracted coal using the KBr pellet method. After applying baseline correction and smoothing the spectra the data were compared, as shown in Figure 5.

From Figure 5a, it is apparent that the spectra of the extracted coal samples from the various groups were quite similar to that of the coal immersed in water as a control, with some differences in intensity at certain wavelengths. These spectra exhibit peaks related to C-H bonds at 819 and 876 cm^{-1}, C-O bonds at 1045 and 1084 cm^{-1}, -CH_3 groups at 1370 cm^{-1}, C=C bonds between 1447 and 1617 cm^{-1}, -COOH groups at 2338 and 2361 cm^{-1}, -CH_2 groups at 2925 cm^{-1}, -CH_3 groups at 2968 cm^{-1}, -CH groups at 3060 cm^{-1}, and -OH groups at 3385 cm^{-1}. These spectra indicate that coal molecules containing these groups were readily extracted [38,39]. The spectrum obtained from the raw coal is presented in Figure 5b and the functional groups identified in these materials are summarized in Table 4.

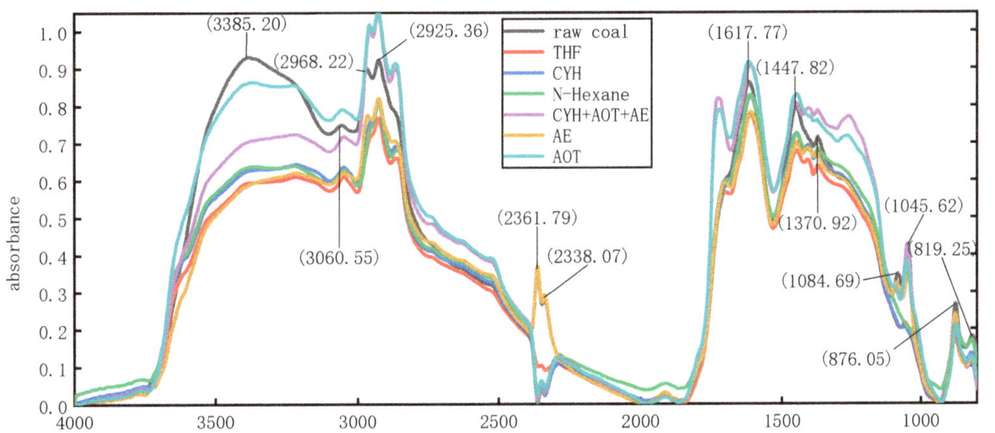

(**a**)

Figure 5. *Cont.*

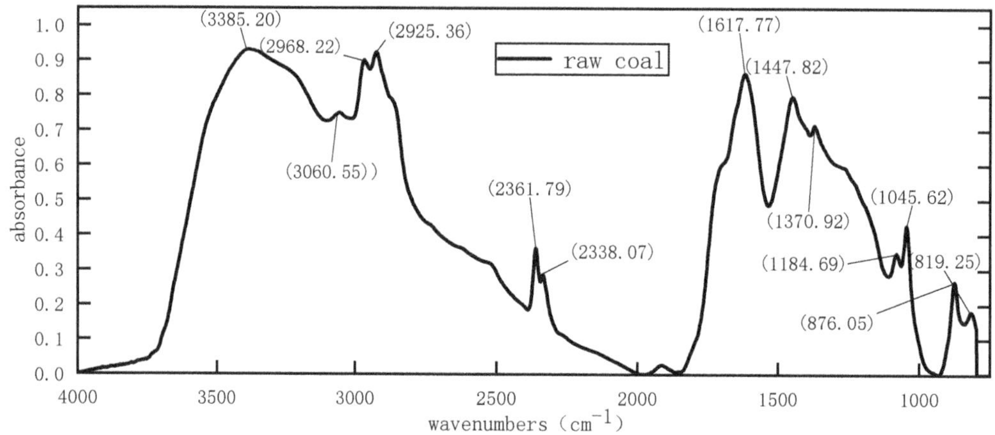

(b)

Figure 5. FTIR spectra acquired from (**a**) various extracted coal specimens and (**b**) raw coal.

Table 4. Infrared absorption peaks generated by the raw coal and associated functional groups.

Spectral Peak Number	Peak Position/cm^{-1}	Absorption Peak Band/cm^{-1}	Spectrum Peaks Belonging	The Group Type
1	819	798–840	C-H external bending vibration of ectopic substituted benzene	substituted benzene
2	876	840–931	Substituted benzene C-H out-of-plane bending vibration	substituted benzene
3	1045	931–1064	Saturated fat ether C-O-C symmetric stretching	C-O
4	1084	1064–1108	C-O stretching vibration	C-O
5	1370	1311–1388	$-CH_3$ symmetric variable Angle vibration	$-CH_3$
6	1447	1388–1533	C=C frame stretching vibration in aromatic ring/thick ring	C=C
7	1617	1533–1847		C=C
8	2338	2300–2344	-OH stretching vibration in -COOH	-COOH
9	2361	2344–2389		-COOH
10	2925	2883–2946	Methylene C-H stretching vibration	$-CH_2$
11	2968	2946–3012	Antisymmetric stretching vibration of methyl group	$-CH_3$
12	3060	3012–3100	Aromatics C-H stretching vibration	-CH
13	3385	3100–3737	-OH in phenols, alcohols, carboxylic acids	-OH

These spectra provide evidence for several peaks related to aromatic structures. Specifically, these peaks comprised the C-H bending vibration of substituted benzene at 819 cm^{-1}, C-H out-of-plane bending vibration of substituted benzene at 876 cm^{-1}, C=C stretching vibrations at 1447 and 1617 cm^{-1}, and C-H stretching vibration at 3060 cm^{-1}. The peaks related to alkyl groups were the $-CH_3$ symmetric stretching vibration at 1370 cm^{-1}, -CH_2- stretching vibration at 2925 cm^{-1}, and $-CH_3$ antisymmetric stretching vibration at 2968 cm^{-1}. The main absorption peaks related to oxygen-containing functional groups were attributed to the symmetric expansion of C-O-C in saturated ethers at 1045 cm^{-1}, the C-O stretching vibration at 1084 cm^{-1}, the -OH stretching vibrations of -COOH groups at 2338, and 2361 cm^{-1}, together with an -OH peak at 3385 cm^{-1}. The peak observed at approximately 2400 cm^{-1} was determined to result from CO_2 gas and so can be ignored. It should be noted that only qualitative assessments were possible based on these spectra because of the possibility of overlapping peaks and other issues, and so additional analysis was required.

3.3. Infrared Data Processing

To accurately assess the functional groups in each specimen and evaluate the extraction effects of the solvents, seven groups of spectra were processed by peak fitting to allow quantitative analysis. This peak fitting was performed using the Origin software Peak Fit module with Gaussian fitting and manual selection of peaks. The first derivative of each spectrum was calculated and Savitzky–Golay smoothing was applied, after which suitable functions were obtained for the relevant fitting parameters and sub-peak fitting was performed. Figure 6 presents an example of this process using the spectra obtained from the raw coal. The resulting peak fitting data are summarized in Table 5. The fitting results for the other six groups of FTIR data are shown in Figure 7.

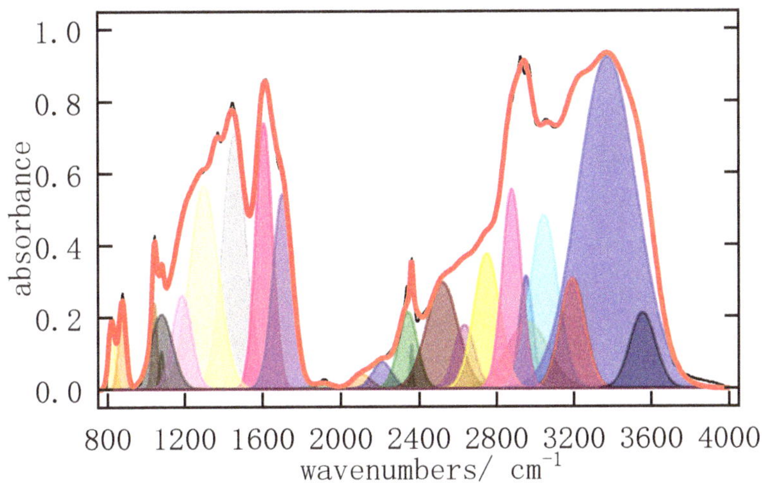

Figure 6. Peak fitting results for the raw coal FTIR spectrum.

Table 5. Peak fitting results for the raw coal FTIR spectrum.

Type	The Average Center	Absorption Peak Attribution	Peak Area
aromatic hydrocarbon	819.8679	substituted benzene	7.73811
	874.0013	substituted benzene	11.41824
	1456.101	C=C	106.9726
	1606.582	C=C	78.49535
	1906.579	C-H	2.10692
	3045.912	aromatic hydrocarbon	86.86852
aliphatic hydrocarbon	1362.781	$-CH_3$	7.33262
	2878.893	$-CH_2$	72.19418
	2953.148	$-CH_3$	35.99535
	2970.776	$-CH_3$	54.26141

Table 5. *Cont.*

Type	The Average Center	Absorption Peak Attribution	Peak Area
oxygen-containing functional group	1043.967	C-O	8.10314
	1080.765	C-O	2.86622
	1083.765	C-O	29.71096
	1186.923	C-O	32.11245
	1299.023	C-O	104.4787
	1702.144	C=O	66.69472
	2104.148	-COOH	4.98769
	2210.091	-COOH	9.21649
	2341.694	-COOH	26.25363
	2361.334	-COOH	2.15573
	2518.659	-COOH	59.63693
	2634.681	-COOH	23.09355
	2749.189	-COOH	62.38078
	3192.25	-OH	50.10662
	3374.61	-OH	349.8566
	3557.065	-OH	35.27509

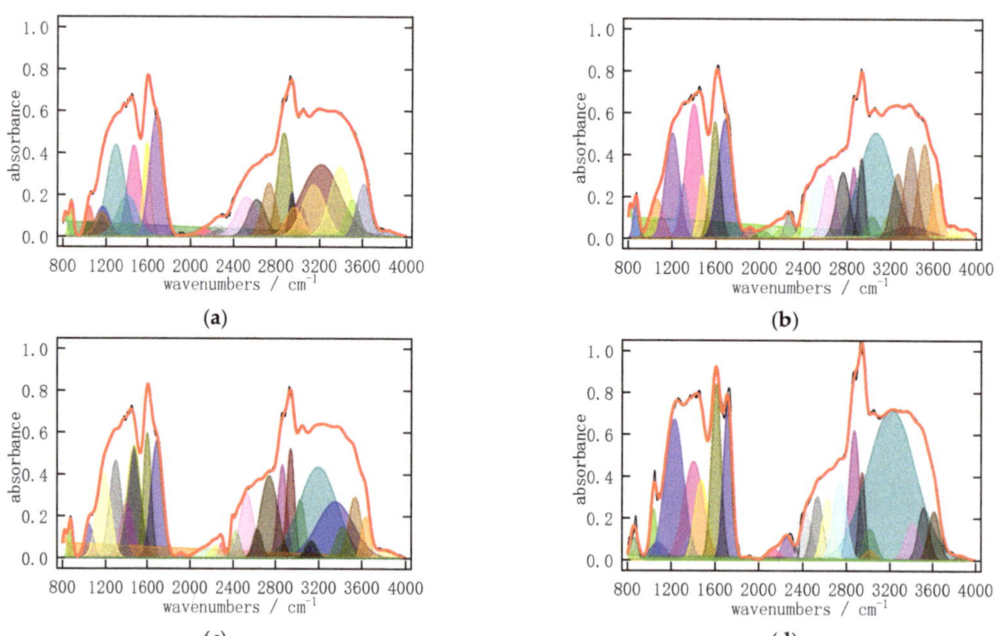

Figure 7. *Cont.*

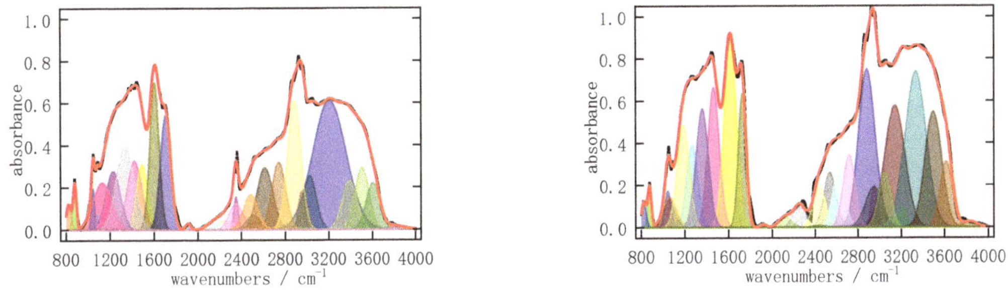

Figure 7. Peak fitting results for FTIR spectra acquired from coal samples extracted using (**a**) MT, (**b**) CYH, (**c**) n-hexane, (**d**) CYH + AOT + AE, (**e**) AE and (**f**) AOT.

This peak fitting indicated that several of the original peaks represented two or more overlapping peaks. Therefore, this analytical process was used to assess the functional groups in the coal specimens.

The area of each fitted peak was calculated and the peaks were assigned to oxygen-containing, aromatic, or alkyl structures.

3.4. Quantitative Analysis of Functional Groups

The data in Figures 5–7 allow an assessment of the functional groups in the various extracted coal samples. The results of this assessment are provided in Figure 8.

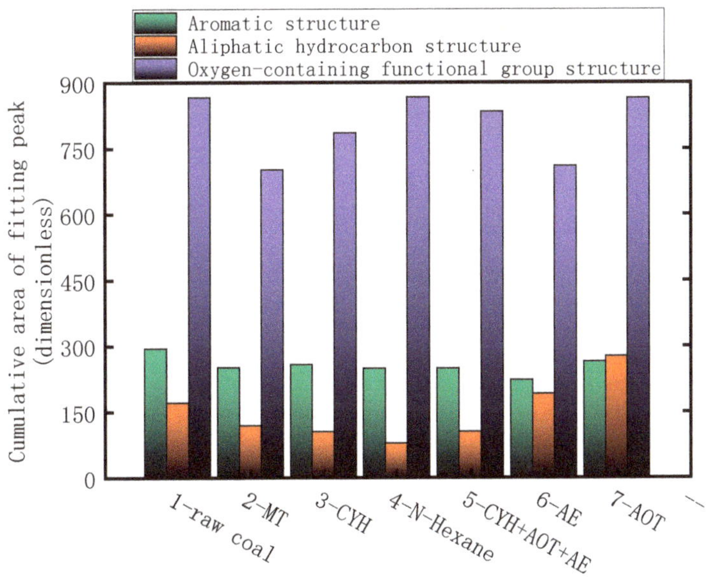

Figure 8. Total areas of FTIR peaks related to group structures in extracted coal samples.

The extent of extraction was determined from the cumulative FTIR peak areas obtained from each extracted coal sample. These cumulative areas were found to decrease in the order of AOT (cumulative area of the peak: 1403) > YM (1330) > n-hexane (1191) > CYH + AOT + AE (1184) > CYH (1145) > AE (1121) > MT (1069) (the later the order, the better the extraction effect). Thus, it was confirmed that all the solvents exhibited relatively effective extraction of the coal. Figure 8 demonstrates that each solvent effectively extracted aromatic structures, although the extraction efficiency decreased in the order of YM > AOT > CYH > MT > n-hexane > CYH + AOT + AE > AE. The best-performing solvents for aromatic groups were therefore n-hexane, CYH + AOT + AE and AE. All solvents showed a good extraction effect for C-H bonds based on the peak at 1920 $\pm$ 10 cm^{-1}. The relative proportion of aromatic C=C-H bonds in each of the extracted coal samples was lower than that in the raw coal sample. The material extracted using MT showed the lowest C=C bond content (close to 84% of that in the raw coal). Compared with the other types of functional groups, C=C bonds were found to be more stable and more difficult to degrade during the extraction process.

Figure 8 shows that the various solvents effectively extracted aliphatic structures and this effect decreased in the order of AOT > AE > YM > MT > CYH > CYH + AOT + AE > n-hexane. Thus, the best solvents for this purpose were n-hexane, CYH and CYH + AOT + AE. With the exception of the sample extracted using AOT, the total areas of the peaks at 2875 $\pm$ 10 cm^{-1} (the methylene C-H stretching vibration) and 2950 $\pm$ 15 cm^{-1} (the methyl group antisymmetric stretching vibration) of all the extracted coal samples were lower than those of the raw coal. The -CH_3 content of the material extracted with AOT was higher than that in the raw coal. Figure 8 demonstrates that the various solvents were able to effectively extract coal molecules with oxygen-containing structures. This effectiveness decreased in the order of YM > n-hexane > AOT > CYH + AOT + AE > CYH > AE > MT, and CYH, and AE and MT were the optimal solvents in this regard.

The cumulative areas of the peaks related to C-O bonds between 1100 $\pm$ 10 and 1300 $\pm$ 20 cm^{-1} in the spectra of the extracted coal samples were lower than those obtained from the raw coal. A comparison of the areas of peaks related to hydroxyl structures between 3250 $\pm$ 20 and 3850 $\pm$ 5 cm^{-1} indicated that, although the peak area of the CYH + AOT + AE sample was lower than that of the raw coal, the extraction effect of this combination was far lower than those obtained using solely CYH or AE. The peak area of the AOT specimen was closer to that of the raw coal. The area of the peak related to methyl structures obtained from the coal extracted with AOT was approximately 160.49% that of the raw coal sample. Based on the single-phase fluidity of the reverse micelle layer formed by the AOT, it is evident that this surfactant effectively extracted methyl group structures while excluding structures containing -CO and -COOH groups.

4. Conclusions

Methanol, cyclohexane, n-hexane, ethanol, and AOT were used to extract long-flame coal from the Caojiatan Coal Mine. The extraction rate was found to decrease in the order of MT (4.457%) > AE (2.749%) > n-hexane (2.449%) > CYH (2.359%). The cumulative FTIR peak areas obtained from the various trials decreased in the order of AOT (with a value of 1403) > YM (1330) > n-hexane (1191) > CYH + AOT + AE (1184) > CYH (1145) > AE (1121) > MT (1069). The optimal solvents for the extraction of aromatic structures were determined to be n-hexane, CYH + AOT + AE and AE, while those for aliphatic structures were n-hexane, CYH and CYH + AOT + AE. Finally, CYH, AE, and MT showed the most efficient extraction of oxygen-containing structures.

Author Contributions: J.G.: Conceptualization, Supervision, Formal analysis. Y.Q.: Writing (original draft), Formal analysis. H.W.: Methodology, Investigation. X.Z.: Methodology, Investigation. G.C.: Supervision, Formal analysis. Y.J.: Conceptualization, Supervision. All authors have read and agreed to the published version of the manuscript.

Funding: This research was funded by [National Natural Science Foundation of China] grant number [52004209 and 52174198 and 52174197]; Shaanxi Province Qin Chuangyuan "scientist + engineer" innovation team. Mine Fire Intelligent Monitoring, Early Warning and Prevention (2022KXJ-166).

Institutional Review Board Statement: Not applicable.

Informed Consent Statement: Not applicable.

Data Availability Statement: Not applicable.

Conflicts of Interest: The authors declare no conflict of interest.

References

1. Wang, L.C.; Liang, Y.T.; Luo, H.Z. Research progress and prospect of mine thermal dynamic disaster theory in China. *Coal Sci. Technol.* **2018**, *46*, 1–9.
2. Guo, J.; Liu, H.; Jin, Y.; Cai, G.; Liu, Y.; Yang, P. Review of underground coal spontaneous combustion hidden fire detection methods and new technology prospects. *Chin. J. Saf. Sci.* **2022**, *32*, 111–119.
3. Guo, J.; Cai, G.B.; Jin, Y.; Zheng, X. Research progress and trend of coal spontaneous combustion fire prevention technology. *Saf. Coal Mines* **2020**, *51*, 180–184.
4. Zheng, X.Z.; Jia, Y.X.; Guo, J.; Wen, H.; Wang, B. Research status and prospect of coal field fire monitoring technology. *Ind. Mine Autom.* **2019**, *45*, 6–10+61.
5. Wang, D.M.; Shao, Z.L.; Zhu, Y.F. Some scientific problems in coal mine thermal power disaster. *J. China Coal Soc.* **2021**, *46*, 57–64.
6. Tan, B.; Zhu, H.Q.; Wang, H.Y. Structure and key technologies of mine fire intelligent prevention and control. *Saf. Secur.* **2022**, *43*, 1–7+9+89.
7. Li, X.X.; Ma, Y.L.; Wang, D.; Shi, C.Z.; Li, J.; Wang, J.L.; Wang, W.M.; Zhang, S.J.; Zhao, M. Study on chemical constituents of ETOAC extract from corn whisker. *J. Qiqihar Univ. (Nat. Sci. Ed.)* **2020**, *36*, 54–56.
8. Wang, S.; Tao, D.Y.; Meiheriguli, A.; Liu, Q.; Jiang, X.M. Mechanism of antifungal effect of n-butanol extract from Paniculata saliculata on Staphylococcus aureus. *Prog. Vet. Med.* **2019**, *40*, 44–47.
9. Hu, J.P.; Peng, Y.; Liu, F.; Zhang, M.; Peng, C.; Li, B. Screening of antioxidant and antibacterial substances from hawthorn kernel and analysis of chemical constituents by UPLC-Q-TOF/MS. *Food Sci. Technol.* **2020**, *45*, 334–340.
10. Wang, X. Application of supercritical extraction technology in the extraction of anti-aging components from Dandelion. *Chem. Ind. Times* **2022**, *36*, 14–16+39.
11. Deng, J.; Li, Y.Q.; Zhang, Y.T.; Yang, C.; Zhang, J.; Shi, Q. Effect of hydroxyl (-OH) on oxidation characteristics of side chain active groups of coal spontaneous combustion. *J. China Coal Soc.* **2020**, *45*, 232–240.
12. Wang, J.R.; Jin, Z.X.; Deng, C.B. *Quantum Chemistry Theory of Coal Spontaneous Combustion*; Science Press: Beijing, China, 2007; pp. 195–207.
13. Deng, C.B. Study on Spontaneous Combustion Mechanism and Spontaneous Combustion Risk Index of Coal. Ph.D. Thesis, Liaoning Technical University, Fuxin, China, 2006; pp. 12–35.
14. Li, K.J.; Rita, K.; Zhang, J.L.; Barati, M.; Liu, Z.; Xu, T.; Yang, T.; Sahajwalla, V. Comprehensive investigation of various structural features of bituminous coals using advanced analytical techniques. *Energy Fuels* **2015**, *29*, 7187–7189. [CrossRef]
15. Ndaji, F.E.; Butterfield, I.M.; Mark, T.K. Changes in the macromolecular structure of coals with pyrolysis temperature. *Fuel* **1997**, *76*, 169–177. [CrossRef]
16. Zhang, Y.T.; Yang, C.P.; Li, Y.Q.; Huang, Y.; Zhang, J.; Zhang, Y.; Li, Q. Ultrasonic extraction and oxidation characteristics of functional groups during coal spontaneous combustion. *Fuel* **2019**, *242*, 288–294. [CrossRef]
17. Ma, Y.Y.; Ma, F.Y.; Mo, W.L.; Wang, Q. Five-stage sequential extraction of Hefeng coal and direct liquefaction performance of the extraction residue. *Fuel* **2020**, *266*, 10. [CrossRef]
18. Jovanovicalinovska, R.; Kuzmanova, S.; Winkelhausen, E. Application of ultrasound for enhanced extraction of prebiotic oligosaccharides from selected fruits and vegetable. *Ultrason. Sonochem.* **2015**, *22*, 446–453. [CrossRef]
19. Deng, J.; Chen, W.L.; Xiao, Y.; Lu, H.; Wang, H. Experimental study on influence of imidazolium-based ionic liquids on thermo-physical parameters of coal. *J. Xi'an Univ. Sci. Technol.* **2018**, *38*, 523–529.
20. Gorden, J.; Zeiner, T.; Sadowski, G.; Brandenbusch, C. Recovery of cis, cis-muconic acid from organic phase after reactive extraction. *Sep. Purif. Technol.* **2016**, *169*, 1–8. [CrossRef]
21. Marek, S.; Jerzy, S.; Marek, K. Extraction of brown coals with alcohols under supercritical conditions. *J. Energy Inst.* **2020**, *939*, 1933–1998.
22. Liu, C.H. Thermal reaction model construction and oxidation process research of Jurassic coal. Master's Thesis, Xi'an University of Science and Technology, Xi'an, China, 2021.
23. Bai, Z.J. Microscopical characteristics of ionic liquid inhibiting low temperature oxidation of lignite. Master's Thesis, Xi'an University of Science and Technology, Xi'an, China, 2018.
24. Zhang, Y.N.; Liu, C.H.; Shu, P.; Hou, Y.; Yang, J.; Li, L. Study on active group and thermal effect of low temperature oxidation of weak viscose coal. *J. Saf. Sci. Technol.* **2021**, *17*, 98–104.
25. Fan, X.; Xu, H. Research progress of mass spectrometry method in coal molecular structure analysis. *Clean Coal Technol.* **2022**, *28*, 23–30.
26. Niu, Z.Y.; Liu, G.J.; Hao, Y.; Zhou, C.; Wu, D.; Yousaf, B.; Wang, C. Effect of pyridine extraction on the pyrolysis of a perhydrous coal based on in-situ FTIR analysis. *J. Energy Inst.* **2019**, *92*, 428–437. [CrossRef]
27. Hao, Z.C.; Zhang, X.D.; Yang, Y.Q.; Sun, F.Y. Chemical composition and kinetic mechanism of tetrahydrofuran extracts from different coal grades. *J. China Coal Soc.* **2018**, *43*, 2904–2910.
28. Xiao, Y.; Lv, H.F.; Ren, S.J.; Deng, J.; Wang, C. Study on the characteristics of imidazole ionic liquids in inhibiting coal spontaneous combustion. *J. China Univ. Min. Technol.* **2019**, *48*, 175–181.

29. Rahman, M.; Samanta, A.; Gupta, R. Production and characterization of ash-free coal from low-rank Canadian coal by solvent extraction. *Fuel Process. Technol.* **2013**, *115*, 88–98. [CrossRef]
30. Alcalde, R.; Garcia, G.; Atilhan, M.; Aparicio, S. Systematic study on the viscosity of ionic liquids: Measurement and prediction. *Ind. Eng. Chem. Res.* **2015**, *54*, 10918–10924. [CrossRef]
31. Zhang, Q.; Chen, F.S. Effect of reverse micellar extraction on structure and properties of soybean protein isolate. *Food Sci.* **2019**, *40*, 108–113.
32. Yang, W.Q.; Mao, K.M.; Mo, W.L.; Ma, F.Y.; Wei, X.Y.; Fan, X.; Ren, T.Z. Analysis of methanolysis mechanism of extraction residue of Zhumaohu lignite based on reaction path of model compounds. *J. Fuel Chem. Technol.* **2022**, *50*, 396–407. [CrossRef]
33. Zhang, H.L.; Wu, N.Y.; Qiao, X.X.; Huo, R.; Gao, W.; Zhao, X.J. Study on crystallization thermodynamics of lycopene crystal in hexane-acetone mixture. *China Food Addit.* **2021**, *32*, 135–140.
34. Zhang, Q.; Chen, F.S.; Sun, Q. Study on the relationship between microstructure of reverse micelles and pre-extraction rate. *Food Res. Dev.* **2018**, *39*, 4–10.
35. Costi, E.M.; Sicilia, M.D.; Rubio, S. Multiresidue analysis of sulfonamides in meat by supramolecular solvent microextraction, liquid chromatography and fluorescence detection and method validation according to the 2002/657/EC decision. *J. Chromatogr. A* **2010**, *1217*, 6250–6257. [CrossRef] [PubMed]
36. *GB 474-2008*; Method for Preparation of Coal Sample. Standardization Administration of China: Beijing, China, 2008.
37. Sun, Z.Y.; Zhang, X.D.; Zhang, S.; Liu, X.; Zhang, S. Mechanism of composition and chemical structure change of THF soluble low molecular compounds during coal evolution. *J. China Coal Soc.* **2021**, *46*, 3962–3973.
38. Weng, S.F. *Fourier Transform Infrared Spectroscopy*; Chemical Industry Press: Beijing, China, 2010; pp. 377–388.
39. Lu, M.M. Experimental Study on Low-Temperature Oxidation Spontaneous Combustion Characteristics of Jurassic Coal and Its Indicator Gas. Master's Thesis, Xi'an University of Science and Technology, Xi'an, China, 2018; pp. 20–22.

MDPI AG
Grosspeteranlage 5
4052 Basel
Switzerland
Tel.: +41 61 683 77 34

Fire Editorial Office
E-mail: fire@mdpi.com
www.mdpi.com/journal/fire

www.ingramcontent.com/pod-product-compliance
Lightning Source LLC
Chambersburg PA
CBHW040758250926
51850CB00007BA/1029
* 9 7 8 3 7 2 5 8 6 1 6 7 5 *